T0201067

Parlay/OSA

Parlay/OSA

From Standards to Reality

Musa Unmehopa, *Lucent Technologies, Bell Labs Innovations, The Netherlands*
Kumar Vemuri, *Lucent Technologies, Bell Labs Innovations, USA*
Andy Bennett, *Lucent Technologies, Bell Labs Innovations, UK*

John Wiley & Sons, Ltd

Published by John Wiley & Sons Ltd, The Atrium, Southern Gate, Chichester, West Sussex PO19 8SQ, England

Telephone (+44) 1243 779777

Email (for orders and customer service enquiries): cs-books@wiley.co.uk
Visit our Home Page on www.wiley.com

This publication is designed to provide accurate and authoritative information in regard to the subject matter
covered. It is sold on the understanding that the Publisher is not engaged in rendering professional services. If
professional advice or other expert assistance is required, the services of a competent professional should be sought.

Other Wiley Editorial Offices

John Wiley & Sons Inc., 111 River Street, Hoboken, NJ 07030, USA

Jossey-Bass, 989 Market Street, San Francisco, CA 94103-1741, USA

Wiley-VCH Verlag GmbH, Boschstr. 12, D-69469 Weinheim, Germany

John Wiley & Sons Australia Ltd, 42 McDougall Street, Milton, Queensland 4064, Australia

John Wiley & Sons (Asia) Pte Ltd, 2 Clementi Loop #02-01, Jin Xing Distripark, Singapore 129809

John Wiley & Sons Canada Ltd, 22 Worcester Road, Etobicoke, Ontario, Canada M9W 1L1

Wiley also publishes its books in a variety of electronic formats. Some content that appears
in print may not be available in electronic books.

Library of Congress Cataloging-in-Publication Data:

Unmehopa, Musa.
  Parlay/OSA : from standards to reality / Musa Unmehopa, Kumar
Vemuri, Andy Bennett.
    p. cm.
  Includes bibliographical references and index.
  ISBN-13: 978-0-470-02595-6 (cloth : alk. paper)
  ISBN-10: 0-470-02595-6 (cloth : alk. paper)
  1. Telecommunication systems–Management. I. Vemuri, Kumar.
II. Bennett, Andy.   III. Title.
TK5102.5.U55 2006
621.382′1–dc22
2005032767

British Library Cataloguing in Publication Data

A catalogue record for this book is available from the British Library

ISBN-13: 978-0-470-02595-6
ISBN-10: 0-470-02595-6

Typeset in 10/12pt Times by Laserwords Private Limited, Chennai, India
Printed and bound in Great Britain by Antony Rowe Ltd, Chippenham, Wiltshire
This book is printed on acid-free paper responsibly manufactured from sustainable forestry
in which at least two trees are planted for each one used for paper production.

Trademarks and Permissions

To Odette and Aron M.U.
To Sai and Family K.V.
To Katie, Eleanor, Ewan and Matthew A.B.

Contents

About the Authors

Musa Unmehopa is a Distinguished Member of Technical Staff in the Wireless Standards department of Bell Labs within Lucent Technologies, responsible for service mediation architectures. He has nine years experience in telecommunications, most recently as a standards manager active in consortia like 3GPP, 3GPP2, ETSI, IETF, the Open Mobile Alliance™ (OMA), and the Parlay Group. Prior to his standards activities, Musa worked as a lead development engineer and systems engineer of large scale, carrier-grade mobile communications software projects. He has been involved with Parlay standards since the early inception of OSA technology within 3GPP in 1999 and has held numerous editor and rapporteur positions for the Parlay/OSA specifications. Most notably, Musa has served on the Joint Working Group management team as the vice chairman of the 3GPP TSG CN-5 working group, and as such held a seat on the Parlay Technical Advisory Committee. Musa was part of the early Parlay prototyping activities within Lucent Technologies and acts as standards consultant to the architecture team of the Lucent MiLife™ Intelligent Services Gateway product. Currently, he acts as his company's alternate member on the Parlay Board of Directors. In addition, Musa presently serves as the chairman of the Architecture Working Group within the Open Mobile Alliance, and previously held the position of vice-chairman of the Mobile Web Services working group of OMA. Musa received his M.Sc in Computer Science from the Technical University of Twente, the Netherlands, in 1996, has published several papers in the area of service delivery and service mediation, and has several patents pending.

Kumar Vemuri is a Member of Technical Staff in the CTO Organization of the Applications Solutions Business Unit of Lucent Technologies. He has several years experience in the telecommunications industry, and has been involved in the research, architecture, systems engineering, prototyping, and design phases of several projects during this time. Most recently, he has worked on the architecture, analysis, and design of products and standards relating to service mediation. Kumar has authored several papers in the areas of service mediation, service delivery, network architectures and converged networking. He holds two patents and has several pending. Kumar received his M.S. in Computer Science from the University of Cincinnati. He currently resides at Naperville, Illinois, USA.

Andy Bennett is a Distinguished Member of Technical Staff working in the Wireless Standards organization within Lucent Technologies. Andy has worked in the telecommunications industry for 15 years, gaining experience in both wireline and wireless technologies. He has held the position of Parlay Framework Working Group Chair and has worked extensively in the 3GPP CN5 OSA and SA2 Working Groups. Andy has authored a number of technical papers and presentations for journals and conferences and has several patents pending on service delivery and mediation.

A Note to the Reader

Books are written for specific target audiences. Sometimes, the material within the books is widely usable. Some books we keep as references on our desks, while others we enjoy, critique, discuss, and then move on to other things. A book, any book, that educates, makes one stop and think, and promotes productive discussion, can be considered successful. We sincerely hope to have written one of these.

We hope the reader will enjoy the structure and content – the technical and the abstract, the easy and the difficult parts, the elementary as well as advanced sections. We envisioned, while writing, that this book would be useful to managers in telecommunications companies trying to keep up with new technologies, to engineers in the same companies who need to know the gory details, to product, solutions, and offer managers as well as those that make purchasing decisions in service provider companies, and their supporting technical staff. And last, but not least, we expect that application developers who want to learn about Parlay and OSA with a view to acquiring skills to enable them to build applications utilizing these technologies will be helped by reading this book.

Part I of this book will be useful to all readers without exception. This part gives you a good grounding of Parlay fundamentals. Part II will be of greater interest to those who need to understand the evolution of standards, how Parlay has arrived at where it is today, and (the nuts and bolts of) how Parlay solutions are expected to work. Parts III and IV are intended primarily for technical readers, the engineers and their management, who are more focused on implementation-related considerations. These provide a lot of food for thought, though they do not always provide all the answers. Often, asking the right questions is more important than answering others. We hope to help the reader make wise design choices through these discussions. Finally, Part V studies some of the more advanced topics, both with deployments, and with the standards, and looks forward into the future.

To keep the size of the book reasonable, some discussions are a little succinct, though still complete. To help readers who may not have all the required background to still follow along without frustration, some appendices are provided with supporting and background material. These can be located at the web page accompanying this book: http://www.wiley.com/go/parlay. In these appendices, we have also attempted to cover supplementary topics related to the book, but not directly woven into the Parlay story.

We encourage readers to read the entire book. However, we realize some sections are rather rigorous and there is a little bit of math in the book which may not be everyone's cup of tea. Advanced topics that may be skipped on a first reading are therefore helpfully marked with asterisks in the table of contents. These can be omitted in a first reading with no loss of continuity.

The Authors

Acknowledgments

The journey of a thousand miles begins with a single step.
Ancient Chinese proverb

Writing a book, any book, is a lot of work – even more so when this is a first for some of the authors. The scope and magnitude of the task is daunting, but as they say, 'no guts, no glory'; 'nothing ventured, nothing gained'. Still, it was a long path from concept to completion – from dreaming about holding a copy of the published work in one's hand, to seeing the day when the book is actually available.

The process of writing has been an interesting journey in and of itself, but the learnings, discussions and interactions that took place along the way were at least equally, if not more, valuable. No man is an island, and our work here has benefited immensely from suggestions, comments and constructive criticism from our colleagues. We would like to thank them for their help and acknowledge their contributions here. We apologize in advance for any people we may have missed. We would like to thank Michel Grech and Igor Faynberg for a careful review of the entire draft of the book, and the following people for comments, discussions, insights shared, constructive criticism, corrections, and for their support of our efforts in particular ways: Shehryar Qutub (the Policy Management part of Chapter 6), Nick Landsberg (Chapter 10), Ram Batni (for discussions relating to Chapter 12), Dirk-Jaap Plas (Chapter 13), Ramesh Pattabhiraman (Chapters 16 and additional website material), Jeroen van Bemmel (Chapter 16 and the Presence appendix), and Rick Hull and Maarten Wegdam (Chapter 17). For her tireless support with software packages and manuscript preparation, we are grateful to Viv Weir. We would also like to thank John Stanaway, Ransom Murphy, Julian Santander, Rick Lewis, John Reid, Doug Varney, Jack Kozik, Fran O'Brien and Ajit Rudran for their general help and book-related discussions. Any residual errors of course, remain our very own.

Apart from our colleagues at Lucent Technologies, our work has benefited deeply from our interactions with our peers at other companies during Parlay and OSA standards meetings. The Joint Working Group over the past couple of years has provided a challenging environment for technical debates. Specifically, we would like to thank Chelo Abarca, John-Luc Bakker, and Ultan Mulligan.

We similarly, and tacitly, also acknowledge the help, understanding, and support of our families as we spent hours in front of a computer somewhere working on endless drafts of the book – for missed appointments, and meals eaten alone. A big Thank You! for putting up with this. And we would like to thank our respective management chains at work for seeing us through some hectic times as the book was being put together.

No worthwhile book can be published without the help of the editing staff and other support from the publishing house. We would, last but not least, like to thank Birgit Gruber, Joanna Tootill and Richard Davies at John Wiley & Sons for the excellent job they did helping and guiding some novice authors through the intricacies of the book publication process. Copyediting is a thankless task and we would like to thank Andy Finch for carrying this out.

Portions of this text pertaining to Parlay APIs and specifications are reprinted with permission of The Parlay Group, Inc.

The Authors

End-user Scenarios

In this preparatory chapter, we present some end-user scenarios. These are meant to be illustrative of some of the end-user or subscriber and service provider needs, wants and frustrations. They will be used to motivate the discussions in the first few chapters of the book as we seek to understand first the requirements of a solution that might better help meet their needs, and later, as we demonstrate how Parlay/OSA technologies fulfill these requirements.

Scenario 1: The Operator's Perspective

October 23rd, 2003

Liz 'Why don't they just call me Elizabeth' Montgomery was having one of those days. She was a senior network engineer for a large Wireless Service Provider, Freedom Wireless. She had just finished meeting with a third-party application developer who was building a new application for the Freedom Wireless network.

'Every time we have to add a new application to our network', she thought, 'we go through this same agonizing process with the third parties who build our applications. Sure, they've got bright ideas, but there is certainly a downside: the painful hand-holding as they try to build, test and integrate their applications with our billing, management, and provisioning systems (and the millions of dollars spent to achieve the same). Not to mention the at least six to eight month window before the service can be rolled out to subscribers. It looked great in Powerpoint so why did it take so long to roll out field carrier-grade telecommunications applications? There has to be a better way to do this'.

She had heard some of her colleagues mention how Parlay and OSA technologies could help alleviate this situation. She was skeptical however. In her 20-odd years of networking experience she had seen even genuinely exciting technologies fail to deliver on their promises. Would Parlay really solve her problems? WDTJCME let out a long, drawn-out sigh. A cup of strong coffee was what she needed.

Scenario 2: The Application Developer's Perspective

October 25th, 2003

Joe Friday worked for the Acme Computer Applications company. They were systems integrators and applications providers, and made their revenues by being contracted by large service providers like Freedom Wireless to systems engineer, build, integrate, test, and trial applications. In fact, Joe had just come from a meeting with Liz Montgomery, who was not too happy with his proposed schedule and cost structure, though she seemed excited by the initial idea.

They had deployed applications in several service provider networks before, but each integration exercise presented its own unique challenges. 'They're never clear on requirements,' Joe said to himself, 'They keep changing their minds on what systems we need to integrate with and how we ought to do it. Feature creep is a big problem. And they have changed some of their legacy equipment, which means it will take longer to integrate and test some application aspects. And why do they all insist on sourcing their kit from multiple vendors? Each telecom equipment vendor adds their own small tweaks to the standard protocols, and if Freedom wants me to use these custom extensions, that will take more time and cost more money...and yet, they seem to think it already takes too long to design, build and test new applications, and that our rates are too expensive. Our margins are low enough as it is. There has to be a better way of doing this'.

Scenario 3: End-user Perspective

December 11th 2003

Allie Dunning was a financial adviser at a securities firm. She had lots of meetings with clients and prospective clients, and spent most of her time on the phone talking to them, or researching stocks and company backgrounds on her computer. Time was always an issue and she came to depend heavily on her calendar. She routinely struggled with getting her calendar aligned with that of her co-workers, and with using her telephone-based features in synergy with applications available to her on her computing platform...She also often wished there was a means whereby her computer and telecommunications infrastructure could work together collaboratively to provide her with a single unified user environment, to enable more effective communications with colleagues. She did not really care about the details of the underlying networks and the associated difficulties relating to inter-networking. She just wanted to get her job done as easily and efficiently as possible.

Wouldn't it be nice, she mused, if colleagues could stop sending me Instant Messages when they saw I was on the phone and did not want to be disturbed? Or if I could set up a conference call on the fly by scheduling in their calendars via some kind of shared interface? Or use information like their location, current mood and mode of accessibility (cell-phone vs. at the desk at work) to determine what kind of mechanism I can best use to share information with them? Surely, there has to be a better way?

Scenario 4: Yet more perspectives

May 5th, 2004

Now that he could take his number with him even after switching service providers, thanks to the magic of number portability, Tom Anderson decided to switch carriers from Freedom Wireless to Utopia. It's not really freedom if the poor network coverage meant he could use his cell phone in the city and at home, but not during his commute or on the road, he thought grimly. He had heard that Utopia's coverage was one of the best, and the price plans were almost the same.

Down the street, Eleanor Alsace was doing the exact opposite. Her three-year contract with Utopia had finally expired, and she could not wait to get on to the Freedom Wireless network. Coverage was indeed good with Utopia, she admitted to herself, but there was a distinct lack of exciting end-user services. She had seen friends do some really cool things with their 'Freedom phones', including picture messaging, instant messaging, gaming, and streaming video. She longed to be able to do the same. Well, now she would.

Jim Singleton had used both Utopia and Freedom Wireless in the past. He liked Utopia's coverage and Freedom Wireless' services. He was currently with Utopia, but wondered if it would be unreasonable for a subscriber to want the best of both networks.

News Flash (Sometime During 2005–2006)

At some point in 2003–2004, Freedom, Utopia, and many other wireless, wireline and high-speed Internet service providers started supporting Parlay technologies in their networks. All users saw drastic improvements in their ability to access new services. Service providers were able to develop, test and deploy new applications much more quickly and cheaply, and grow their revenue bases with sticky services and by increasing subscriber reliance on the underlying network. Freedom Wireless and Utopia Networks now competed on coverage and the offered application set. Application developers were happy too; they could build and integrate telecommunications services more easily, adding new feature support (within reason) took less time, and they could build an application once, and sell it to multiple service providers without significant changes, thanks to standardized interface specifications. The telecommunications industry was finally recovering from the slump of the last three years, and Parlay/OSA were acting as drivers for growth.

Scenario 5: The Future

December 12th, 2007

Alan Friedman ('Alfie' to his friends) woke just in time to catch the tail end of the captain's announcement – 'we'll be landing at London Heathrow in 20 minutes'. He did not usually sleep on planes, but he was tired, and the continual monotonous hum of the engines helped him drift into unconsciousness. Or was it the hours on end of inhaling recycled kerosene exhaust? At any rate, he was back home now…and he loved getting back. What was it that people said? 'Be it ever so humble, there's no place like home'.

Fully awake now, he freshened up, and checked the programming on his cell-phone. Good. It was ready. A well-maintained schedule of tasks that needed to be done, so he would not have to key in each one separately.

He had already typed in all his email messages before he took his unplanned nap, and marked them for immediate dispatch (once a connection was available). His phone was programmed with his security preferences (end-to-end encryption – his employer had insisted on that), and connection information.

The limousine company number was also programmed in, along with his reservation code, destination information and other details. Last, but not least, he had programmed in a call to his Internet Service Provider (ISP) to enable him to get the latest news and weather reports, stock quotes for his portfolio, and other information of interest to him. His computer would organize information gleaned from the various news sources into the format he liked. Now this was what he called coming home.

He wondered briefly how far things had come technology-wise in the past three years. Wireless connectivity was almost ubiquitous now, and people seldom used wires for interconnecting devices. In fact, his laptop case now had a cradle for his cell-phone, and he had a wireless earpiece as well. True, this sometimes made it difficult to work out whether people were talking to you or into their phones when you heard them speak, but that was the price of convenience. Now people could talk to each other at all hours of the day thanks to technology improvements. It was arguable whether their ability to communicate increased significantly – there were still the same old misunderstandings, misinterpretations and mis-representations of information, but don't blame technology for that. At least now people had higher bandwidth, instant presence updates and media-rich content – this helped reduce some of the communication gap.

As he left the plane Alfie turned on his cell-phone and dropped it next to the laptop in his carry-on case. Then he turned on his computer and affixed the earpiece in his ear. Now he was ready to face the world.

His Automatic Personal Assistant (APA) program was immediately activated. He called the voice 'Sandy'. The program was pretty flexible, and he had a choice of settings for her personality, modes of interaction, etc. There was a second part of the program that ran on his computer – this enabled the cell-phone and computer to work together in new and interesting ways, leveraging each other's capabilities. A third part, which ran on the service provider network, gave him some of the context-sensitive features that factored in his location, presence, and other information as he got his services.

He still recalled how long it had taken to provision all these details and configure this application (some things never change). But it had been worth the trouble. Talking to Sandy almost felt like conversing with real human being. 'How was that for a Turing test?' he thought, smiling to himself. But of course her domain of expertise was limited, though sufficient for the tasks she was required to perform.

He told her to execute the script he'd programmed in. She said there was no signal yet, but told him he was connected to the network as soon as he exited the plane. She had him connected to the limo company, and was telling the operator that took the call that Mr. Friedman had arrived at Heathrow, his reservation number was #89423231, he was headed to Hyde Park in London, and where would the limo pick him up? Alfie clearly heard that he was supposed to wait outside door 5E (as in 'Edward' – he still wondered why they just wouldn't use the standard phonetic system 'Alpha', 'Bravo', 'Charlie', . . .), and his pickup would be in around 18 minutes. Sandy thanked the operator and hung up. Of course, he had the option of barging into that conversation at any time, overriding Sandy if he felt the need to do so – the program was pretty flexible and let the user assume control anytime he wanted – he liked that. At least Sandy would not be cross, when interrupted.

Next, 'she' called into his corporate network. She always told him what she was doing. A couple of minutes later, she told him his email messages were on their way (he could read them in the limo as he was being driven back home). The transfer complete, she informed Alfie, terminated that connection and set about gathering the news stories and other information based on the configured options. She used his personal ISP account for this. Then she read him the weather forecast, and breaking news stories – he liked listening to the news, and she was pretty good at reading summaries. He interrupted her after the weather and asked her to call home so he could speak to his wife, which she did with the practiced ease of a seasoned secretary after looking through his wife's presence and availability profile, and current location information.

Sandy would screen calls for him too, unless it was from preprogrammed family numbers. The caller would be asked what the call was regarding, or to enter the APA-override code if Alfie had given it to them (in which case they were directly connected to him, or to his voice mail). She would then courteously take their message, ask them to hold briefly while she checked Alfie's presence profile preferences, and, if he indicated an interest in taking the call, would connect them to him. She could be asked to save the conversation for later replay, or even take voice commands during the call when he asked her by name for assistance with certain tasks (e.g. Sandy, Katie is not on our conference call yet. Can you check her availability? Can you call her and conference her in? Thank you.)

Multi-modal services were where personal agents came in truly handy however. He remembered how, just a couple of days ago, as he took a walking tour of downtown Chicago, she had directed him along the shortest path which passed by various landmarks, and had told him about each in turn, while displaying pictures on his cell-phone screen with interesting facts about each of them. She had led him to the museum of Science and Industry, as she had 'remembered' that he had browsed their web site three times in the past couple of weeks. She even explained how steam engines worked as he lingered in front of the 'History of Transport' display. He had purchased tickets to the museum, and for the train ride back to his hotel also, through the same interface.

Another day in the life of an average end-user. . .and a powerful reminder of what an effective confluence of speech, artificial intelligence, and Parlay/OSA technologies could achieve.

Part I

Background and Introduction

Fasten your seat belts it is going to be a thrilling ride! In this first part of the book, we examine the current ways of doing things, ecosystems of networks and services, their associated value-chains, and study possible improvements that could be effected in the present mode of operation. We then distill a set of requirements from this analysis, synthesize a generic solution, trace the evolution of standards to Parlay technologies, and discuss how Parlay meets the requirements so derived. After this, an overview of Parlay operation is presented from a standards perspective.

Chapters 1 to 4 are intended for all readers who want to understand Parlay – technical, business and marketing alike. Chapters 5 and 6 will be of greater interest to technical readers, though still accessible to non-technical ones.

1

The Internet is Calling – Today's Network Ecosystems and Their Evolution

1.1 Introduction

In this chapter we briefly discuss, at a high level of abstraction, the different kinds of telecommunications networks, the technologies in common use in them today, and then explore why newer technologies continue to remain attractive to users, application developers, and service providers.

Time restrictions, and the need to provide a book that can be carried without the aid of a wheelbarrow, mean we cannot start at the 'very beginning' (since every story starts somewhere after the 'Big Bang'). But we will plunge into the history of telecommunications at a point that will provide readers with some background for the chapters that follow. We have included the key wireline and wireless network technologies in use today and have tried to give a flavor of each, emphasizing what makes them unique and picking out those aspects that are most relevant to the book. This chapter is not intended to be a thorough or complete tutorial and so interested readers are referred to e.g. [Faynberg 1996, Miller 2002, WAP] for more complete discussions of individual topics covered in this chapter.

So what is a network? At the highest level, any communications network may be visualized by the reader as being conformant to the abstract diagram in Figure 1.1. First, there is an access technology – this gives the user access to the network itself, and the services it hosts. Second, there is the core network infrastructure. Multiple access networks may 'hook into' the same core – this typically happens as networks evolve, and a strong need is felt to share services across them either for feature parity (within limits of reason, of course), or for reuse of deployed core network infrastructure and services across different terminal types or for other different reasons altogether. Third, there is a services layer that spans the core network as an overlay. This layer provides the real intelligence and value-add to the core that performs switching and related functions. The services layer also contributes directly to the end-user experience. Finally, we have gateways to other networks, for, as we will see in the main body of the chapter, no network can afford to be an island, and that interconnectedness increases value.

In this chapter, we first explore traditional wired telecommunications networks as this provides a natural lead into the discussions of wireless, WAP and other network technologies.

Parlay/OSA: From Standards to Reality   Musa Unmehopa, Kumar Vemuri, Andy Bennett
Copyright © 2006 Lucent Technologies Inc. All Rights Reserved

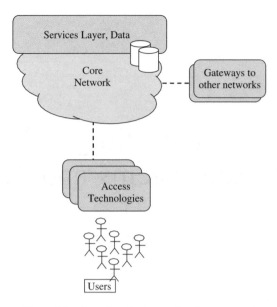

Figure 1.1   Overall logical network reference model

1.2   Traditional Telephony and Intelligent Networks

Since the dawn of time, man has been a gregarious creature with the need to communicate with others of his species. Communication has enabled us to transmit beliefs, traditions and inventions down through the generations and so escape the limits of evolution. People talk, gesture, whisper or find any means available to communicate ideas, feelings, warnings and secrets, and do so for a large part of their waking hours (and sometimes when sleeping too). When we have found a need to communicate over distance we have solved it, using sound (drums), light (beacons along the Great Wall of China) or electricity (the telegraph and the telephone).

However, it was the last of these, the telephone, that fundamentally changed the way people communicate. There was magical quality to hearing a person's voice over hills and valleys, oceans and seas that separated two people talking to each other. The world hasn't been the same since telephony took off in a big way in the third and fourth decades of the last century and things are only getting better, as newer capabilities to share text, documents, video or other media become more widely available.

To set up calls between two parties interested in communicating, one needs an element called a switch. Since not all phones are connected directly to all other phones in the world (the 'two cans and string' model is not very scalable), lines could be connected to a switch, which could link them together whenever the parties tied to those lines wanted to talk to each other. As the number of phones grew, so did the number of switches, and switches had to be connected together as well, to permit users connected to one switch to talk to users connected to others. This led to the birth of telecommunications networks.

Telecommunications networks started out as interconnected networks of switches that permitted users to make and receive simple voice calls (this was referred to as POTS – the Plain Old Telephone Service). A network is an ecosystem, defined in Webster's dictionary as 'the complex of a community of organisms and its environment functioning as an ecological unit'. Here, the network equipment (switches etc.), the user equipment, and interactions between these define the complex environment of interest. Just as in real world ecosystems, changes need to be carefully handled in the network.

For two parties to communicate using the network as a medium, network elements between the calling and the called party (sometimes termed the caller and the callee), need to somehow propagate first the desire to set up the communication path, and then the content of the communication itself, between the two entities. Note that the path is both physical in terms of the trunks tied up to carry the conversation, and logical in that the conversation between the two communicating parties is carried across it. The path between the caller and the callee is not permanent, but needs to be maintained for the duration of their conversation. Thus, the network elements supporting this interaction need to track some state associated with this call. This process is referred to as 'call processing'.

Every phone call between two parties engaging in a conversation was represented in the network in terms of a call model or a state machine on each switching element between the source and the destination[1]. In other words, each switch in a call path would execute a call model or state machine as call processing progressed and the two parties trying to communicate were connected.

Gradually, the need for services became more pronounced, due both to end-users maturing in their use of telephony related technology and in operators' desire to stabilize and expand their subscriber bases. To meet this need, additional features were introduced within call models[2] [Dobrowolski 2001] whereby special code was executed within the context of individual states in the call processing state machine, and new capabilities were provided to the parties involved in the call. These features, since they executed on the switching elements themselves, were typically referred to as 'switch side features'.

But there were issues with this architecture. For one thing, as each new feature was introduced, all switches that needed to support that feature had to be upgraded with this capability. Since there were many switches in the fabric, this kind of an upgrade was not easy to carry out transparently. Also, with the greater proliferation of telecommunications, people began to rely rather heavily on the network, and service outages (both planned outages for the service retrofits, as well as unplanned outages due to the vulnerability of having to update the entire complex fabric) became unacceptable.

Another problem was the degree of difficulty involved in making additions to existing switching logic, and then, testing these additions to ensure that new features did not interact with each other, and with the already deployed features in strange and undesirable ways. This was a far from trivial thing to do. Also, the software architecture of switches rarely allowed a sufficient degree of functional separation of services from the 'normal' logic flow and data structures, etc. relating to call processing. This led to serious issues with switch performance, as well as compromises in the design of the new feature itself.

To alleviate some of these concerns, and to provide a more flexible environment that could change more rapidly with the times as new feature capabilities were added, the Intelligent Network (IN) paradigm was introduced.

In the new (IN) model, switches are no longer merely simple executors of state machines. Service logic is separated from basic call control logic. Service features that were heretofore limited to being hosted on, and executed by, switches, are extracted from the switching elements and collocated into a separate physical element dedicated to execute the enhanced service logic for the newly introduced feature. Such a physical element is called a Service Control Point or SCP. The call model state machines at switches were enhanced to support the capability to query this SCP element (using a well-defined message set) and receive instructions that could be factored into their call processing operations.

Thus, switches now perform two functions – one (called the Call Control Function or CCF in the IN Distributed Functional Plane or DFP) that deals with the execution of the Basic Call State

---

[1] For readers unfamiliar with call processing and state machines, a very gentle and accessible introduction to these topics is provided in Appendix A [Parlay@Wiley].

[2] The concept of call models is only very briefly introduced here. Later chapters will expand upon this concept as they present more details relating to Call Control in the Parlay space.

Model (BCSM) that implements the call processing logic, and two (called the Service Switching Function or SSF in the IN DFP) that is concerned with the ability of the switch to interact with the SCP, request instructions, receive responses and so on. IN-capable switches are also referred to as SSPs or Service Switching Points. In Wireless Networks, these are sometimes also called MSCs or Mobile Switching Centers. But more about Wireless Networks later.

The SCP itself was a physical manifestation of two logical elements from the IN DFP – the SCF or the Service Control Function, and the SDF or Service Data Function. The former of these refers to the service logic that executes a relevant feature at the particular point in the call at which the switch sought SCP assistance, using the data that the switch provided in the request message to generate a suitable response. The latter refers to the capability whereby subscriber data or other data pertaining to numbers, translations etc., are hosted in a large database and made accessible to the service logic for use as appropriate as features execute[3].

With the proliferation of IN, one can still build in switch side features, but one has added flexibility in deploying new features and capabilities to better the end-user experience, through use of SCPs. Introducing new IN-based services in the network no longer necessitates updating all switching elements in the fabric. Rather, only the physically separated SCPs needed retrofitting. Basic service for call connectivity remains unaffected throughout such an update.

These concepts are illustrated in Figure 1.2. The reader is also referred to [Faynberg 1996, Chapter 5] for more details.

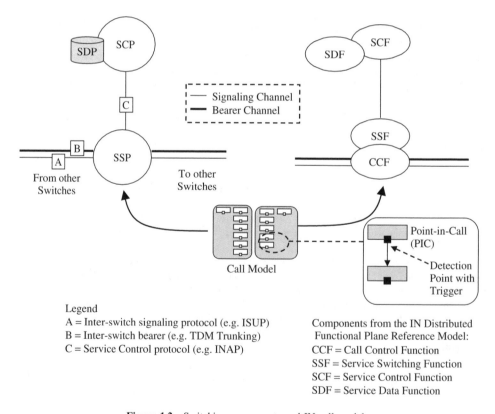

Figure 1.2   Switching components and IN call models

---

[3] This latter function may also be supported by a dedicated physical element standalone, in which case it would be called a 'Service Data Point' or SDP.

1.3 Signaling

When any two elements communicate they need to use a medium (such as a wire connecting them, for example), and a language they both understand, called a protocol. Communication may involve the transmission of data ('Watson, come here, I need you') or information pertaining to the data transmission (end-user Alexander G. Bell wants to connect a call to end-user Dr. Watson). The former is sometimes referred to as bearer (or payload) information, while the latter is called signaling.

Several signaling protocols are in existence today. Different networks use different signaling protocols. Different protocols are used between different types of network elements, and between the same network elements when they are involved in performing different functions.

Good signaling protocols are designed to be flexible and extensible for the addition of new parameters, messages or functionality, efficient in the number of bits of information that need to be transmitted between two communicating elements to share state or other information, and easy to process with minimal overhead.

Another characteristic of such well-designed protocols is that they are layered, such that each layer provides specific functional capabilities to the protocol as a whole, and the upper layers build on capabilities offered by the lower layers. Accessing these lower layer capabilities takes place through connect points called Service Access Points (SAPs), so they can be used in performing the tasks of the upper layer. The data unit supported by the protocol, and more specifically at each layer, is referred to as a Protocol Data Unit (PDU).

A layered architecture that is widely used in the design and operation of protocol stacks, called the Open Services Interconnect (OSI) data model, was developed by the ITU. This model, as shown in Figure 1.3, is composed of seven layers, and most protocols in use today adhere closely to it. The OSI model is discussed in greater detail in [Tanenbaum 2003].

A complete discussion of the design of good signaling protocols merits a book in itself. The interested reader is referred to [Holzmann 1991].

Signaling can be of different types, depending on where and how it is used. It can be classified in different ways, and In what follows, we study some of these ways.

One way of classifying signaling considers whether the signaling stream touches any end-user equipment (e.g. the phone on your desktop). The signaling link between end-user equipment and the network element (such as a switch) is commonly referred to as UNI or the User-to-Network Interface. Signaling links between network elements are referred to as NNI or Network-to-Network Interface.

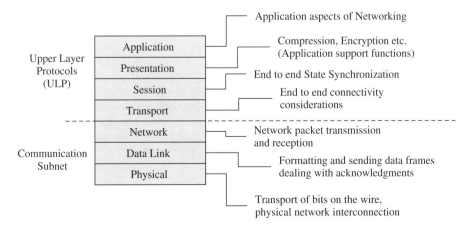

Figure 1.3 The Open Systems Interconnection (OSI) model layered protocol reference architecture

Another categorization considers the role particular signaling protocols play in the overall call flow. For example, user equipment to switch signaling, or switch to switch signaling during call setup, is referred to as call control signaling, while the communication that takes place between SSP and SCP elements is called service control signaling. Typically, different protocols are used in networks to fulfill each of these roles.

There are several other ways of categorizing signaling protocols, such as in-band vs out-of-band, etc. However, we do not study those distinctions for they are best left to books dedicated to signaling. The categorizations we cover above suffice for the purposes of the concepts we intend to develop later in this book, i.e. UNI and NNI, and Call Control and Service Control.

1.3.1 Signaling and Standards Bodies

As signaling pertains to communication among disparate elements in a complex networked environment, some form of agreement on the definition of these protocols is desired. Enter standards. Some standards are developed in bodies focused almost exclusively on data networks, while others are focused on voice communications, and some support working groups (WGs) fall into the gray area in between the two. In this section, we briefly look at some standards bodies of interest to this discussion in an attempt to give the reader a better feel for where and how the various signaling protocols are developed.

1.3.1.1 Telecommunications-oriented Standards Bodies

ITU – The ITU (International Telecommunication Union) is a specialized agency of the United Nations. With telecommunications networks spanning the globe, there is a need for standardization and regulation of such networks on the same scale, that is, globally. The mission of the ITU is to ensure efficient and smooth development and operation of telecommunications technology worldwide, and the general availability of this technology to the global population. As these globe-spanning networks were made up of an enormous mixture of national networks, interconnected by countless, often very specific signaling protocols, the Open Services Interconnect (OSI) data model, referred to earlier in this chapter, was developed by the ITU as a reference model for communications networks and their protocols. The development by the ITU of ISUP as the signaling standard for bearer traffic and INAP as the signaling standard for service control served as a major catalyst for the global proliferation of digital circuit switched telephony networks. With IP networks reaching the same ubiquity, the ITU developed H323 as the international standard for session oriented communication over the Internet [ITU].

3GPP™ – the 3rd Generation Partnership Project (3GPP) is a partnership of regional standards bodies that defines the standards for GSM-based wireless networks and for their evolution into a third-generation UMTS architecture. 3GPP provides several technical specifications aimed at addressing specific interfaces, services and network elements from within its reference architecture [3GPP].

3GPP2 – the 3rd Generation Partnership Project2 (3GPP2) is a partnership of regional standards bodies that defines the standards for CDMA-based wireless networks just as 3GPP performs similar functions for GSM technologies. Given the large overlap in technical directions and architecture between 3GPP and 3GPP2, the latter has agreed to reuse the specifications issued by the former body wherever applicable. In addition, most recently a harmonized reference architecture (called IMS or the IP Multimedia Subsystem) that melds both the 3GPP and the 3GPP2 models has been adopted to further drive convergence in the work being done in these two bodies [3GPP2].

1.3.1.2 Data Network-oriented Standards Bodies

IETF[4] – The Internet Engineering Task Force (IETF) is an organization that hosts numerous working groups dedicated to developing protocols and standards that govern network element

---

[4] There is also a research wing that parallels the work done by the IETF, called the IRTF (Internet Research Task Force) and also run by the same body, the ISOC. This body does more of the 'forward looking' work,

communications within the Internet, and other Internet Protocol (IP)-based networks. In fact, IP was itself designed by this body [IETF].

Among the numerous IETF WGs, the following are of immediate interest and relevance to our current discussion[5]. A brief summary of the work carried out in each of these groups is provided below:

1. *Iptel* – The Iptel working group designs standards for use in supporting telephony over the Internet Protocol, specifically (inter-domain) routing of voice calls over the Internet.
2. *SIP* – This WG is focused on developing the base Session Initiation Protocol and extensions that enable it to be efficiently used in setting up and tearing down multimedia sessions. This WG was spawned off earlier work accomplished under the charter of the MMUSIC (Multiparty Multimedia Session Control) WG of the IETF.
3. *SIPPING* – Session Initiation Protocol Project INvestiGation is dedicated to studying the applications of SIP and non-base-protocol extensions in support of SIP applications.
4. *PINT* – The PSTN/Internet Interworking WG deals with scenarios where an end-user connected to an IP network such as the Internet can request services from an SCP in the PSTN network. Examples of such services include Click-To-Dial (CTD), where a user clicks on a link or submits an HTML form and causes a call to be set up between herself and a customer service representative representing a business.
5. *SPIRITS* – The Services in PSTN/IN Requesting InTernet Services WG addresses scenarios that are an exact converse of PINT scenarios. So the focus here is on services in the PSTN/IN that require IP-host based feature assist capabilities. Internet Call Waiting (ICW) is an example of such a service. If a user is connected to the Internet via his phone line through a modem, incoming call notifications can be piped to the user via that Internet connection even though his phone line is busy at the time. Both PINT and SPIRITS recommend the use of SIP as a signaling protocol.
6. *Sigtran* – The Signaling Transport WG has produced several protocols including SCTP (Stream Control Transmission Protocol) and others that define the lower layers of a protocol stack to enable the transparent transport of SS7-based protocol payloads over IP. The intent here is to promote seamless convergence where possible through use of upper layer protocols (OSI layers four and above) across network types.
7. *Megaco* – The Media Gateway Control WG developed, in concert with the ITU, the Megaco protocol (also referred to as H.248.1) that defines the communications between Media Gateway Controllers and Media Gateways. (See Section 1.4.2 on 'Converged Networks' in this chapter for more details.)

1.3.2 Some Examples of Signaling Protocols

In traditional telephony networks (also called the Public Switched Telephone Network (PSTN) in wired contexts or Public Land Mobile Network (PLMN) in wireless contexts), switches communicate with each other over SS7 (Signaling System #7)-based signaling protocols [Russell 2002].

For example, switches in the PSTN utilize different protocols for user equipment to switch signaling (e.g. Ear & Mouth (E&M) Protocol, Telephone User Part (TUP)), switch-to-switch signaling (e.g. ISDN User Part (ISUP)), and switch to SCP signaling (e.g. IN Application Protocol (INAP)). As explained previously, the first two of these are typically called call processing signaling, while the last of these is referred to as service control signaling. All these are SS7-based.

---

and has made significant contributions to protocols in the area of AAA, SPAM-filtering etc. [IRTF]. The AAA work has since been absorbed into the IETF AAA WG.

[5] In later chapters, work being done in other IETF WGs may also be introduced as appropriate. This listing is merely intended to give the reader a taste for the kind of work the IETF undertakes.

The Internet, the largest, most widely prevalent, almost ubiquitous network today utilizes the Internet Protocol or IP as the basis for communication between computers. Various application level protocols ride atop IP to provide a range of functional capabilities between communicating applications on computers connected to this network. Examples of these protocols include Simple Mail Transfer Protocol (SMTP) for email, File Transfer Protocol (FTP) for file transfers, Hyper Text Transfer Protocol (HTTP) for Browser to Web Server interactions, etc. Most IP-based protocols use either Transmission Control Protocol (TCP) or User Datagram Protocol (UDP) as the layer-4 protocol of choice. The reader is referred to [Comer 1999, Comer 2000] for more details on IP.

Both the SS7 and the IP protocol suites are compliant with the OSI model.

1.4   A Foray into Other Network and Service Architectures

In this section, we discuss some other network and service architectures of interest. A good understanding of some of these will be useful in later chapters as we address how Parlay and OSA technologies relate to them. Others are introduced to give the reader an appreciation of how network evolution takes place, and the different generations of related technologies as new standards are defined.

1.4.1   Voice over the Internet Protocol (VoIP)

Metcalfe's law states, 'The usefulness, or utility, of a network equals the square of the number of users'. The very ubiquity of the Internet, low barriers to the entry of new endpoints, and the overwhelmingly large number of users, along with its ability to carry various protocols that perform different functions leads to increased value per Metcalfe's law, and a positive feedback loop that continually contributes ever more to its growth.

This, combined with a widespread interest in utilizing the Internet for voice communications, has led to voice becoming one of the most widely transmitted payloads on the Internet today[6]. The use of IP to transport voice is referred to as Voice over IP (VoIP). The Internet's inherent ability to transport data, pictures and other visual media such as video in concert, and potentially interleaved with, voice, leads to a truly powerful multi-media user experience.

As with PSTN/PLMN networks, voice-, or more generally multimedia- session setup requires some session setup, processing and teardown signaling, in addition to the ability to transport bearer information over IP. This support signaling can be provided using various protocols. H.323 (developed in the ITU [H.323 2003]) and SIP (Session Initiation Protocol [RFC 3261], defined by the IETF) are popular IP-based protocols for this today. The former is still widely used, while the latter, widely acknowledged to be the protocol of choice for the future, continues to gain in popularity.

IP-based telephony does not have as clear-cut a partitioning between service-related and call-related signaling. The traditional telephone network supports almost all the user services needed, with a minimal set actually supported by the user handset in a manner independent of the network. In contrast, the IP-based telephony model supports a near equal, if not skewed in favor of the handset, distribution of services between the network and the handset domains. This means end-users have greater flexibility in the kinds of services they can access (since this is handset dependent), but also means user reliance on the terminal is greater than in the PSTN world (e.g. user data are hosted more on the user terminal than on the network, so if the user has to initiate a session from a

---

[6] In fact, with the prevalence of high-speed Internet connections such as those using cable or DSL lines, VoIP technology is now really taking off in a big way. VoIP service companies like Skype [Skype] and Vonage [Vonage] are seeing a sharp uptake in their subscriptions over a high-speed Internet access infrastructure. What seems really interesting with some of these services is that one gets a real phone number assigned to the 'always on' high-speed Internet connection, and this number is not 'geographically bound' – the user can get a local phone number in the New York area, while residing in London, and can use this transparently, without the caller knowing his or her actual physical location. Judicious choice of phone number can cut down on long-distance bills, especially if one makes more calls within a particular area code.

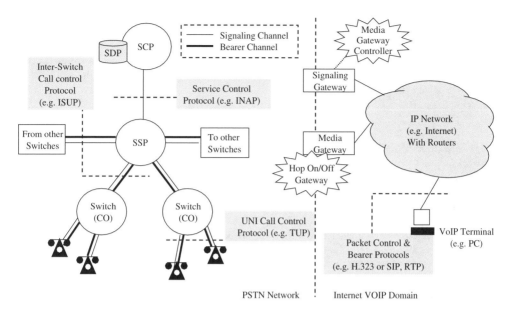

Figure 1.4   The PSTN and IP networks today, and VoIP

different terminal, the experience may be less pleasant than if the data were stored on the network and available to him transparently).

Lately, some protocols, even in the IP-domain have started adding mechanisms to provide support for service control related signaling. But rather than define new protocols aimed specifically at service control, in most cases they have relied on extensions to the base signaling protocol to fulfill these needs as well. However, it is still possible to draw rough analogies between the traditional IN architecture in the PSTN and the IP architecture for multimedia call setup (Figure 1.4). Some of these details for a specific protocol (namely SIP) will be covered in later chapters.

1.4.2 Converged Networks[7]

Isolated networks of users who cannot communicate with users of other networks still feel isolated, though not necessarily alone. Ubiquity, ease of network access, and interconnectedness, contribute towards a feeling of community. The PSTN is useful because it permits a user at any phone connected to it to call another user at any other phone. Connectivity contributes to value. Recall Metcalfe's law.

Definition: Gateway

Where network elements belonging to different networks, and using different protocols, but providing similar functions within their own network contexts need to communicate, an element called a gateway, that speaks both protocols, is used to mediate between these two elements. The gateway element, in its simplest form, functions as a protocol translator, and enables the

---

[7] Networks today include Cable and DSL access technologies as well. These are used, for instance, to support high-speed broadband Internet Access. In late 2004, in the US, broadband Internet access, for the first time in networking history, surpassed dialup access to the network. For the sake of simplicity, these Cable and DSL aspects are not depicted nor explained in any detail in this chapter.

two elements, one from each network, to talk to each other. Gateways form the basis for most convergence in networks today.

When viewed from the perspective of a signaling flow, a gateway through which a flow enters a given network is typically referred to as a 'hop on' gateway, while one through which it exits is called a 'hop off gateway' (Figure 1.4).

As VoIP took off, service providers gradually came to view the Internet, or other managed IP networks, as a means to offload some of the voice traffic to a more cost efficient, less resource constrained environment that supported more optimized routing (and tied up less resources during session setup). In addition, Internet users wanted the ability to call telephones connected to the traditional PSTN or PLMN. In order to support these and other similar needs, convergent architectures came into being.

Convergence may be achieved in a variety of areas. Convergence in terms of signaling transformations, as call processing signaling transits an IP to PSTN or PSTN to IP network boundary, may be carried out at network elements called Signaling Gateways.

Bearer stream transformation as it transits a network boundary of the kind described above is carried out at network elements called Media Gateways (MG), and is commonly referred to as transcoding[8].

If the signaling stream controls a media or bearer stream associated with it, the Signaling Gateway is also referred to as a Media Gateway Controller (MGC), for not only is the signaling transformed as the network boundary is crossed, but the associated media characteristics are also controlled as this transformation takes place, through interaction with the Media Gateway element where the bearer stream is being transcoded. Megaco ([H.248.1], jointly developed by the ITU with the IETF) is the IP-based protocol of choice for MGC to/from MG communication.

Convergence can also be achieved in the services domain. This is of great interest to service providers and also to end-users. If services originally developed for one network could be transparently used in another, then this offers great benefits. For one thing, it saves money while promoting complete and immediate feature parity. And the immediate availability of all existing services in a new network context does wonders for the end-user experience and in meeting user expectations.

The specific signaling protocols supported both for call/session control and for service control may vary based on the specific domains being inter-worked. The degree or ease of inter-working may also vary depending on how closely the call/session and service state models align between the two types of networks in question.

IP-based telephony might want to reuse IN elements in support of providing deployed features to VoIP users. This could be supported by carrying the IN service control protocols over IP for example. This forms the basis of the Sigtran work in the IETF.

IN SCPs could be enhanced to interact with IP-based application servers for new feature logic that is shared between the IN and IP domains. Work in this area has been done in the IETF PINT [RFC 2848] and SPIRITS working groups [RFC 3910] to support end-user capabilities such as Click-To-Dial (CTD) or Internet Call Waiting (ICW). For more on PINT and SPIRITS, the reader is referred to [Kozik 2000, Gurbani 2003].

These are but two examples of service reuse. Several other elegant models [Vemuri 2000] have been discussed in the literature. For more information on services for converged networks, the reader is referred to [Faynberg 2000].

---

[8] Media is typically encoded in some bit format for transport across the network. This encoding is done using software called a codec. Since voice is encoded using one set of codecs in the Internet domain, and another different set of codecs in the Telecommunications domain, the operation of switching the encoding scheme or codecs at the media gateways is referred to as 'transcoding'.

1.4.3 Internet Access via the PSTN

It will be useful for the reader to have a basic understanding of how one may access the Internet via the PSTN – the so-called 'dial-up' Internet access. In this section, we provide a high-level summary description of the same. Internet access has also evolved with time. Initially, this was achieved through the use of modems that enabled users to dial in via analog phone lines – this was somewhat slower, and offered speeds in the 56–128 Kbps range depending on the specifics of the modems in use. More recently, broadband Internet access has seen widespread growth where the use of cable modems and digital subscriber lines (DSL) enables subscribers to achieve speeds in the megabits per second range. Again, in order to keep the discussion simple, and since we merely aim to give the reader a flavor for how the basic technology works, we explain only the dial-up access.

An ISP (Internet Service Provider) typically supports modem pools at several geographic locations called POPs (Points of Presence). When the user dials the phone number associated with a POP (and if there is one in your area, you may not have to pay long-distance charges) by running the appropriate dialer software on her computer that talks to her local modem, the modem from the pool answers the phone. Once the basic call is set up, the PPP protocol (Point to Point protocol, developed by, you guessed it, the IETF) runs across the link and sets up the data connection.

The modem pool is collocated with a NAS or Network Access Server, which also functions as the AAA client. This element interacts with a AAA server hosted within the service provider network (typically over protocols such as TACACS [RFC 1492], or RADIUS [RFC 2138] or, more recently, DIAMETER [RFC 3588]) to perform end-user authentication, authorization and accounting procedures, and sets up filters for IP traffic that transit the user connection (PPP link) and IP network for that session. An IP address may be assigned to the computer for the duration of the session (for packets to flow back to it), using either a statically assigned pre-configured ISP-owned address (relatively rare), or a dynamic address obtained through protocols such as DHCP [RFC 2131] or IPCP [RFC 1332].

Once the session is established, the end-user can transmit and receive data from her computer over this link. Once done, the user simply hangs up, and the IP address (if dynamically assigned), becomes free for reuse for other user sessions, as does the port on the modem pool. Figure 1.5 illustrates dial-up access to the Internet.

1.5 Wireless Networks and Generations of Technology

We started this chapter by looking at very abstract network architectures and introduced call/session control signaling and service control signaling, after which we explored how wireline networks have evolved specifically in terms of their protocols and interfaces and the network elements processing these. Now it is time to look at wireless networks.

So far, we have used the term 'wireless network' in a rather generic sense to refer to networks of mobile terminals built to support cellular technology. In this section, we examine these kinds of networks in a little more detail. We shall briefly introduce the concept of cellular communication and then describe wireless networks in terms of their network elements, their signaling protocols, and the service data they store for subscriber services. We shall then look at how the circuit switched core of wireless telephony networks has been expanded with a packet switched domain to support mobile access to services residing in data networks. After that, we will describe how third generation wireless networks are evolving from current mobile communication systems by introducing a new radio access technology and by further evolving the core network.

Generally speaking, wireless and cellular networks are not strictly the same. Every cellular network is a wireless network, but not all wireless networks need necessarily operate using cellular technology. WiFi (Wireless Fidelity or IEEE 802.11b wireless networking) is an example of localized wireless networking that does not operate on cellular technology. For the purposes of this section, we shall use the term 'wireless' to mean cellular in a generic sense.

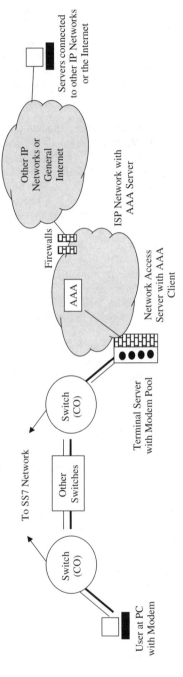

Figure 1.5 Dial-up access to the Internet

1.5.1 Cellular Communication

Wireless (or, to be more precise, cellular) networks are assigned a certain frequency band to use for setting up radio links to the mobile devices of their subscribers to complete the communication path. When engaged in a phone call, the mobile device is allocated a certain frequency in the available spectrum. As radio waves form a shared medium, for the duration of the call, this frequency (or a timeslot in a specific frequency, depending on the details of the wireless radio technology in use) is uniquely assigned to that specific mobile device, in order to avoid interference. This means that the capacity of a mobile network is confined by the number of unique frequencies one can assign within the available spectrum. Cellular systems address this issue of limited capacity by dividing the coverage region of a network in largely non-overlapping areas, called cells. Frequencies are then reused in non-neighboring cells, to increase the overall network capacity.

Cell sizes may vary depending on the area they cover, the technology in use, and in the frequency spectrum utilized by the technology in question. For example, in rural environments a typical cell size may be larger than in the city, as the total number of concurrent mobile phone calls can be safely expected to be lower and hence less reuse of frequencies is required. Also, cells may vary in shape. In dense urban areas the cells may be evenly shaped and arranged like roof tiles or the scales of a fish collectively to cover an entire downtown area. Along major highways or subway and train lines the cells may be stretched in length to offer travelers and commuters continuous radio coverage, whereas on either side of the route coverage may drop quickly.

Although one tries to achieve ubiquitous coverage with cellular technology, sometimes, in the interiors of large buildings or such hard to reach places (for the radio signal), no coverage may be available to make or receive cell-phone calls. Such areas are termed 'urban canyons'.

1.5.2 Wireless Networks and their Elements

Wireless networks, irrespective of the specific technology deployed, all share a similar network architecture. This wireless network architecture is depicted in Figure 1.6, in which we also recognize of course the overall logical network reference model introduced in Figure 1.1, with the separation of access, core, and services.

One of the most successful wireless network technologies, in terms of global deployment, is GSM. We will draw on GSM to introduce and further define wireless networks, and their network elements and signaling protocols. Interested readers are referred to [Mouly 1992] for an excellent

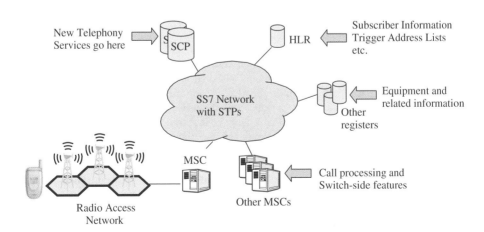

Figure 1.6 Sample wireless network architecture

coverage of GSM technology, in its full breadth and depth. Wireless networks based on CDMA technology will be covered later in the chapter.

GSM networks reuse much of the PSTN and are SS7 networks at their core. Inter-switch signaling is based on the same SS7 protocols deployed in PSTN networks, allowing for large scale reuse whilst smoothly facilitating fixed-to-mobile and mobile-to-fixed calls.

Mobile devices communicate with the network via a radio link to a Base Station System (BSS), consisting of a Base Station Transceiver (BST), or the 'antenna', and a Base Station Controller (BSC). The BSC communicates with the Mobile Switching Center (MSC), connecting the radio part of the network with the SS7 core of the network. The MSC, which is the telephony exchange in GSM networks, performs the basic call processing procedures and interconnects with other MSCs or with the PSTN or ISDN exchanges for network connectivity, via the SS7 core. So mobile telephony systems like GSM only make use of radio resources at the edge of the network, when completing the last step of the communication path to the mobile device.

An MSC differs from a PSTN switch in that it serves mobile devices rather than fixed phones. As mobile subscribers have the freedom of picking up their phone and moving about, contrary to fixed phones and a PTSN switch, there is no static relationship between a mobile device and a specific MSC. Depending on the location, a mobile device is served by a given MSC, which is referred to as the serving MSC. As the serving MSC may alter when the subscriber is changing location, all information pertaining to mobile subscribers is located in a centralized database, called the Home Location Register (HLR) – this is a fundamental difference with the PSTN: as terminal mobility is supported, there is a registry like the HLR that is maintained in wireless architectures. An MSC may query the HLR to obtain service subscription profiles for a given subscriber, or routing information required to locate the subscriber in the network in order to complete an incoming call destined for that subscriber. Originating services are also deployed at the HLR. If for instance the subscriber is not allowed to receive calls while she is registered with another network in a foreign country[9], this information will be stored in the service subscription data of the subscriber. So in the case where the HLR will be queried for routing information with the intention of terminating a call to the subscriber, the HLR will return a decision not to allow further processing of the call and the attempt will be rejected (or barred).

For the purpose of minimizing the need to perform database queries to the centralized HLR database, a temporary local copy of the subscriber data is stored in the Visitor Location Register (VLR) associated with the serving MSC. The VLR record includes information required to page the mobile device and perform call setup procedures. Information relating to so-called terminating services is stored in the VLR as well. An example of a terminating service is 'Call Forwarding on Not Reachable', e.g. when a terminal is switched off. This is a terminating service as only after paging a mobile device, is the not-reachable status for the device established. It is the serving MSC, using service subscription information from the VLR record, that will perform the service logic involved with terminating services, without having to interrogate the HLR.

As service subscription data may change over time, the data stored in the VLR need to be maintained in synchronization with the data kept in the HLR record. Whenever changes occur in the HLR record, the VLR record gets updated. Also, as subscribers move around, they may cross MSC boundaries. As VLR records are associated with the serving MSC, such a crossover (or inter-MSC hand-off) will result in the creation of a new VLR record and the deletion of the old one. The signaling protocol for HLR to VLR communication is the MAP protocol (Mobile Application Part), which is an SS7 based protocol.

As is the case in PSTN networks, IN-based services can be applied in GSM networks as well. In this case, the IN system is referred to as CAMEL (Customized Application for Mobile Enhanced Logic). The service control protocol between MSC and SCP is the CAP protocol (CAMEL

---

[9] Such a service may serve to protect the subscriber for incurring the additional costs associated with receiving incoming calls when roaming, or it may be applied by the operator for subscribers who have overdrawn their user account.

Application Part), which, and this will not be a surprise by now, is SS7 based[10]. Similar considerations also apply to CDMA architectures, which we will see later on.

In addition to the HLR and the VLR, the SCP now introduces a third location for service data pertaining to the mobile subscriber, and a third location for service logic execution. A MAP (Mobile Application Part) interface is introduced between the SCP and the HLR to ensure service data does not conflict and undesired feature interactions are avoided. Also, the trigger address lists for the CAMEL services of a given subscriber are stored in the HLR.

1.5.3 Evolution of 2nd Generation Wireless Systems

Wireless networks as introduced above are referred to as second generation wireless networks, as they embody the progression from analog technology (the first generation) to digital communication. The second generation GSM network is a circuit switched communication system, seeing that a fixed route through the network is established between the parties, for the entire duration of a call. With the advent of packet switched technologies, and the type of always-on, IP-based services that are facilitated by these technologies, GSM networks evolved by adding a packet domain to the circuit switched core network. The packet domain is used to transport packet data efficiently across the GSM network, from a mobile device to external packet networks. This new GSM bearer service is called General Packet Radio Service, or GPRS.

The first order of business in realizing the packet domain is the introduction of packet switches required to route packet streams. These packet switches are called Serving GPRS Support Nodes, or SGSNs, and their main function is to route the packets to the mobile device and vice versa. As with MSCs, a notion of serving SGSN applies and a VLR record is associated with the serving SGSN. The HLR continues to be the centralized place where subscriber data and service profiles are stored.

In order for the SGSN to transport the data packets to external packet data networks, a Gateway GPRS Support Node (GGSN) is introduced. One of the functions of a GGSN is to perform the translation of GPRS data packets into the data protocol in use within the external packet network. Similarly, an address scheme conversion is required in order to deliver packets originated in an external packet network to a mobile device in the GSM network.

Within the GSM network, a GPRS backbone network is in place between the SGSNs and the GGSNs to carry the data packets. As there may be several external packet networks, e.g. IP or X.25, packet gateways (GGSNs) are required for each such external network. However on the GPRS backbone all packets look alike, as external packets are encapsulated and tunneled across the backbone[11]. A specific session that may exist within a tunnel on the GPRS backbone, established between a GPRS-capable mobile device and a specific address in a given external packet network, is called a Packet Data Protocol Context, or PDP Context. With a PDP Context, a GPRS-capable mobile device in the GSM network is now addressable by entities in the external packet network, and payload packets can be exchanged to and fro.

CAMEL capabilities are in place to allow for IN-based service control of PDP Contexts. To support such service control, a CAP interface exists between the SGSN (or the gprsSSF to be exact) and the SCP.

GPRS itself evolves to EDGE (Enhanced Data-rates for GPRS Evolution), which is sometimes informally called 2.75G. This evolved form of GPRS technology results in increased data rates without any changes to the underlying core network. EDGE is not further discussed in this book.

---

[10] The reader should note that 3GPP Release 4 is also tending towards including support for TCAP/IP type scenarios as the network continues to evolve. Such work has been in progress for a while in other standards bodies like the IETF for a few years now, where the underlying transport mechanisms for carrying SS7 protocols were being developed. The interested reader is referred to [Sigtran] for more details.

[11] The signaling protocol used on the GPRS backbone is called the GPRS Tunneling Protocol, or GTP. For the remainder of the material addressed in this book, GTP is not important.

1.5.4 Third Generation Wireless Systems

Two developments characterize the dawning of the third generation in wireless networks. The first improvement is the launch of a new radio technology introducing higher data rates, advances in DSP technology and more efficient use of radio spectrum. The second advancement is the establishment of an all-IP core network.

The radio technologies in use in second generation wireless networks are based on frequency division multiplexing, where each connection uses its own dedicated radio frequency, or time division multiplexing, where each connection uses a dedicated frequency only part of the time, in fixed time slots. There are generally two drawbacks with FDMA (Frequency Division Multiple Access) and TDMA (Time Division Multiple Access), and those are that adjacent or nearby frequencies interfere with each other and the fact that each frequency can only be used for one connection (in any given time slot).

Determined to condemn such drawbacks to history, spread spectrum technology emerged that allows multiple mobile devices to use the same time slots and frequencies at the same time. Interference is avoided by cutting up the speech payload of all active mobile devices into tiny fragments and transmitting all of them simultaneously over the radio link. A unique code is assigned to the speech segments of each individual connection. So even though all communication data are shared over the airwaves, any given mobile device will be able to distinguish and identify the speech payload destined for it, by means of the unique code for its speech connection. This radio technology is termed CDMA (Code Division Multiple Access).

The second advancement in third generation wireless networks is the establishment of an all-IP core network. With the evolution of 2G (second generation) networks we have seen that GPRS adds the possibility for a mobile device to connect to external packet networks through the GSM network, and obtain services residing and executed in those external networks. To facilitate the exploitation of increased efficiency and enriched service capabilities made possible by IP technology, wireless networks need to advance beyond the capability of offering access to external packet networks. This trend is visible as an evolution of the nucleus of wireless networks into an IP core. In 3GPP this core is called the IP Multimedia Subsystem (IMS). The principal objective of IMS is the realization of an integrated voice and data network infrastructure, capable of delivering multimedia capabilities, be it real-time or otherwise. IMS is gaining wider industry acceptance and is likely to see widespread deployment by the time this book is published. Later sections in this chapter introduce IMS in a bit more detail.

1.5.5 CDMA Network Evolution

Broadly speaking, CDMA networks evolve along similar lines, though the details are somewhat different (and a discussion of these finer points merits a book in itself). A high-level summary view is presented here. The 2G CDMA networks evolve forward to support CDMA 1X-RTT (Radio Transmission Technology) – a technology that provides for more efficient over the air interfaces and higher bandwidths. An overlay network, called CDMA 1X-EVDO (Evolution for Data Optimized, sometimes also called Data Only), may also be deployed to support packet traffic as the evolution continues forward. EVDO was not designed to support voice[12], and so CDMA 1X-RTT evolves forward into an integrated packet infrastructure with CDMA 1X-EVDV (Evolution for Data and Voice). CDMA evolution, with support for CDMA 1X-EVDO is depicted in Figure 1.7. The reader interested in learning more is referred to [Viterbi 1995].

CDMA 1X-EVDO supports nodes such as the PCF (Packet Control Function), and the PDSN (Packet Data Serving Node), and these roughly translate, at the highest layer of abstraction, to elements similar to the SGSN and GGSN from GPRS networks. While the GPRS networks support

---

[12] Strictly speaking, EVDO evolves towards EVDO Rev A also sometimes referred to as DOrA (read 'Dora') that can in fact support VoIP. EVDV, the next phase of the evolution, provides for higher bandwidth and increased data rates over and above EVDO.

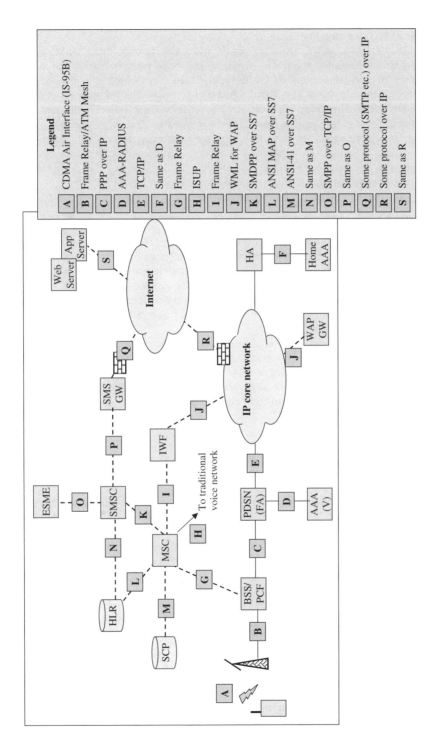

Figure 1.7 An evolved CDMA network with 1X-EVDO

the establishment of PDP contexts for handling end-user sessions, CDMA makes use of PPP (recall this was used in 'dial-up' scenarios). Also, CDMA 1X-EVDO utilizes Mobile IP (designed by the IETF Mobile IP WG) for mobility management.

The astute reader can conclude from the above sections that conceptually GSM and CDMA networks operate on the same principles – signaling and radio protocols are different, but at a high level, they are very similar indeed. So the reader may draw a high level generic model of mobile networks in her mind. Both networks can be viewed as more detailed instances of the generic wireless network architecture shown in Figure 1.6. The reader will see, however, that understanding of some of the differences will help in later chapters as we study Parlay/OSA service capabilities and mappings of service capability APIs to underlying networking technology details. For now though, we continue to focus on the similarities by recognizing that whilst the radio access network technology between 3GPP and 3GPP2 networks differ, the IP core networks of both are harmonized. Both organizations partner in the development and standardization of the IMS.

1.6   The IP Multimedia Subsystem (IMS)

So far, we have looked at various network ecosystems in place today, and have studied the evolution of cellular networks, from the current 2G incarnations to the future 3G evolved forms. At times, a reference was made to an all-IP manifestation of these 3G networks, called the IMS or IP Multimedia Subsystem in 3GPP. Mention was also made that a similar evolved architecture is supported by 3GPP2 as it describes CDMA evolution into an all-IP environment, and that in the latter case, it also goes by the same appellation in addition to sometimes being called the Multi-Media Domain or MMD. We shall use IMS to refer to both.

The IMS architecture is poised to enable the dream of anywhere, anytime communication. What this means is that IMS will enable every networked device, and the people using them, to communicate with any other device, over any network – be it wireline or wireless - with any service, in any media. IMS creates a common core network that can span both wireless and wireline networks, thereby providing seamless service control and delivery across these two types of networks.

1.6.1 A Standards View

In these sections, we study IMS in some detail. 3GPP defines most of the architecture, requirements, and call flows for the IMS in documents such as [3GPP 2002a, 3GPP 2004a, 3GPP 2004b, 3GPP 2005a, 3GPP 2005b] among others[13], and 3GPP2 also utilizes these documents as a basis for its own standards (this enables quicker convergence and reuse) which include the following documents [3GPP2 2003a, 3GPP2 2003b, 3GPP2 2003c, 3GPP2 2003d, 3GPP2 2003e, 3GPP2 2003f, 3GPP2 2003g, 3GPP2 2003h, 3GPP2 2003i, 3GPP2 2003j]. These documents, just like the 3GPP documents previously indicated, contain overviews of the IMS architecture, descriptions of reference points, reference point operational descriptions, and finally, protocol mappings and functional call flows in support of particular required capabilities. Last but not least, the OMA or Open Mobile Alliance™ [OMA] also talks about how the IMS architecture can be supported, albeit more from a services perspective, in a manner that promotes seamless access and use of IMS capabilities in both 3GPP and 3GPP2 contexts. The last of these (i.e. the OMA documents on IMS) are covered by an OMA Enabler Release, called 'IMSinOMA' [IMSinOMA 2005]. [Brenner 2005] provides an introduction into OMA and some of its activities.

In what follows, we shall explore the IMS architecture. This is admittedly a simplified view of IMS – for a more comprehensive treatment, the reader is referred to the standards documents

---

[13] The interested reader who reads through one or more of these standards documents will soon see how the documents reference one another, and how one quickly gets drawn in, with greater understanding, into more and more other standards documents that explain more of the esoteric details. For help in locating standards documents, the reader is referred to Appendix B, which is included as advanced reading in [Parlay@Wiley].

listed above, which total several hundred pages together. But the lightweight treatment of IMS concepts here shall suffice for most readers to provide a clear view of this all-IP architecture, and its relevance and relationship to Parlay and OSA technologies.

1.6.2 Simplified View of the IMS Architecture

Figure 1.8 depicts a simplified view of the IMS architecture – a view that covers the most important service-related aspects and is sufficient for our purposes. As was alluded to earlier in this chapter, the HLR or Home Location Registry is the centralized repository for service subscription data and service profiles. With the evolution of cellular networks into their all-IP 3G form, the HLR element evolves forward into the HSS or Home Subscriber Server. This HSS element stores information pertaining to subscribers and their subscribed services, among other things, in the 3G environment, and is accessible to call control elements (called CSCFs or Call Session Control Functions), application servers (analogous to SCPs from the traditional IN model), and other authorized network entities that require this information in processing end-user requests.

The IMS supports SIP as the protocol of choice for all signaling, for call control and for most service control interactions between the CSCFs and application servers. Generic SIP (as defined in [RFC 3261]) is used as is, for the most part. The one exception to this is the reference point between the CSCF and the AS, called the ISC (IMS Service Control) reference point. Along this interface, the ISC protocol (SIP with some special private header extensions) is used. Interfaces to the HSS component are normally implemented using the DIAMETER protocol. The main reference points of interest along with the associated protocols are indicated in Figure 1.8.

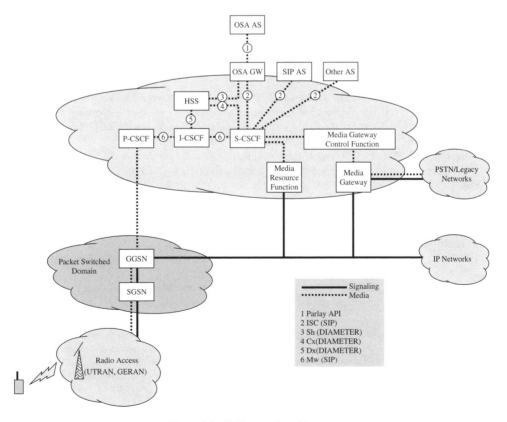

Figure 1.8   IMS network architecture

1.6.2.1 Application Servers in the IMS Architecture

The IMS architecture defines a service layer that supports different kinds of application servers, most prominent among them being the SIP AS and the OSA Gateway.[14]

- The SIP AS supports SIP-based applications and receives SIP (or ISC) messages from the network and responds with messages in the same protocol to enable further processing of user requests at the CSCF or to otherwise be able to provide an enhanced end-user experience[15]. SIP ASs may use any SIP-based technology (e.g. SIP CGI [RFC 3050], SIP CPL [RFC 3880], SIP Servlets [JSR 116], etc.) as they support the value-added application logic.
- The OSA Gateway is the Service Mediation Gateway or SMG that is referred to in later chapters in the book. This is a gateway component that implements the standards defined by the Parlay and OSA specifications. Since 3GPP defines IMS and 3GPP defines OSA, it is logical that references are made to OSA (and not Parlay) in this context. However, as we will see in Chapter 4, the two technologies are similar to the point of being virtually indistinguishable.

The OSA gateway serves as a gateway element (as the name suggests), enabling different OSA-compliant applications that are themselves hosted on application servers (called OSA ASs), access to network capabilities via the OSA-defined SCF APIs. Since most of the book is dedicated to the topic of the OSA Gateway (or Service Mediation Gateway, SMG, as we will call it), we do not discuss that in any more detail here.

1.6.2.2 The Different Types of CSCFs

The IMS architecture classifies CSCFs into three types based on their location and the logical function they perform in call flows. These are as below:

- Proxy-CSCF or P-CSCF: The Proxy CSCF is the contact point into the IMS for an end-user's terminal. The P-CSCF may reside in a visited network, in case the user is roaming. In case the IMS network is realized as an overlay on top of a GPRS network, the P-CSCF is the first point of contact after the GGSN that routes to the user's home IMS network.
- Interrogating-CSCF or I-CSCF: The Interrogating CSCF is the contact point into the user's home IMS network from other networks. Its job is to locate the right S-CSCF for the user after querying the HSS, and then to forward the SIP Registration request to the S-CSCF. Once registration is completed, and the S-CSCF is known, the I-CSCF is no longer involved, and SIP Invite messages are forwarded directly from the P-CSCF to the S-CSCF for outgoing calls and vice versa for incoming calls. There is one notable exception. If for some reason the network operator wishes to keep their network configuration hidden[16], the I-CSCF remains in the path between the P-CSCF and the S-CSCF. In this case, the I-CSCF performs the function of a Topology Hiding Inter-network Gateway (THIG).
- Serving CSCF or S-CSCF: The Serving CSCF serves as the SIP session control point for the end-user's terminal device and, like the I-CSCF, always resides in the user's home IMS network.

---

[14] The OSA Gateway will be explained in later chapters in its full breadth and depth. For the moment, we shall just focus on its position in the overall IMS architecture, and its relation to the HSS and S-CSCF.

[15] Unlike the SCPs in traditional IN domains that are limited to providing (rather critical) services to call processing, ASs in IMS, which can support capabilities in areas other than just call control (think Presence for example), may be able to provide enhanced end-user experiences even outside immediate call control contexts. Hence the 'or' in this statement.

[16] An example for one such reason could be to hide capacity information like the exact number of S-CSCFs from other networks for competitive motivations.

The S-CSCF maintains state information required for the support of services, however, the S-CSCF does not contain service logic itself. For service logic execution, the S-CSCF refers to application servers (ASs) using the SIP-based IMS Service Control (ISC) interface. All calls and sessions go through the S-CSCF and the S-CSCF controls all services, irrespective whether the end-user is roaming or not, thus ensuring continuous and consistent end-user experience.

1.6.3 Service Control in IMS

Service control in IMS takes place entirely on SIP ASs, as we have seen that the various CSCFs do not contain any service logic themselves. Determining the sequence and invocation of applications running on these SIP ASs for a given call may be done in two places: the S-CSCF (service filtering) and SCIM (service brokering). The procedures for service filtering have been standardized in detail [3GPP 2005a], whereas the mechanisms for service brokering are largely under-standardized. Both mechanisms will be explained below.

1.6.3.1 Filter Criteria

For execution of service logic, ASs are involved by the S-CSCF through the ISC interface. So, like in Intelligent Networking, we see a separation of call or session processing logic and service control logic. Unlike IN however, delegation of service logic execution to ASs is not based on a call model or state machine with detection points. Rather, the S-CSCF may decide to forward a certain SIP message to a particular AS based on a variety of criteria. These criteria include:

- the type of SIP message received (e.g. an INVITE or a REGISTER message);
- whether or not some specific header element is present in the SIP message;
- the content of certain header elements;
- whether the SIP message pertains to an incoming or outgoing call.

These criteria are referred to as filter criteria. Based on the filter criteria, the S-CSCF decides to forward certain SIP messages to a specific AS for service logic execution, whereas other SIP messages are processed by the S-CSCF itself for call or session processing. We can distinguish two types of filter criteria.

1. Initial Filter Criteria (iFC) are part of the subscription and service profile of the end-user and are stored in the HSS. The iFC are downloaded from the HSS into the S-CSCF over the Cx interface, upon registration of the end-user device in the network.
2. Subsequent Filter Criteria (sFC) are determined by the AS, once it has been involved in service control by the S-CSCF as a result of the iFC. SFC are determined dynamically based on service logic execution and signaled back to the S-CSCF over the ISC interface.

The iFC are specific for a given Application Server. Hence if the end-user is subscribed to more than one service, multiple iFCs can be part of the subscriber profile. As part of the iFC a priority is defined which allows the S-CSCF to determine the order in which to contact the various Application Servers. Default behavior is also part of the iFC definition in case the AS in question cannot be contacted.

1.6.3.2 The Service Capability Interaction Manager

The Service Capability Interaction Manager (SCIM) is defined as part of the Application Servers that provide service control in IMS networks. As any aspect within an Application Server is left unspecified in 3GPP, providing it handles any SIP exchange appropriately according to ISC definitions, the SCIM is left unspecified as well. The role of the SCIM is that of service broker in more complex service interaction scenarios than can be supported through the service filtering mechanism; for example, feature interaction management that provides intricate intelligence, such as the ability to blend Presence and other network information, or more complex and dynamic application sequencing scenarios. This added complexity may provide another reason why SCIM is not fully standardized.

Whereas the service filtering mechanism can be used to manage application interaction and straightforward sequencing, the SCIM may provide a more enhanced end-user experience by blending applications with each other and with context-sensitive information like Presence and Location, and Policy functions. In addition, the SCIM may incorporate multi-session awareness, with a session context that can comprise multiple sub-sessions for example, for voice, video and data streams.

The SCIM is mentioned here, though underspecified in standards and hence mostly proprietary, because the Parlay Gateway, in its capacity as Service Mediation Gateway, can be deployed to fulfill the role and function of the SCIM in IMS networks.

1.7   Related Technologies

Now that we have familiarized ourselves with various networks and their service architectures, a number of related technologies are introduced here as they provide service capabilities in the network that Parlay can provide access to. Later chapters in the book will elaborate on the programmatic interfaces defined by Parlay to make use of these capabilities when building end-user applications.

1.7.1 WAP Technology

WAP or the Wireless Application Protocol, defined initially by the WAP Forum™[17], really came to the fore around 1997, and was the precursor to some of the more exciting 'data to mobile handset' applications of which we will see more as network evolution to 3G continues. A very high-level, simplified view of WAP operation is presented here. As was previously stated in the section that discussed Internet access technology, for WAP as well, multiple alternative access paths to WAP-based services are afforded by the networks of today. In particular, GPRS and EDGE networks (and their CDMA equivalents) provide for the notion of data-session supporting nodes such as SGSN/GGSNs and PCF/PDSNs for WAP sessions. To keep explanations simple, and since we are using the PSTN network and its evolution to drive our discussions, we shall focus on the circuit-switched based data access path for WAP in this section. The interested reader is referred to [WAP] for more details.

WAP enables the user to browse the Internet from her mobile terminal. WAP, in concept, is independent of the radio access or core network technology in use, and can be deployed just as effectively in CDMA and GSM networks. There are several million users of WAP today. Since the screen-size of wireless handsets is usually small, and other limitations exist (such as the thin pipe to the handset over the air interface, etc.), an element called the WAP gateway is introduced into service provider networks where WAP is deployed, to perform conversions of accessed web page contents for suitable rendering on handsets, and for transport over the air.

The WAP standards define a complete protocol stack for use between the handset and the WAP gateway including layers for session control (WSP), security (WTLS), etc., as well as content encoding related aspects. The latter includes a WAP binary format for over the air transmission of accessed

---

[17] This body has since been subsumed under the OMA [OMA]. As Andrew Tanenbaum once remarked, 'The one good thing about standards is that there are so many to choose from.' Lately, market forces have caused a kind of consolidation of some of these distinct bodies, thereby contributing greater stability, and enabling vendors to make more judicious choices of which protocols to implement in their products.

data, and an encoding format called WML or the Wireless Markup Language – derived from HTML, which most web pages are written in today – for easy rendering to handset screens and so on.

Figure 1.9 provides a view of the network infrastructure needed to support WAP (recall that our focus here is primarily on network technologies). The digital switch or MSC is provided with a connection to an Interworking Function (IWF). All WAP data calls (or dialed calls where the destination is a WAP service) transit this link. The IWF connects on its other interface to a wireless service provider hosted IP network (LAN or WAN), to which a WAP gateway is connected. Some kind of simple handshake takes place between the IWF (representing the user device) and the WAP gateway as this connection is set up, and user credentials such as the subscriber phone number and other information are exchanged across this interface at that time.

The WAP gateway maintains the association between the user identity, and the IP address assigned to this connection, and then works to forward on user requests for web content to web or WAP servers (also called Origin Servers) either within, or outside, the service provider network. The WAP gateway then performs the required conversions on the data returned, and forwards it on along the same path, but in the reverse direction, back to the handset.

The reader should note that here data are being carried over the circuit call established between the handset and the digital switch. When the session ends, the user simply disconnects the call.

1.7.2 Location Based Services

Of late, there has been an upsurge in location technology and its use particularly in mobile networks, but sadly, the uptake here in terms of real-world networks has been somewhat sluggish. Location has been used in wired networks for many years now. The E-911 system in the US, has for example, relied on reverse directory lookups in databases to advise emergency operators and dispatchers of the location and routing information from the nearest police/fire station or hospital, so as to better assist people in distress in more timely a fashion. But use of location technology in networks with wireless handsets is somehow more appealing, primarily due to the mobility of the terminals in question.

In their simplest form, location-based services may be classified into two types. One is where the user himself provides his location while requesting location-specific information from a server. An example of this is where Bob enters his zip code into an HTML form to obtain local weather information, and possibly a Doppler radar image of his vicinity. A second, more enhanced service experience could result if Bob simply asked for location specific information, and his location were transparently obtained by the server in question (factoring in his preferences and privacy

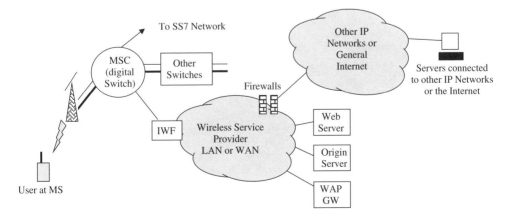

Figure 1.9 The WAP access model

permissions of course), and he were provided with context sensitive information without having to provide his location explicitly to the service.

The latter could be achieved in several different ways in cellular networks today. Recently, phones are coming equipped with GPS receivers, thereby enabling them to provide a fairly accurate location fix that can then be passed on (again, with end-user permission) to network services that require it. Alternatively, network elements such as MPCs (Mobile Positioning Centers) in CDMA networks, and GMLCs (Gateway Mobile Location Centers) in GSM networks, which talk to other Position Determining Equipment (PDE) in the network, are able to obtain location information (Figure 1.10). Such location information could consist of the cell-ID and cell-sector the handset is currently in, or even the latitude and longitude (sometimes even altitude) co-ordinates (sometimes called lat/long or X/Y/Z) determined using various algorithms and triangulation mechanisms such as AFLT, EFLT or Network Assisted GPS (this uses network information in concert with GPS information to locate more accurately a handset).

These MPC and GMLC servers can be made accessible to applications either directly, over protocols such as MLP (Mobile Location Protocol, an XML-based protocol defined by the Location Interoperability Forum or LIF, now subsumed by the OMA standards body), or indirectly, through OSA/Parlay capable service mediation gateway elements via the User Location interfaces supported by such gateways. Regardless of what mechanisms are used, once this information is obtained by the location-based services, specific context sensitive content can be served to users more transparently. Furthermore, this technology may be used very effectively to respond to emergency calls made from cell-phones where the caller either does not know, or is otherwise unable to specify, his or her location.

1.7.3 Short Message Service and Multi-media Messaging

Messaging capabilities are an intrinsic part of the wireless networks of today. Its most well known exponent is the Short Message Service (SMS). Given the popularity of SMS and the resulting high volumes and thus revenues this service generates, it is interesting to consider that the success of the service was really a fluke. In early GSM deployments, part of the available network capacity remained unused. Taking advantage of the characteristics of digital technology available in second generation networks, SMS was introduced as a low-bandwidth, packet-based message exchange mechanism, mostly bundled by equipment vendors at a discount with GSM voice service as part of a package deal. Adding SMS messaging services to the more sparsely used frequency bands in the network allowed network operators to make more use of their available bandwidth, and potentially increase average revenue per user. So, born as a capacity optimization feature, a killer application has emerged blinking into the daylight.

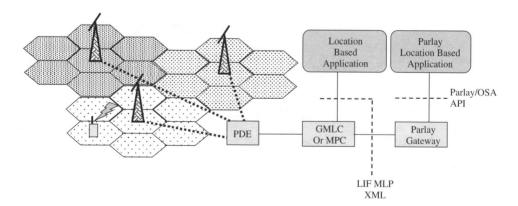

Figure 1.10   Logical architecture schematic of location-based services

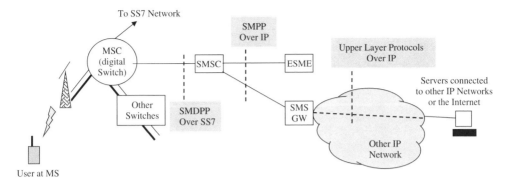

Figure 1.11   SMS network view

The Short Message Service is a store-and-forward message delivery technology, where a Short Message Service Center is introduced in the GSM network as the message store (Figure 1.11). SMS uses the wireless network for message transport and delivery. Because of the store-and-forward nature, SMS basically consists of two point-to-point services, from the originator to the SMSC (Mobile-originated short message, or MO-SM), and from the SMSC to the destination (Mobile-terminated short message, or MT-SM).

The SMSC uses the HLR to locate the destination party for an SMS message. The HLR is also used for supplementary services applicable to the SMS bearer, such as for example the barring of incoming SMS messages. Short messages are short, as they are transmitted out-of-band, over a low bandwidth medium. The messages are limited to 160 alphanumeric characters, although messages may be concatenated. The signaling protocol between the SMSC and the HLR (e.g. to obtain the location of the destination for the short message) is the SS7-based MAP protocol.

Given its enormous popularity, the basic SMS service has been enhanced in many ways. Simple examples include the ability to concatenate the short messages, and the addition of point-to-broadcast to the basic point-to-point capabilities. Enhanced Messaging Service (EMS) adds the capability to send formatted text messages (including bold and italic fonts), simple pictures and animations, and ring tones and logos.

The latest step in this process of enhancing the SMS capabilities and building on the success of the service is the Multimedia Messaging Service (MMS). MMS messages may be used to stream audio or video to the mobile device, or to exchange photos and download games.

In order to support MMS in the network, the basic SMSC does no longer suffice. An MMS Relay/Server is introduced to support MMS capabilities. The basic functionality of the MMS Relay/Server is still the storing and forwarding of messages, MMS messages in this case, but given the much richer content involved, interfaces are introduced to value added service applications, content stores, and external networks.

1.8   Summary

In this chapter, we have covered, with a broad brush, many of the networking technologies in use today. The intent here is to provide the reader with a background and a little more appreciation of the complexity involved in network architectures, and also to introduce, albeit at a high level, the kinds of interfaces in existence, and the reference points where programmatic interfaces could be introduced (this latter point will become more apparent in later chapters). This chapter serves as the basis for the discussions in the rest of the book – we scatter some magic idea seeds here, and these grow into a forest of beanstalks in the pages to come. Next, we look at some marketing, business, and technology drivers for change.

2

The Need for New Technologies

2.1 Introduction

In the last chapter, we studied some of the different kinds of communications networks and the associated technologies and protocols in common use today, albeit at a high level of abstraction. Here, utilizing this knowledge along with an understanding of the scenarios presented at the start of the book, we try to distill a reasonably comprehensive set of requirements that new solutions should satisfy in order to meet the expectations of various parties that are as yet unfulfilled by the technologies deployed commercially today.

Note that although the previous chapter covered the various types of networks from a standards or a 'reference architecture' perspective, each deployed network is subtly different from every other, even where the same technology is used. Each has its own nature if you will, similar in a sense to how people have different natures and react differently in different situations. The make-up, or psyche, of the network is determined in large part by the protocols that are used, which vendors' equipment is deployed in support of which functions, how the various elements are deployed, what paths are open for inter-element communications, etc. In other words, even where two networks appear remarkably similar, there may be subtle differences between them. The reader is encouraged to keep this in mind while reading through the various chapters. The principles covered in this book are however expected to be applicable in the vast majority of situations.

2.2 Issues with Networks Today or The Drive to Improve

Humankind has always focused on self-improvement. Communications is no exception. In this space, as in all others, whatever is newer, faster, better, more appealing and meets compelling user needs while being cost-effective for mass deployment still catches on, and gains wide acceptance.

Let us delve a little bit deeper to study some issues with the networks of today, examining issues from three perspectives – the network operator, the application provider and the end-user. Please refer to Figure 2.1 while reading the sections that follow, for additional context.

2.2.1 Network Operators

Network operators are those corporate entities that provide end-users with network connectivity. These are companies that own, operate and manage communications infrastructure (and are licensed to perform these functions, an idea that is significant from a regulatory perspective), and are commonly referred to as 'phone companies'. Some network operators also double as service providers in that they not only provide the 'dumb' infrastructure, but also the set of services that execute atop that base.

Parlay/OSA: From Standards to Reality   Musa Unmehopa, Kumar Vemuri, Andy Bennett
Copyright © 2006 Lucent Technologies Inc. All Rights Reserved

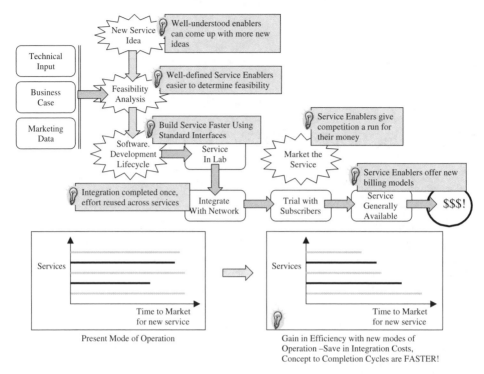

Figure 2.1   Telecom services – potential impact of technological improvements

The traditional telephony features such as call forwarding or call waiting are examples of such services, but other categories of services such as Click To Dial capabilities, auto-Attendants, etc. also exist, since operators are not restricted merely to supporting traditional telecommunications infrastructure any more. For the purposes of this book, unless otherwise specified, we treat service providers and network operators as a single kind of entity, and shall use the terms network operator, and service provider, interchangeably.

One instance where they may NOT be the same is in the case of the Virtual Network Operator (VNO). Your network operator may or may not be the same as the service provider whose name appears on your telephone bill each month. This is because VNOs (called MVNOs or Mobile Virtual Network Operators in Wireless contexts) resell access to another network operator's infrastructure to end-users. This is supported by a business agreement reached between the VNO and the ultimate network operator whose infrastructure is used in supporting your telecommunications services. We leave VNOs and MVNOs out of the picture until later chapters, where some advanced concepts relating to these entities are presented.

We refer the reader back to Scenario 1. Service providers would like to stabilize and grow their subscriber bases, thus bringing in more revenue. Doing so requires that they increase end-users' reliance on the network, by storing profile and preference data within the network context and by providing compelling user experiences the end-user cannot do without. In addition, they would like to offer newer, more exciting services more quickly, to attract subscribers from competitors' networks, and make these services 'sticky'. In other words, make subscribers reluctant to move to another service provider. Of late, with the advent of service provider portability, being mandated by governments in certain countries, it has become very easy for subscribers to switch from one operator to another (while maintaining their current phone number), and service providers need to do what they can to maintain and grow their subscriber bases. Services are thus a life-blood of any network and a critical area for differentiation.

True, one has to spend money to upgrade the services deployed, or deployable (for new technologies may enable you to deploy services not previously possible) in one's network. However, if the right technology, business model[1], and related choices are made, one can reap rich rewards. New services can increase ARPU as well as AMPU[2], if the services offer seamless integration of new capabilities with existing supported features subscribers have come to know and love.

Another aspect of importance here is the fact that in certain markets the opportunity to grow the subscriber base is limited, even physically constrained, due to the already large uptake in terms of members of the population being subscribed (i.e. penetration). Penetration is close to 100% or even larger than 100% in Western Europe. For instance countries like Italy have a penetration of larger than 100%, indicating that some people own more than one subscription.

When deploying new applications, one common hurdle service providers face is the large number of communicating legacy systems already deployed. Inevitably each new service or application needs to interface with a number of them. These could include elements like billing systems, provisioning systems, operations support systems, network management systems, etc. Each integration point costs money, time and other resources (engineers, who could be assigned other tasks, have to engage in hand-holding the application developers). Typically this has to be repeated with each new application that is added since there is very limited reuse of the functionality. This raises costs, and makes the deployment process slower and more expensive than it could otherwise be. This phenomenon of limited reuse of common functionality across various services (e.g. a billing solution for location services and a billing solution for presence services) is often termed 'vertical integration'.

Subscribers that sign up for individual services expect the service provider to provide customer care support to help them with provisioning, configuration and other related aspects of their accounts when they get started (and periodically thereafter if and when they have issues with the service). Unless new applications are closely integrated with the rest of the customer care infrastructure, support for each new service or application adds to the service provider's operating expenses.

Typically, the design of telecommunications services and applications requires a detailed working knowledge of all the various protocols and interfaces connecting to the element hosting the service logic, and such knowledge is relatively arcane, making trained engineers available to build such carrier grade applications hard to find. This contributes further to the costs involved as new telecom applications are built.

Sidebar
Telecommunications services and hardware are typically referenced with adjectives such as 'carrier-grade'. This means that the service or hardware element has characteristics such as high availability and reliability (robust, deterministic, and fails very infrequently). It is also used to indicate that the failure characteristics and recovery or repair times and strategies are well known and that procedures for these are well defined. Traditionally these characteristics are a prerequisite of being a telecommunications carrier.

---

[1] Business models are important too. Some revenue- (and risk-) sharing models may involve the service provider agreeing to deploy a service but giving the application developer a cut of each transaction. Others may involve the application developer (or enterprise) hosting the new application and paying the service provider a flat fee per month for access to the latter's subscribers. Yet other transaction based models may be used with online or offline revenue reconciliation mechanisms put in place.

[2] ARPU and AMPU stand for Average Revenue Per User and Average Minutes (of usage) Per User respectively. Any service that raises ARPU across the user base and brings in more revenue than the amount expended on providing the service, contributes directly towards profits for the service provider. Similarly, any service that generates more minutes of use or drives user behavior that results in more usage of existing services is a significant moneymaker for the service provider.

Contrast the shortage of trained engineers with the Internet domain where there are thousands of qualified programmers who can fairly competently build applications to requirements through the use of widely liked and deployed technologies such as Java and C++, using Internet Toolkits for software development. If only there were a way for telecom service providers to tap into this resource pool, it would alleviate some of the cost related issues with regard to new service development.

To summarize, what service providers really want is a cheaper, more flexible environment, with faster 'concept to completion' cycles, that leverages the untapped pool of Internet resources and technologies to provide services more efficiently, and which provides effective dynamic feedback on new services and applications deployed. If there were also a means by which these services or applications could be hosted in other domains (with suitable agreements on sharing of responsibility for service outages), thus enabling them to reduce customer care costs, it would be even better.

2.2.2 Application Provider

Applications are blocks of logic that utilize the service provider supplied service-enablers to provide value-added capabilities to an end-user context in interactions with other users connected to the network, or with the network itself. In the IN context, applications such as Voice-VPN (Virtual Private Network with closed user groups and private numbering plans) are supported using a base set of Service Independent Building Blocks called SIBBs. Application developers or providers build these applications, either under contract with service providers, or independently for resale across service provider customers.

The reader is referred to Scenario 2. There is a small pool of available talent in the pure telecom applications space, and the stringency of requirements tied to telecommunications applications (very high availability and reliability are examples of these), coupled with the detailed low-level binary protocol encoding aspects of application building and one-off repeated integration of each new application with legacy systems, makes the development procedure extremely involved, less agile, time-consuming, and very expensive. Application providers also have to build potentially the same application over and over again, for each new service provider customer, due to network differences (though this may not be such a bad thing, since each integration job results in a revenue influx).

What applications providers would like is a development environment where they can seamlessly utilize Internet technologies and toolkits as they build telecommunications grade applications, and do so more quickly. If they could build an application just once, and then customize it to fit into various service provider network contexts, this would be even better (not to mention a single off-the-shelf product contributing to multiple revenue-generating transactions).

Let us illustrate the points from the preceding paragraphs by means of some examples from the history of telecommunications and networking.

1. The IN model was so successful and saw widespread deployment because now suddenly it gave service providers (who were previously constrained to developing services within switches and having to deploy them throughout the network in all switches that had to provide the capability) the option of building and deploying services in centralized nodes called SCPs. Telecom-specific development environments, protocols, and technologies were used to effect this, but it was still a radical improvement from the previous state of the art, so there was widespread acceptance of IN relatively quickly. Besides, if one operator deployed

IN and started reaping rewards, others would have to follow suit pretty quickly to remain competitive, and this drove market acceptance. Now, operators whose network properties span multiple countries simply develop the service once and deploy the same in different networks subject to cultural and regulatory differences.

2. The Internet, where the number of new nodes grows exponentially, with a burgeoning talent pool, saw the adoption of new technologies, toolkits and programming languages like Java, and now. NET, which enable almost anyone to develop programs, applications and services, and communicate with others. Of late, this has even been facilitated through mechanisms such as Peer-to-Peer sharing and communication, thereby further overlaying a sense of community across geographic boundaries, over a physically widely distributed network. The success of the Internet can be widely attributed to the low barriers of entry.

3. Services like i-mode[3] took off quickly in Japan. Introduced via proprietary mechanisms and protocols, these services were still successful because a movement snowballed where some 'cool' services saw a wider following, followed by a wider interest in a larger application development population to reach out to a larger interested audience, which led to yet more subscribers and more applications and so on. This positive growth spiral provided service providers in Japan that came up with this technology with stable and growing subscriber and revenue bases, provided subscribers with newer, 'cooler' services, and provided application developers with markets found ready for intelligent new concepts and more revenue for the services they built – a true win-win situation for everyone involved. For service providers, there is no necessity to come up with the elusive 'killer application'. Because of the ease of development and the large developer population, the resulting abundance of new applications will increase the likelihood that a few winners emerge, while less successful attempts will falter. It is no longer the burden of the service provider to find the killer application and it is no longer the risk of the service provider that some applications will not make the grade. Therefore, opening up their network, their most prized and hence most protected asset, offers them the ability to reap great rewards.

Parlay/OSA technologies are targeted at exploiting the strong points of all three examples above:

A. Use of standard interfaces, so application developer risk is reduced, and new services, once developed, can be sold to a multitude of service provider or enterprise customers.

B. Enabling the large talent pool from the Internet domain to build telecommunications applications or applications that leverage telecom capabilities.

C. Provide a rapid feedback environment in which newer, more exciting, services could be built, tested, trialed and deployed, much more quickly than through traditional paradigms available today.

Additionally, third-party application developers may have some really new and exciting ideas for converged services that span telecommunications networks and other environments. In today's prevailing conditions they are frustrated due to their inability to access or leverage any of the telecommunications network capabilities in a wider integrated communications services context.

---

[3] i-mode is the mobile Internet access system developed by NTT DoCoMo. For more information, the reader is referred to [Natsuno 2003].

2.2.3 End-users or Subscribers

An end-user or subscriber is a person who owns, leases, or operates end-user equipment, and has a contract with a Network Operator[4]. This end-user equipment could be a phone (wired or wireless), a computer, a PDA, any device that is capable of interacting with any infrastructure-based network. We exclude ad-hoc networks of roaming wireless endpoints from our discussion since these do not require service provider hosted equipment, typically (though not always) rely on low range technologies such as BlueTooth[5], and do not generally assume support for network hosted services accessible via standard protocols.

It is ultimately the end-users who, as a community, decide which of the services and applications the service provider hosts are successful, and which are not. Of course, lots of factors play into this dynamic, including the charging or billing model offered, the ease of use of the service in question, etc., and the technology in question may or may not be a factor in whether a particular service or application is to the users' liking.

There are also regulator-mandated services – for instance, there is a government requirement in many countries today that all wireless networks implement device or handset tracking by a certain date to provide support for Emergency Services such as 911. Another such requirement might be the support some governments' need for wiretaps (also called 'legal intercept') and other communication interception once subpoenas or other legal documentation is provided to authorize such activity in the interests of national security.

The reader is encouraged to re-read Scenarios 3 and 4. Services, as we have stressed before, are a service provider's main avenue for competitive differentiation. However, users would like to get as much as they can from their communications experiences while spending as little as possible. This imposes a cap on the amount that service providers can charge for particular features or packages of features (either per-use or subscription based). Given take-rates and related assumptions this implies a certain limit is already pre-imposed on what can be spent on a new service or application even before it is designed and deployed.

There are forces that require that costs be kept low, but also that new services be periodically deployed to give the end-user a better experience than competitors. What subscribers really want is a comfortable access to well-integrated services, good customer care facilities, and cheap service that meets their needs more easily.

2.3   Summary: Required Characteristics of a Desirable Solution Technology

The last paragraphs in each of the three preceding sub-sections outline the requirements that network operators, end-users and application providers would like to see resolved. Figure 2.1 indicates some of the areas in a services definition, development and deployment life-cycle where potential improvements can be made, and the impact these small changes may have in the overall gains (both in terms of revenue and stability or growth of the subscriber base) that service providers may be able to reap. These are briefly summarized below:

• Network Operators – looking for increased revenue from a stable or growing subscriber base. Attractive services are key, hence Network Operators require a flexible environment based on Internet technologies, yielding to faster 'concept to completion' cycles. Network Operators want more innovative and attractive services. They want the ability to brand these services to show "ownership" and differentiation, and they want to leverage capabilities and services already

---

[4] Technically, there is a difference between end-users and subscribers. A subscriber is the one who engages in a contract with a service provider, i.e. subscribes to the service. An end-user however can be a subscriber, or for instance an employee of a large enterprise, where the enterprise owns the subscriptions for all its employees. For the purpose if this book, the distinction between end-users and subscribers is not significant.

[5] BlueTooth is a low-range radio technology connecting mobile and handheld devices. Interested readers are referred to [BlueTooth]

deployed in their network by exposing them in a secure and well-regulated manner, to applications that can generate new revenue at reduced cost.

- Application Providers – Looking for ways to leverage network service capabilities to develop revenue-generating services, using Internet technologies and toolkits. Application Providers want more innovative and attractive services. They want toolkits that enable them to create new services and applications while allowing them to leverage the knowledge and skills they already have; they do not want proprietary toolkits with interfaces and capabilities forced by particular IN vendors.
- End-users – looking for low cost, low maintenance services to enhance their communications means and enrich their experience. End-users want more innovative and attractive services.

We take these as input requirements, which we then use to construct a suitable solution in the next chapter. Subsequent chapters will flesh out more details within the solution space as they show how Parlay/OSA technologies could be employed to meet these and other needs of the three groups of technology users described above.

3

Follow the Yellow Brick Road

3.1 Introduction

In the previous chapter, we identified some important needs of three different kinds of user communities that are as yet unmet by technologies in common use today. These three kinds of user communities were Network Operators, Application Providers, and End-users. In this chapter, the focus is on building a generic solution that may be implemented to fulfill some of these needs in the cheapest or simplest way possible, so that an elegant, cost-effective and working solution can be implemented. Attempts are made to address as many of the requirements as possible, while presenting the reader with cogent arguments as architecture choices are discussed, and either selected (added to the solution set) or discarded from consideration. The solution architecture is no doubt interesting, but the discussion is important in and of itself, for it illustrates some of the principles involved in systems architecture and design.

Sidebar: The Most Efficient Solution – of Square Pegs and Round Holes

In many engineering schools, a required course in the first year is one that is sometimes called 'Engineering Drawing'. Essentially, it is intended to teach visualization, three-dimensional geometry and the ability to think 'outside the box' – this experience can be very rewarding, or very frightening, for new students.

One problem students are often presented with in class is where the teacher challenges them to come up with the simplest, most elegant, and efficient solution to what is typically called a 'single plug' problem. 'There are two holes in the wall,' the teacher says, 'and you must fill them both with the simplest plug you can devise. One hole is circular in cross section, the other a square or rectangle. You can imagine them to be of whatever dimensions you like, but the key is to come up with the most efficient solution.'

Students think long and hard, and come up with various possibilities – a sphere or cylinder fused to a square peg etc., but the truly astute among them realize the correct answer is simply a cylinder. Look at a cylinder from two different perspectives. It has a circular cross section when viewed from one angle, and a square or rectangular cross section when viewed from another 90 degrees away. See Figure 3.1.

Problems have different possible solutions. If we look at problems from different angles or perspectives, we are able to come up with, and evaluate these, and then perhaps pick the most efficient one to use. Sometimes, we may stumble upon the most efficient solution serendipitously. In other cases, we may algorithmically or through an extensive search, come to the right

---

Parlay/OSA: From Standards to Reality   Musa Unmehopa, Kumar Vemuri, Andy Bennett

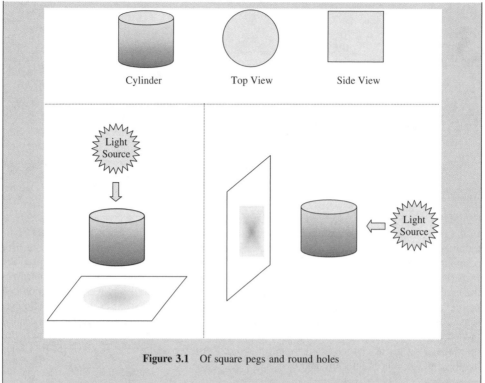

Figure 3.1 Of square pegs and round holes

conclusion. And in some cases, as in the above example, a square peg may in fact be the right plug for a round hole.

3.2 Of 'Smoke-Stacks', Value-Chains, and Service Layers

In Chapter 1, we studied various network ecosystems, but did so primarily from a technical perspective so as to give the reader an appreciation for what happens from a network viewpoint as she communicates with friends and family or surfs the web, using telephony and related infrastructure. In this section, we look at the associated value chain involved. We use this discussion as a spring-board into how network evolution and support for service enablers, instead of vertically integrated smoke-stacks of applications, will help drive a wider proliferation of new applications more quickly into the networks of today. As also previously indicated, the more new services, the more the subscriber reliance on network hosted capabilities, and hence the greater the ability for service providers to generate new revenue. But let us examine these statements in a little more detail.

In the previous chapter, we identified three stakeholders in the communications value chain: Network Operators (or Service Providers), Application Providers, and End-users. For us to be able to understand the entire value chain involved, in addition to those corporations that provide the network and associated services (these are companies such as Sprint, Verizon, Cingular, Vodafone, NTT DoCoMo, T-Mobile etc.) we need to also recognize the corporations that provide these Service Providers[1] with their switching infrastructure (companies like Alcatel, Ericsson, Lucent Technologies, Motorola, Nokia, Nortel, and the like). The latter are called Telecom Equipment Vendors. They architect, design and develop the various network components such as switches,

---

[1] Those providing cellular wireless service are called Wireless Service Providers.

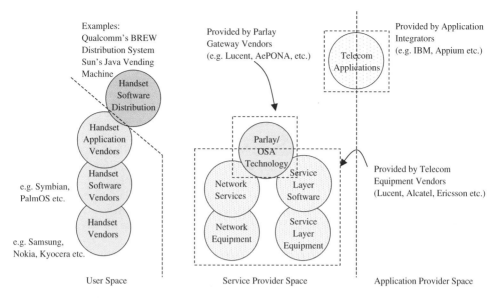

Note: All trademarks are the properties of their respective companies.

Figure 3.2   Smoke-Stacks and Value-Chains and how Parlay fits in

HLRs, SCPs, Parlay gateways and so on. Figure 3.2 shows the stakeholders in the value chain and demonstrates how Parlay fits in.

Service providers purchase and deploy equipment from telecom equipment providers into their networks. The larger ecosystem thus encourages creativity and innovation on the part of the equipment providers, enabling those with more competitive price and performance to achieve greater market penetration. Similarly, competition between the Service Providers themselves, for subscribers and subscriber dollars, requires that the service providers promote more attractive offerings, and lure more end-users to their networks either by providing more services, more minutes, better end-user experiences, etc. Increasing ARPU and AMPU helps service providers make more money. And since this is a service provider goal, telecom equipment vendors need to factor in such thinking as they develop their own products, solutions, services and offerings, in order for them to be successful[2].

To put it succinctly, revenue flows from subscribers to service providers to telecom equipment vendors, with associated value flowing in the opposite direction (value has to be paid for) with equipment vendors providing useful new capabilities to service providers who in turn leverage these to deliver better user experiences. The perceived value of new capabilities is therefore magnified several fold at each step. This is a value-chain.

Thus, the ability to more quickly, cheaply and effectively deploy newer and more attractive services into networks is of paramount importance. Many attempts have been made to make legacy networks more programmable, since this would lead to shorter 'concept to completion' cycles for new deployments. Most, if not all of these, were based on proprietary technologies in the past, and sought primarily to tweak existing processes with improvements, not cause paradigm shifts in the way things were done.

---

[2] Admittedly, there is some difference between customer success and customer satisfaction, but regardless, one can only succeed if one's customer succeeds.

3.3   The Programmable Network

With the advent of IN, the concept of a service layer gained prominence. However, new services and applications were still developed in a vertically integrated manner in the form of what is typically referred to as the 'smoke-stacks' model. This is explained in greater detail in the sections that follow. Here, there is a rat's nest of interfaces, and repeated costly integration needs to occur each time a new service has to be deployed. The idea of a service layer was a good one though, and is being leveraged more and more in network evolution. Enter the service enabler.

Parlay, the latest[3], established, new technology in the domain of enhanced programmability of network-hosted services, attempts to leverage the network capabilities in the form of a service layer, but does so through a set of standards-defined interfaces designed specifically to support the existing value chain, but also simultaneously open up the network to a whole new domain of programming expertise – the Internet. This is what makes this paradigm shift so interesting and different from previous attempts. The focus here is on positively impacting the programmability aspects of services and applications through new enablers that can be seamlessly leveraged by them, while not disturbing other aspects of network operation where investments already made are still being leveraged to generate revenue.

The Parlay service enablers are the new service layer – more resilient to change, more future-proof, and capable of supplementing the existing service layer already deployed. Gradual displacement of old services with new technologies with a seamless mechanism for service providers to tap into new revenue streams is what makes this so attractive. These aspects are covered in much greater depth from a technical standpoint in later chapters.

3.4   Services and Applications

Before proceeding much further, it is important that the reader understands the distinction between the terms 'service' and 'application' that we have, so far, been using synonymously.

A service is (in the context of this book) a core-network supported enabler that permits the easier development of applications. Examples of services supported by typical telecommunications networks include user terminal location (X, Y co-ordinates for the cell-phone with number 312-555-1212), user terminal status (is that phone connected to the mobile network, turned off, or engaged in an active conversation at this time?), and call control (set up a call between the phone with number 312-555-1212 and an application that reads out a weather forecast at 312-555-0800). Services are also sometimes referred to by the term 'service capability'.

An application makes use of services in providing subscribers with better communications experiences. For instance, Billy could connect via his cell-phone to a weather application, which, without his providing any additional information, could find his current terminal location via the user terminal location service supported by the service provider, and then perform a dip in a weather database to generate context-specific content that is then downloaded to his handset. Similarly, a presence application could let Alice determine Bob's availability by tracking his handset status and location, and vice versa, thereby engendering a feeling of community among the users connected to the network.

3.5   Developing a Satisfactory Solution Architecture

Before trying to design the solution, let us first review some important concepts from Chapter 1 that we will use here. We have previously discussed how signaling protocols may be broadly classified as call/session control protocols and service control protocols, with the latter being used between

---

[3] Terms such as 'latest' may be subject to some debate. As of this writing, there is a movement afoot in standards to deploy Parlay (or 'Parlay-like') services (and make them accessible) over Web Services or other XML technologies. This is sometimes referred to as Parlay-WSDL or Parlay-X. Later chapters discuss this in greater detail.

core network elements (such as switches) and the service control elements that host service logic or enablers.

Since most of the requirements under consideration, from what we shall refer to as the 'problem domain' as per Chapter 2, deal with services, it stands to reason that any solution we come up with must factor in the service control protocols and service control elements.

The requirements for a solution to our problem domain, in Chapter 2, were identified from the perspective of three user communities. We have seen that these communities may have different requirements, unique to their environment and situation, yet principally yielding the same ultimate desire: more innovative and attractive services. We therefore generalize the requirements of Chapter 2 into the following four categories:

a) Reducing Integration Costs, Faster Development Cycles
b) More Efficient Application Development, Reuse across Network Types
c) Lowered OPEX, Shared-hosting Models
d) More Effective Use of Deployed Legacy Systems, Evolution Independence

These four categories are investigated in more details in what follows, where we load up our shopping cart with architectural components when we shop for a solution to our 'problem domain'.

3.5.1 Reducing Integration Costs, Faster Development Cycles

In an attempt to reduce integration costs, new service logic can either be hosted on existing service elements, or on new service elements that can support existing protocols. These two options provide us with least cost alternatives (in terms of monetary amount required to implement it) in trying to meet the requirements in question. They are also minimally intrusive in terms of the kinds of changes that need to be made to networks already in existence.

Let us look at each of these in turn:

1. The former of these requires that new service logic be hosted on existing service elements or legacy platforms, the obvious benefit being that no new network element has to be integrated in the network. This could entail however some of the same integration related issues that application providers and service providers see today in terms of cost, time to market, and inter-operation with deployed billing, network management, and operations support systems (OSSs). Besides, more often than not, the integration has to be repeated each time, per newly introduced application, given the way these legacy platforms are structured, and this still does not permit the transparent reuse of Internet Toolkits and technologies in service and application design and deployment. So we ignore this option for now as not really viable for long term evolution.
2. The latter option, i.e. new service logic be hosted on new service elements that can support existing protocols in communicating with existing network element, is interesting. It may be possible to build a new platform that factors out the integration aspects into the basic infrastructure, thereby 'opening up' the service logic development and execution environment to third-party developers via Internet Toolkits. This would permit developers to build applications while not necessarily having in-depth knowledge of the underlying network protocols in use. Such an approach would also reduce some of the service provider and application developer concerns with costs and 'concept to completion' schedules for applications.

The second approach seems like it might work, if it meets the other requirements against which we evaluate our solutions. We therefore retain this option for further consideration, and in subsequent sections assess whether it indeed meets the other requirements.

We can steal a page from the IN model, and apply it to this design. Core service capabilities are provided in legacy equipment that is already deployed in today's networks (Figure 3.3). Telecom

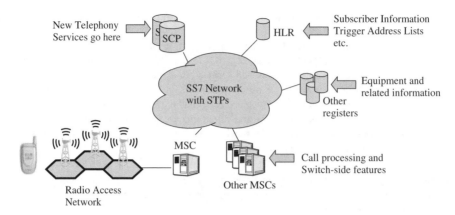

Figure 3.3 Sample wireless network architecture

applications provide value in being able to leverage these core services (such as those that provide a user's terminal location, the terminal status – active, inactive or busy, the user's presence and mood, etc., with the ability to charge a prepaid account for service rendered) and support an integrated user experience that appears seamlessly to factor these in. The applications provide the controlling logic for these end-user transactions in truly multimedia domains (voice, video, data etc.) just like SCPs provide service control for calls in the IN arena. The gateway option we are discussing may thus be thought of as an SCP-analog for evolving networks (Figure 3.4).

The element in option (2) above is a kind of 'gateway' in that it serves on the one hand to support a flexible services environment, and abstracts away low level protocol details from an application development perspective, while implementing conversion from the programmatic to the protocol interfaces, and a common integration with legacy systems on the other[4].

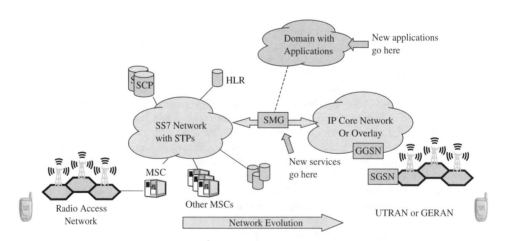

Figure 3.4 Evolved network architecture including a Service Mediation Gateway (SMG)

---

[4] We must emphasize that the problem of integration does not really go away. It is merely reduced. Rather than integrate each new application and its associated platform with the underlying network each time, the gateway is integrated once with the rest of the network, and applications are permitted to connect via the gateway to other network elements on an as needed basis.

The architecture component we have added to our shopping cart is:

- a Gateway that bridges between the application domain (hosting the new service logic) and the existing network elements.

3.5.2 More Efficient Application Development, Reuse across Network Types

The reuse of Internet Toolkits in the development environment also implies that the heretofore-untapped resources and talent from the Internet space can be easily brought to bear in the area of telecommunications services. More people can now build applications more quickly using tools that are widely available, to interfaces that they can easily understand, with technologies they know and like.

Prior to the advent of this model, telecommunications service and application development required detailed knowledge of the network, of the sometimes rather arcane set of tools, and of very complicated (many-a-time binary) interfaces. Now, one only requires knowledge of the tools and the interfaces, and these are now rooted in the much more widely accepted IT (Information Technology) world.

Application providers may still want to build their applications just once, and sell them to multiple telecommunications service provider customers to realize greater revenue from their investments. This can be easily achieved if the programmatic interfaces (also called Application Programming Interfaces or APIs) supported by the gateway in question were based on some open standard interface definition. Please see Figure 3.5.

Service providers also benefit from this support for open standards, since it means they now have a variety of off-the-shelf applications developed by various application providers that they can choose from to best meet the perceived needs of their subscriber bases.

The architecture components we have added to our shopping cart are:

- programmatic interfaces at the northbound of the gateway that need to play well with Internet technologies and toolkits;
- programmatic interfaces at the northbound of the gateway that need to conform to open standards;
- abstraction of the details of telecommunications protocols at the southbound of the gateway, through the programmatic interfaces at the northbound.

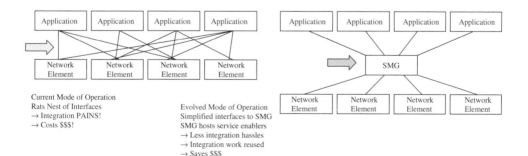

Figure 3.5 Evolution of services architecture

But having a gateway element with support for open, standards-defined APIs is insufficient. What about costs associated with customer care? Would not the rapid development environment for new applications require a proportionate increase in the service provider's operating expenses (OPEX) budget and outlay? How can this be addressed?

3.5.3 Lowered OPEX, Shared-hosting Models

These issues can be addressed by making the API accessible from remote locations or across an IP network, through the use of suitable middleware and an appropriate communications protocol between the applications and the gateway. Applications could be hosted in enterprise or corporate domains, and still leverage the core service capabilities provided by the service provider network in their respective applications contexts as they provide service to their end-users[5].

A major concern with this approach is security, especially for applications hosted outside the secure bounds of the service provider network. Service providers would like to make certain that only authorized applications from external domains were able to access the core network supported service capabilities, that suitable policies be enforced to ensure that this is so, that external applications can be billed for service usage, and that the service level agreements between the various domains are honored as inter-network service usage is supported. We will address this concern of secure access in later chapters.

A major advantage is that all application management issues, and related expenses, are now the concern of the enterprise or corporate domains that host the applications in support of subsets of the service provider subscriber population, and some revenue sharing agreement can be worked between the service provider and the enterprise to enable the former to realize some income for enterprise application use of its core capabilities and assets in processing transactions[6]. Some amount of self-care can still be provided, if the service provider so desires, at the gateway level for subscribers that want to control the low-level aspects of service delivery or update their service profiles within the service provider domain itself. These may not be too extensive, thereby contributing to lowered OPEX.

The responsibility for providing a good quality application experience now rests, in part, on both the enterprise domain hosting the applications, and on the service provider that provides the underlying service components that the application uses – though users may consider the enterprise ultimately responsible for the quality of their application interaction.

The architecture components we have added to our shopping cart are:

- suitable middleware and an appropriate communication protocol between the gateway and the application domain;
- security for applications hosted outside the secure boundary of the service provider network;
- means for revenue-sharing between the service provider domain, and the enterprise domain hosting the applications.

3.5.4 More Effective Use of Deployed Legacy Systems, Evolution Independence

A beneficial by-product of the architecture we have so far come up with is that legacy systems that could previously only be used by a small number of new applications can now be accessed by a

---

[5] Note that nothing prevents the service provider from simultaneously hosting other applications locally within the trusted service provider domain as well. We architect the solution for greatest flexibility.

[6] It must be noted that the service mediation gateway model provides a great deal of flexibility in that it still permits service providers to host critical or important applications that service providers want to retain firm control over, within their networks even as the other benefits that stem from the use of this new paradigm are realized.

much larger set of the same, leading to better reuse of capital well-spent in revenue captured from the application space.

In addition, if the APIs were well designed, and protocol mappings could be easily developed to the underlying network elements for different network contexts, or to different vendors' implementations of the same core network components, the use of a gateway-based architecture enables the core network to evolve independently of the application layer. This low degree of coupling is another point in favor of this architecture. Decoupling applications from the specifics of the underlying network for instance allows for application portability, i.e. reuse of applications across various underlying networks.

The architecture components we have added to our shopping cart are:

• means to apply the new application domain to legacy systems;
• means to maintain the applications domain independently from underlying network evolution.

3.6 Service Mediation and Mediation Gateways

If we now take our shopping cart and proceed to the checkout counter, we may quickly summarize our solution architecture as follows:

It provides a gateway that enables applications (located either within the service provider domain or outside) built to standard Internet protocols and toolkits, to leverage service provider hosted service capabilities exposed via standards-defined APIs in a secure, policy regulated manner to provide a better, more integrated, end-user experience.

Such a component is commonly referred to as a 'Mediation Gateway' or a 'Service Mediation Gateway' (SMG), and the carefully regulated access it provides to services (using appropriate security and policy management mechanisms) is by a process called 'Service Mediation'. So effectively, the architecture we have defined in this chapter is a Service Mediation Gateway (SMG) architecture.

Figure 3.6 shows the Service Mediation Gateway architecture that we have derived thus far, supporting all architecture components needed to meet all the requirements for all three user communities.

3.7 Service Mediation Example

Now that we have come up with what appears to be a working architecture, let us work through an example scenario to verify that it does indeed meet the requirements we expect our solution to satisfy.

3.7.1 User Experience

Alice, out on a walk, during a business trip to New York City, exhausted after a long day of business meetings, would like to find the closest ATM to withdraw cash, and then suggestions for good Italian restaurants near her hotel.

Using her fully capable handset, she connects to a locator application, and requests information relating to ATMs. Her handset shows her a street map indicating where she is and ATMs in the vicinity. She clicks on one and the map zooms in to show her the exact streets to take to get to the ATM, and if she desires, tracks her progress as she walks there.

After withdrawing the money, Alice simply selects 'Restaurants – Italian' from the menu. Her phone screen now shows another map with small restaurant icons. She can click on any of those

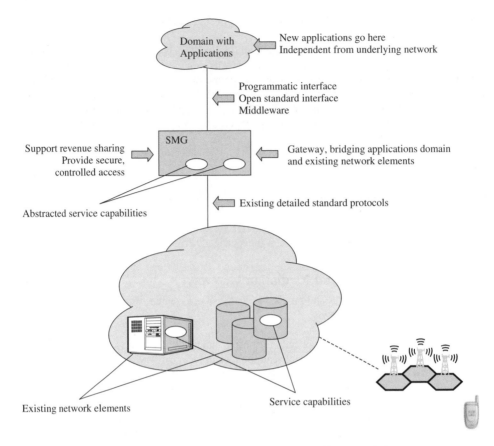

Figure 3.6   Service Mediation Gateway – our architecture solution

icons to get more information on menus, wait time, etc. She could even click on the phone number on the screen to call the restaurant and make reservations for later that night.

3.7.2 Network Operation

The locator application is accessed via WAP. Prior to this scenario taking place, the application has already established a secure connection with the Service Mediation Gateway in the service provider network to which Alice is a subscriber. A pre-negotiated business agreement between the enterprise hosting the application, and the service provider hosting the SMG is in place, and governs the boundaries within which both can interact.

When it receives Alice's request, the locator application looks up her location by invoking the appropriate methods on the service provider hosted Location Capability exposed through the SMG. If Alice has authorized the application in question to receive her location information (authorization is important to alleviate privacy concerns), the application request is processed by the elements in the core of the service provider network and her location returned (either as {X, Y} co-ordinates within a pre-specified range of accuracy, or as cell ID and sector information, depending on technology) to the application via the SMG. The SMG talks to southbound network elements in the core network such as GMLCs or MPCs over suitable protocols to get this information on Alice's terminal location, but this happens in a manner that is completely transparent to the locator application.

The application now factors in Alice's location as it generates a web page to fulfill the 'nearest-X' query from Alice, first for ATMs, and then for Italian restaurants. Her location may be requested more times if necessary, to support additional requests she may make. Subsequent pages can merely use the content store to provide more information relating to the restaurants in question, based on end-user selection.

Once this call flow concludes, and Alice has signed off, the application may debit Alice's account for services rendered (or alternately, this may be a subscribed service where she already pays a steady fee per billing cycle). In the former case, the application would simply request, via the SMG, that her prepaid account be debited for a specified amount, and assuming the application is permitted by Alice to perform this operation (i.e. if Alice is a subscriber to this service), the transaction is carried out in the network.

The above example clearly shows how a Service Mediation Gateway architecture addresses Alice's need to have at her disposal a really cool service and enables the Network Operators to provide this service to Alice. One could argue that a similar user experience could be supported through WAP in conjunction with other interfaces available in telecommunications networks today, and that would be a valid argument for some facets of the call scenario. However, for the complete flow to be supported, or for easy extensibility of existing deployed services and applications into new domains, a service mediation type architecture would provide the most cost-efficient solution since integration performed once can be reused multiple times, thus leading to a reduction of overall cost. Later chapters will elaborate on how the wish of Application Providers to develop portable applications that make use of network service capabilities, while utilizing Internet technologies and toolkits for their development, is granted by the Service Mediation Gateway architecture as well.

3.8 Summary

In this chapter, we have taken the requirements from three user communities and generalized those into a 'problem domain' consisting of four requirement categories. We have assessed and sifted through these four:

a) Reducing Integration Costs, Faster Development Cycles
b) More Efficient Application Development, Reuse across Network Types
c) Lowered OPEX, Shared-hosting Models
d) More Effective Use of Deployed Legacy Systems, Evolution Independence

Based on that assessment we introduced a generic service mediation based architecture that seems to satisfy the requirements that network operators, subscribers, and application providers would like to see met from their communications networks.

The next chapter will explain some of the standards relevant to the context of service mediation, and shall broadly outline standards evolution to Parlay and OSA. Subsequent chapters will pick up this thread on the architecture and develop it more in the context of a standards-defined implementation.

4

Parlay and OSA

4.1 Introduction

Thus far, we have defined a problem space by introducing a set of requirements that network operators, end-users, and application providers wish to see resolved regarding service technologies in use today. We have identified a service mediation based architecture as a solution to this problem space, and hinted at Parlay and OSA as the technology standards fit for the job. It is time to have a closer look at this technology, starting with the standard itself and how it evolved from earlier activities.

The Parlay solution, as captured in the suite of Parlay specifications, is currently being defined in a collaborative effort of various standards organizations and industry fora. Some of the architectural concepts that form the ground works of the Parlay solution are based on the foundations that were provided in other organizations and earlier initiatives. This chapter will start by looking back at how the Parlay work evolved from standards activities in the past in their effort to reuse some of the successful and promising concepts devised elsewhere. We will explore the various seemingly parallel initiatives and introduce the Joint Working Group (JWG) as the main vehicle for ensuring a single harmonized and technology independent solution.

At the end of this chapter the reader will understand that Parlay and OSA are to a large extent one and the same and will appreciate how the various standards activities and industry initiatives, past and present, relate to one another and come together in a single solution.

4.2 The Need for Standards

One may well ask, why do we even need standards? We try to address that issue briefly in this section.

A basic reason for the need for open, standards defined interfaces was presented in Chapter 2, where we talked about how application developers can proactively develop applications reusable across various network types by writing to well-defined interfaces, and can sell these to different service providers with minimal or no changes. We also discussed how service providers benefit from having access to a larger pool of applications, and how the ability to support more applications more cheaply and quickly in their networks enables them to stabilize and grow their subscriber bases and tap into new revenue streams.

There are other reasons why standards are important. They define a common reference architecture, set of interfaces and reference points surrounding the element in question. They enable better interoperability between conformant implementations (as we said before, most networks in existence today are multi-vendor environments, so this is rather important). This contributes to the 'usability' of the technology in question, contributing to its success.

Parlay/OSA: From Standards to Reality   Musa Unmehopa, Kumar Vemuri, Andy Bennett
Copyright © 2006 Lucent Technologies Inc. All Rights Reserved

Thus, standards treat the network elements with which they are concerned as black boxes, and do not typically specify the internal or design level details of particular implementations (or what can be referred to as 'white box' details). This enables the vendors of the element in question to build competitive differentiation into their products, and thereby distinguish their implementation from those that others develop. Similar reasoning applies to applications, and other standards compliant software as well.

While defining standards, one must tread the line between specifying too little information on the one hand and adding too many requirements on the other. The former will lead to implementations that will not be interoperable, leading to problems with deployment and other issues relating to the interpretation on what little is specified, while the latter would stifle innovation, obstruct creativity, and while it would lead to better interoperability, reduce the number of parties interested in building the element(s) in question, thereby becoming anti-competitive, and against common-sense.

4.3   The Parlay Family Tree

In order to grasp a full understanding of Parlay it is important to understand its pedigree, i.e. how and where it originated, how it was influenced, and how it applied influence to others. Writing good standards documents is like good management, and is therefore also part art and part science. Micro-management or excessive laxity and over-delegation seldom yield good results. But under a good manager, a team can perform wonders. Every standard goes through growing pains, but also becomes stronger with every challenge it faces and overcomes. The following section presents how Parlay/OSA have gotten to where they are today, starting from humble beginnings indeed.

4.3.1   The Cradle

Undoubtedly, some aspects of the Parlay solution can be traced back in origin to some of the very basic concepts in communications. For the purpose of this book, however, we will start our genealogy survey at a place where the most fundamental concepts that differentiate Parlay as a solution were being conceived and developed. We will look at the developments that directly contributed to the Parlay technology. From this perspective it is fair to say that the TINA-C initiatives form the birth ground of Parlay.

4.3.1.1   TINA-C

Tiered communication architectures, where one can identify a services layer, a connectivity layer (sometimes called core network layer), and an access layer, are well accepted within the industry to date. The Telecommunications Information Networking Architecture Consortium, or TINA-C, was one of the first organizations to define a formal service architecture based on this tiered principle. TINA-C was founded in 1993, while the first TINA workshop already took place in 1990 [TINA]. The TINA architecture covers a large number of aspects of telecommunications in great breadth and depth. It is predominantly the TINA service architecture that we'll focus on in this section.

TINA defined two main principles for separation, in order to allow for functional distribution and separation of concerns. The first TINA separation principle, one we are all familiar with today, is that of separating service logic from the basic call control or connectivity processing. For instance the concept of Intelligent Networking, as discussed in Chapter 1, makes use of this first separation principle. The second TINA separation principle introduces the distinction between the access to a service and the use of a service. It is this second separation principle that makes one of the fundamental concepts of the Parlay solution.

The service access and service usage concepts are realized in terms of a so-called access session and service session. Service access covers the discovery of a service and the request for usage of the service. It is the access session that implements the business relationship between the application provider and the network operator. Service usage covers control of service behavior and exchange of service content. Within the context of an access session, a prospective user of a service, residing

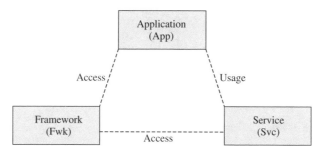

Figure 4.1   Separation of access and usage

in the network, obtains an initial point of access to the network and authenticates and authorizes itself. Once a prospective user's identity and privileges have been established by the network, the user is provided with the means to make use of the service offered by the network, within the boundaries of the agreed business relationship.

The main architectural benefit of introducing an access session is that functionality like AAA (Authentication, Authorization, Accounting) and Integrity Management (e.g. Load Control and Fault Management) can be logically centralized and reused for all services offered by the network.

The main architectural benefit of introducing a service session is that services can be developed independent from the service access aspects. For instance, this enables exposure of various services in a coherent and consistent way, as well as the possible outsourcing of service development to third parties.

Figure 4.1 illustrates the concept of access sessions and service sessions. For more detailed information on TINA, the reader is referred to [TINA 1995,TINA 1997a,Mampaey 2000].

4.3.1.2 OMG TSAS

The output of TINA-C can be classified as a set of architectural specifications and business and information models. The Object Management Group (OMG) – an open, not-for-profit organization [OMG] that among other specifications has published CORBA™, UML™, and IDL – took the work of TINA-C in the area of service access sessions and the Consumer-Retailer relationship [TINA 1997b] and produced an interface specification[1]. This specification is called OMG TSAS (Telecommunications Service Access and Subscription) [OMG 2000a]. The TINA-C work in the area of Consumer-Retailer introduces the concept of subscription management, which deals with the case where the user of a service is not an end-user, but for instance an enterprise that acts as a retailer of the service towards its own customer, i.e. its employees. The enterprise operator (retailer) mediates services on behalf of service providers to its employees (i.e. end-users).

OMG TSAS defined three domains, namely the Consumer Domain, the Retailer Domain, and the Service Provider Domain (Figure 4.2). Subscription in this scenario deals with information about services and contractual relationships between these three domains.

Sessions can exist between these domains. For instance access sessions, in OMG TSAS part of the so-called core segment of interfaces, can exist between a consumer and a retailer and between a retailer and a service provider. Once an access session is established, a subscription segment can be used. The core segment and subscription segment, consisting of a set of interfaces, are defined using OMG IDL.

---

[1] To be precise, the OMG issued an RFI, followed by an RFP, at the end of 1999, inspired by the TINA-C work on the Access Session and Consumer-Retailer Reference Point specifications. Active TINA-C member companies then responded, which resulted in the OMG TSAS specification.

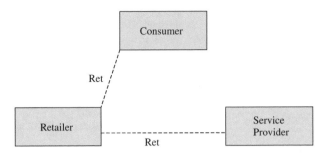

Figure 4.2   Three OMG-TSAS domains

[Mampaey 2000] and [Bakker 2000] provide some information on the relationship between TINA, OMG-TSAS, and Parlay, whereas the entire specification containing all the details can be found in [OMG 2000a].

4.3.2 Early Childhood

The Parlay Group was founded in 1998 [Parlay], after the completion of TINA phase 1 specifications at the end of 1997. Although influenced by the work carried out in TINA-C, the group's objective was not limited to a mere continuation of that body of work. The goal of the Parlay Group was to specify open network application programming interfaces that would bridge capabilities from the IT domain with those of the telecommunications domain. Part of the motivation for the creation of the group was the expectation that regulatory bodies mainly in Europe would mandate equal and open access to service capabilities in the network of incumbent operators to third parties.

The interface specifications in the first Parlay release, Parlay v1.2[2], were rather limited. For example, it was possible to start a service, but not stop it. The interfaces covered only the relationship between the Application and the Framework, i.e. Parlay API Interface 1. The Framework v1.2 specification was based on the TINA Retailer Reference Point specification v1.1 [TINA 1999]. Note that the key concepts of the Parlay architecture, including Framework and Application (Figure 4.3), will be introduced in a lot more detail in Chapter 5.

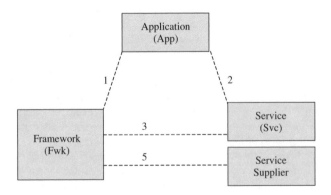

Figure 4.3   TINA-based Parlay architecture

---

[2] The significance of the version number for this particular Parlay release stems from the fact that for this version of the Parlay API specifications IDL descriptions were published. At a later stage, with the introduction of the Joint Working Group, IDL descriptions are automatically generated with each release of the API specifications. Later sections in this chapter will cover IDL and versioning in more detail.

The service interfaces in Parlay v1.2 were limited to Generic Call Control, Generic User Interaction, Mobility, and Connectivity Manager.

The Parlay Group started out as a closed group, consisting of five companies, i.e. British Telecom (BT), Microsoft, Nortel Networks, Siemens, and Ulticom—formerly DGM&S Telecom. Roughly at the same time as Parlay v1.2 was being completed, in the autumn of 1999 3GPP (3rd Generation Partnership Project) created the Open Services Access[3] (OSA) Adhoc group, to start looking at the definition of open network APIs to implement the Virtual Home Environment (VHE) concept. The work was carried out mainly in parallel, as Parlay documents at the time could not be shared openly with non-member companies.

4.3.3 The Wonder Years

We have discussed the Parlay cradle, and seen the technology in its infancy (with all due respect of course). At this stage the concept of open network APIs was becoming more widespread and accepted in the industry. Among other things, this meant that various activities in this area sprang up, in parallel to the continuing activities of Parlay and 3GPP. Parlay was entering adolescence; a defining stage but also sometimes a somewhat confusing stage.

4.3.3.1 Parallel Activities

After publishing Parlay v1.2 a number of events occurred that contributed to a consolidation between the various parallel efforts in this area. The Parlay Group changed to an open organization, and in May 1999 various companies joined the ranks, i.e. AT&T, Cegetel, Cisco, Ericsson, IBM, and Lucent Technologies. Open information sharing between Parlay and, for example, 3GPP was now possible.

In addition 3GPP disbanded the OSA Adhoc and instead created a dedicated OSA working group within the Technical Specification Group for Core Networks (TSG CN). 3GPP TSG CN WG5, or CN5 for short, became responsible for the technical interfaces, or APIs, specific to UMTS OSA.

In parallel, the European Telecommunication Standards Institute (ETSI) had started its own open API activity, for wired networks [ETSI]. This work was carried out within the ETSI Technical Committee for 'Signaling and Protocols for Advanced Networks' (SPAN), working group 3[4]. In SPAN, the focus of SPAN 3 was 'Applications Interfaces for Service Providers and Network Operators'.

The work of OMG TSAS was fed into the Parlay version 2 effort in an attempt to strengthen the Framework specification of Parlay v1.2. The interfaces to allow a service to register itself with the Framework were included to implement Parlay API interface 5 (i.e. between the Service and the Framework). Furthermore, the Parlay v1.2 definition for Parlay API interface 1 (i.e. between the Application and the Framework) was further completed by adopting the remainder of the OMG TSAS interface in this area. With this move, almost all the work of OMG TSAS had found its way into the Parlay specifications.

4.3.3.2 On Route to Harmonization

The developments outlined above meant a great step in the right direction on the path to a single set of harmonized open network APIs, applicable to a multitude of underlying network technologies. 3GPP was developing APIs for UMTS based networks, ETSI did the same for ISUP/IN based wire-line networks, and Parlay continued to address the IT/Enterprise angle. However at this time all

---

[3] Initially OSA was expanded as Open Service Architecture. After 3GPP Release 99 this was changed to Open Service Access.

[4] After the ETSI SPAN reorganization in the autumn of 2000, working group SPAN 3 became SPAN 12. The most recent development in ETSI is the consolidation of technical committee SPAN with technical committee TIPHON™, into the new technical committee TISPAN.

three groups were still working in parallel, having their separate technical working group meetings and producing their own set of specification documents.

The first consolidation step occurred with the decision of ETSI SPAN 3 to collocate all of their meetings with 3GPP CN5. These arrangements were rather straightforward, as ETSI is one of the organizational partners of 3GPP. Hence any company membership issues, and matters related to intellectual property or copyright, did not exist.

The truly significant achievement was the legal agreement between ETSI and the Parlay Group at the end of 2001, to formalize the close cooperation both groups had been having at a technical working group level for the past year and a half. ETSI and Parlay entered into this co-operation agreement with the intent to develop and publish jointly the Application Programming Interface specifications for Open Service Access (OSA)[5].

Through the close family ties between ETSI and 3GPP, and now with the ETSI-Parlay agreement in place, the so-called Joint Working Group (JWG) was created. The JWG is the single technical working group where the API specifications for Parlay and OSA are being developed and maintained, based on a collective set of harmonized requirements. The specifications themselves are jointly published by ETSI and the Parlay Group.

It is important to mention the involvement of the Telecommunication Standardization Section of the International Telecommunication Union (ITU-T), as this is the main organization for setting global telecommunications standards. The technical work takes place in study groups (SGs), which get assigned a group of questions that belong to a broad subject area. SG 11 deals with signaling requirements and protocols. The question relevant to the Parlay and OSA work is 'Question 4/11' on 'API/Object Interface and Architecture for Signaling'. Question 4 of SG 11 requires the definition and specification of API interfaces fulfilling a set of telecommunication requirements. As the ITU-T recognized the activities already ongoing in this area in the organizations involved in the JWG, they decided not to initiate an overlapping initiative, but rather to refer to the work published by the JWG. This decision further added to the consolidation and harmonization in the network API arena.

For a certain period of time, JAIN™ [JAIN] community members also participated in the Joint Working Group in the specification of predominantly the Call Control APIs. This book does not elaborate on the JAIN involvement. The interested reader is referred to [Bennett 2003], which expands on service mediation standards and the role of JAIN with regards to Parlay.

All these relationships towards consolidation and harmonization are pictorially represented in Figure 4.4.

4.3.4 Maturity?

Now that Parlay, through the JWG, had published several versions of the API specification set, and went through various iterations of updates and corrections, as well as requirements extensions, the technology could be considered maturing. Several interesting developments corroborated this growing process of Parlay.

4.3.4.1 New Members of the Family

Halfway through the year 2002, the 3GPP2 organization was showing an increased interest in Parlay technology, for adoption as service architecture in their networks as well. With the publication of the 3GPP2 OSA specification almost a year later in June 2003, in third generation wireless environments there now was a single service mediation architecture, equally applicable to both W-CDMA as well

---

[5] One of the effects of working together was the introduction of terms like Parlay/OSA and OSA/Parlay. The reader should now understand that Parlay and OSA are one and the same. For the purpose of this book, in the remainder we will only use Parlay. Exceptions may be made when referring to the technology in 3GPP or 3GPP2 specific contexts.

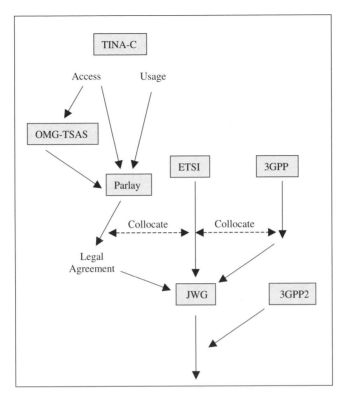

Figure 4.4   Parlay consolidated standards efforts

as cdma2000 networks. An OSA working group was formed within 3GPP2, which joined the ranks with the JWG. In Chapter 8 we will elaborate on the 3GPP2 specifications for OSA.

4.3.4.2 Closing the Generation Gap – Backwards Compatibility

The chapters later on in the book will deal with the details of Parlay version 4, for reasons of stability and availability of this release. At the time of writing Parlay 5 is in its completion phase and Parlay 6 requirements are being considered. All these new generations of Parlay API specifications contain both updates and error corrections to the previous generation, as well as new requirements for additional functionality. It is not likely, and in fact unwanted, that the new generation will instantly take over from older generations. The generation gap needs to be closed.

Enter backwards compatibility. Starting with Parlay version 3.2 the Parlay technology was maturing sufficiently, as evidenced by a number of life network operator trials and initial commercial deployments. The investments associated with these activities needed preserving with the further advancing of the technology. For any mature and deployed technology, one expects older versions of applications to remain operational after upgrades of server equipment, such as a Parlay Gateway.

4.3.4.3 Family Gatherings – Interoperability Testing

A last effort on the way to adulthood for Parlay worth mentioning here are the activities regarding interoperability testing. Interoperability testing is a valuable and effective means in order to ensure the Parlay technology is fit for commercial deployment in a distributed, multi-vendor environment.

ETSI to date has organized two interoperability events where network operators, Parlay application providers and Parlay gateway vendors get together, interconnect their products and perform tests to demonstrate interoperability and standards-compliance. One of the outcomes of these events is the identification of potential ambiguities in the standard specifications where different companies have interpreted specified behavior in different ways. These instances have served as invaluable feedback into the standards process.

Furthermore, the European Union has hosted project OPIUM (Open Platform for Integration of UMTS Middleware) [OPIUM] with the aim to support the accelerated rollout of commercial 3G services within Europe. Parlay technology has played a central role in this project, where network operators, application providers, and gateway vendors came together and participated in a Parlay test bed.

4.3.5 Non-identical Twins

We started this chapter by saying that Parlay and OSA are to a large extent one and the same. In fact, we have just agreed to use the term Parlay to cover both Parlay and OSA. The consolidation efforts described above that have lead to the Joint Working Group provide the foremost explanation for the striking resemblance. However, Parlay and OSA are not identical twins. The distinction is minor and you have to know both twins fairly well to spot the differences, as these do not affect the service mediation architecture introduced by Parlay, or any of the deployment models supported. For reasons of completeness though we will point out the differences in this section, and continue after this section by treating Parlay and OSA as one and the same.

Harmonization and alignment of the APIs is of great importance in the specifications produced by Parlay and OSA, and indeed is one of the most valuable assets. That means that for those functional areas where both organizations publish API specifications, these specifications are the same. However, not all interfaces published by Parlay are applicable to a 3GPP OSA environment. The interfaces in OSA form a proper subset of all the interfaces supported in Parlay. The interfaces that are not part of the OSA subset are Connectivity Management, the Conference Call Control interface, and the Enterprise Operator interfaces in the Framework.

4.4   The Standards Themselves

We have provided you with the history leading up to the definition of the Parlay service mediation architecture and the Parlay APIs (from the cradle to maturity), as well as the environment in which the Parlay specifications are produced and published (the Parlay family, formed by the Joint Working Group). Let us now take a peek in the family album, that is, take a look at the Parlay API specifications themselves.

4.4.1 The Common UML Model

As the Joint Working Group forms the organizational community taking care of alignment and harmonization in producing a common suite of API specifications, the common UML model ensures that all publishable artifacts are aligned as well, from a technical point of view. For each Parlay API there is a formal model specifying the entire interface, its definition, and its behavior. UML (Unified Modeling Language) consists of a collection of modeling techniques including class diagrams, sequence diagrams, and state transition diagrams used to model distributed communication systems. Each of these techniques is used in the definition of Parlay.

The actual textual specification documents are automatically generated from this common Parlay UML model, using software tools. This ensures that the ETSI and Parlay version of the specifications is semantically and syntactically exactly the same as the 3GPP version of the specifications, though each is formatted according to its own conventions and style. ETSI and Parlay publish their specifications jointly, as European Standards (ES), whereas 3GPP publishes their own specifications as 3GPP Technical Specifications (3G TS).

4.4.2 Technology Realizations

The common Parlay UML model is used to generate more than merely the specification documentation. Accompanying each API specification document are three technology realizations. The UML model for Parlay is defined in a technology independent way such that the various technology realizations can all be generated and derived from this common source. The technology realizations consist of three communication technologies in which Parlay products can be potentially deployed.

An added benefit from generating both the technology realization as well as the specification document from a single source is that these two artifacts are aligned as well. There are instances in IN and CAMEL where the specification document does not quite follow the ASN.1 definitions, where one artifact got modified without performing the accompanying changes in the other. Without a single common source, errors will inevitably find their way into the specification document, or the interface definition. By maintaining a single common source, misalignment is physically impossible.

4.4.2.1 CORBA Technology Realization

The first-born son and oldest technology realization is the CORBA realization. CORBA (Common Object Request Broker Architecture) is an object-oriented middleware solution for client server based communications systems. The APIs are defined in IDL, the Interface Definition Language. The IDL files for the Parlay API specifications form the CORBA Technology realization of Parlay.

In the early stages of Parlay (version 1.2) two more middleware based realizations were available, i.e. MIDL and DCOM. At that time the common Parlay UML model was not available yet, and hence all middleware realizations had to be constructed manually. Keeping in mind the fact that the standards process is contribution driven, and since no MIDL (Microsoft Interface Definition Language) and DCOM (Distributed Component Object Model) realizations were contributed for subsequent Parlay versions, CORBA IDL remained for a while the sole technology realization of Parlay.

4.4.2.2 Java™ Technology Realization

Once the common Parlay UML model was available, this opened the door for alternative technology realizations, in addition to CORBA and IDL. Within the Parlay Group, the Java Realization working group was chartered to address the Java developer community and produce a technology realization in support of both the Java2 Enterprise Edition (J2EE™) and Java2 Standard Edition (J2SE™) programming models. This has resulted in two Java realizations, one for J2EE, defining a Java equivalent of the IDL realizations, and one for J2SE, providing a local API on the Parlay Application Server.

4.4.2.3 WSDL Technology Realization

The third technology realization of Parlay is the WSDL (Web Services Definition Language) technology realization, in recognition of the growing interest in XML-based technologies and Web Services deployments. The WSDL realization of Parlay will be covered in more detail in Chapter 16.

With these three technology realizations, all three derived from the common Parlay UML model, organizations wishing to deploy a Parlay solution for their services architecture, are free to pick and choose the particular technology realization that fits best with their installed legacy base and systems, specific deployment requirements, or tool environment and skill set of their particular developer community.

4.4.3 Versioning Schemes and How They Relate

The Parlay history has shown us that the various participating standards organizations joined the party at different stages. As a result, the version numbers for the various releases are misaligned.

Table 4.1 Standards versioning schemes and their relationships

ETSI Version	Parlay Version	3GPP Version
N/A	Parlay 2.1	3GPP OSA Release 99 (3G TS 29.198 v3.x.y)
ETSI OSA Phase 1 (ES 201 915)	Parlay 3	3GPP OSA Release 4 (3G TS 29.198 v4.x.y)
ETSI OSA Phase 2 (ES 202 915)	Parlay 4	3GPP OSA Release 5 (3G TS 29.198 v4.x.y)
ETSI OSA Phase 3 (ES 203 915)	Parlay 5	3GPP OSA Release 6 (3G TS 29.198 v6.x.y)

This can be cause for confusion and hence in this section we will spend some time relating these versioning schemes together. Fortunately, although misaligned, at least the version numbers increase in lock step.

The easiest way to convey this information is in tabular format (Table 4.1).

The legal agreement for joint publication of specifications between ETSI and Parlay was not yet in place at the time of Parlay 2.1, hence there is no equivalent ETSI phase.

For 3GPP specifications, the specification series number (29.198) does not change between releases; instead, the first digit in the version number signifies the release. For instance the digit 4 in version 4.x.y tells you this is a 3GPP Release 4 specification. In the case of ETSI, the specification series number changes from one release to the next.

4.4.4 The Specification Series

In the description of the versioning scheme we referred to specification series, rather than a single specification. The reason for this is that the Parlay specification is really a suite of multiple specification documents, a total of 14 parts within the 29.198 series[6]. There is a general overview part (part 1) and a common data type part (part 2) and twelve parts for all the APIs supported by Parlay[7]. Chapter 6 'Standards Capabilities and Directions', will further elaborate upon each of the APIs.

4.4.5 Specifications and Recommendations

Within 3GPP, besides the API specifications in the 3G TS 29.198 series, protocol mapping recommendations are published in the 3G TR 29.998 series. Whereas the API specifications are normative (hence the TS for Technical Specification), the mapping recommendations are informative (which is why they are contained in a TR, for Technical Report).

A protocol mapping recommendation takes one of the APIs and provides recommendations on how to map API method invocations on specific protocol operations in the network. Mapping recommendations exist for several APIs, and for several standardized signaling protocols. The mappings are informative recommendations, as there is more than one way to implement the support for Parlay in any given network. This includes mappings to standardized signaling protocols in ways not alluded to in the mapping documents, as well as proprietary means to support a specific API method invocation. Another reason for the mappings being informative is that the Parlay APIs are designed in a way independent of the underlying network signaling protocols. The Parlay API provides an abstraction of service capabilities residing in the network. Hence a direct complete mapping to every conceivable network signaling protocol cannot be presented.

---

[6] We will use the 3GPP publications here in this discussion, as the series number remains a constant across releases. However, the ETSI specifications, and hence the Parlay specifications as well, also consist of 14 parts.

[7] Technically, the parts for Connectivity Manager and Generic Messaging are not contained in the 3GPP subset. However, for this discussion we will ignore that fact. Furthermore, in order to preserve the part numbering scheme across organizations, the part numbers for these parts are not re-allocated within 3GPP and are left unused.

Table 4.2 API to protocol mappings

Mapping Recommendation	Title
3GPP TR 29.998-1	Part 1: General Issues on API Mapping
3GPP TR 29.998-04-1	Part 4: Call Control Service; Subpart 1: API to CAP Mapping
3GPP TR 29.998-04-4	Part 4: Call Control Service; Subpart 4: Multiparty Call Control ISC
3GPP TR 29.998-05-1	Part 5: User Interaction Service; Subpart 1: API to CAP Mapping
3GPP TR 29.998-05-4	Part 5: User Interaction Service; Subpart 4: API to SMS Mapping
3GPP TR 29.998-06	Part 6: User Location – User Status Service Mapping to MAP
3GPP TR 29.998-08	Part 8: Data Session Control Service Mapping to CAP

The mapping is performed on a per API method basis, mapping the method to one or more protocol operations, and mapping the method parameters and data types to the protocol operation elements and data types. The value of this exercise is to verify whether an API is sufficiently rich in functionality to make efficient use of network capabilities, as well as ensuring that no API functionality is defined for which there is little or no network support. The exercise has its limitations in that the mapping provides a static view of what the possibilities are with a given API. Dynamic behavior is not described, nor are complete service scenarios included.

The part numbering corresponds to the relevant part number in the 29.198 series. The subpart number indicates different protocols to map to. The subpart numbering is not consecutive. The reason for this is that Parlay initially planned multiple protocol mapping recommendations for each part, and hence subpart number placeholders were put in place. However, not every protocol mapping was completed, although the allocated numbering was preserved. Table 4.2 shows the various supported protocol mappings.

4.5 Summary

In this chapter we have attempted to provide some insight into the pedigree of Parlay, how Parlay arrived at where it is today, and what the organization structure is that ensures the production and publication of a suite of harmonized APIs. We have also presented some insight into the specification structure, and its versioning and release schemes.

5

The Parlay Conceptual Architecture

5.1 Introduction

In this chapter, we will look at the conceptual architecture of Parlay. We cover the key logical elements (Client Application, Service Capability Server and Framework), their roles and the interactions between them. As far as possible we have avoided referencing the methods defined by Parlay that make up the interfaces between these logical elements. These can be found in the Parlay specifications themselves, and will be introduced in some detail in the next chapter. Instead, this chapter provides a more narrative description that provides an introduction to the key concepts.

The goal of Parlay is to provide a means for end-user services (Client Applications) to gain access to the functionality (Service Capabilities) available in telecommunications networks. For example, a tourist may wish to have information about interesting sights delivered to their mobile handset while they walk around a city or other tourist destination. Clearly, the user will be particularly impressed if a description of the history St Paul's Cathedral in London is provided when they are standing in front of its façade rather than while they're having coffee and a doughnut five minutes later. Networks are able to provide the location of the handset (and thus the user, it can be reasonably assumed). This can be combined with the content (pre-recorded audio, or text perhaps) to create a useful service to the user.

To allow the Client Application to get at the network functionality, Parlay defines a number of interfaces, each supporting a different capability of the network. Each defined interface is known as a Service Capability Function (or just "Service") which, when implemented, is known as a Service Capability Server (SCS). A Client Application will make use of functionality from as many or as few SCSs (and of course any non-Parlay functionality) that it needs in order to be useful to an end-user.

In our example of a service (Client Application) that acts as a tourist guide around a city, the SCS being used by the Client Application would be a Location SCS. By adding functionality provided by the Call Control SCS, the Application could set up a call to a recorded description of each place of interest as the user approaches it.

The two-way relationship between Client Application and SCS can be considered the primary one in Parlay since it delivers useful functionality that can be delivered to an end-user. However, as we have seen in previous chapters, network functionality and the integrity of the network is something to be prized and protected. For this reason the relationship between a Client Application and a Service Capability Server must be managed. Thus into the relationship comes a third element: the Framework. This triangle of entities is central to Parlay and so will be central to the rest of this chapter. Figure 5.1 shows the Parlay triangle.

Parlay/OSA: From Standards to Reality   Musa Unmehopa, Kumar Vemuri, Andy Bennett
Copyright © 2006 Lucent Technologies Inc. All Rights Reserved

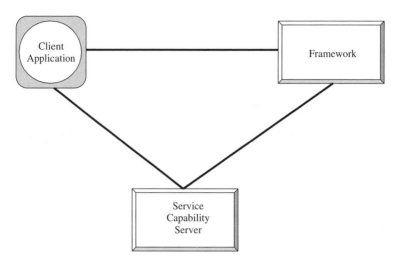

Figure 5.1   The Parlay triangle

Let's now take a brief look at each of the three major entities, examining their roles and responsibilities in a little more detail. This high-level introduction will prepare us for a deeper look at
the Parlay architecture when we put everything together and walk through a scenario. Later in this
chapter, two further entities (the Service Supplier and the Enterprise Operator) will be described
that are part of the Parlay architecture, but for now we need to focus on the triangle.

5.2   The Client Application

The ultimate purpose of the Client Application is to provide a human-machine interface with one
or more end-users. Since network protocols are rarely a good way for humans to communicate, the
Application must ensure that a request from a user is translated into an operation in the underlying
telecoms network. Similarly, it must ensure that a significant event in the network is translated into
some sort of indication to the user. These interactions are illustrated in Figure 5.2.

Take for example a very simple Client Application, designed to allow the user to make and
receive voice calls while out and about. The user might be presented with a screen via which a

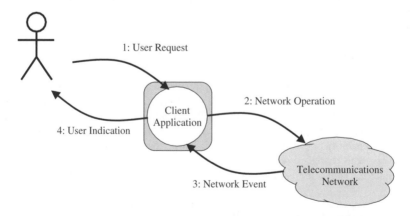

Figure 5.2   Client application interacting with the network, on the end-user's behalf

phone number may be entered and a button to press to initiate the call. When the network has found the destination phone and has started ringing it, the network signals this to the application and the application will inform the user by providing an audible or visual indication.

Sounds familiar? This is, of course, exactly what software in every mobile phone does right now. This software talks to the network using network protocols that have been around for some time. Unfortunately, there are many types of network and corresponding protocols. As we have seen in previous chapters, Parlay hides the nature of the underlying network, so a Parlay Client Application translates Parlay operations into user events, and vice versa. The SCS, which we cover shortly, completes the process by translating Parlay operations into network events.

Of course, since the Client Application interacts with end-users, it needs to manage its relationship with those users. In the case of an application running in a single user's device (phone, PDA, laptop) this relationship is one to one, but an application may have many thousand, or millions, of users. For example, the user may be using a web page as the user interface to an application. Many other users are communicating with the application in the same way. The application will probably want to authenticate each user and will almost certainly have a database of information (subscription data) on each one. The way in which the authentication occurs and the nature of the subscription data are generally outside the scope of Parlay.

A Client Application thus provides an interface to its users and makes use of an interface provided by the SCS to cause things to happen in the network or to be informed of things happening in the network. The interface to the user is not defined by Parlay. It might be the buttons and screen on a phone or it might be a graphical user interface on a PDA, but it is not part of the Parlay world. The interface to the SCS, however, is part of Parlay and the role of the SCS is covered next.

5.3   The SCS

The second of the three entities is the SCS. Essentially the SCS translates from Parlay to the language of an underlying network, and vice versa. There are, of course, other aspects of the SCS's role and some of them will be covered in this section, but that is its purpose in a nutshell so let's consider this first.

5.3.1 Translation

Since the SCS translates requests (method invocations) from the Client Application into operations in the network, the SCS provides the Client Application with an abstracted view of the capabilities in the underlying network available to it. The Client Application therefore only needs to understand how the SCS operates and not how the network operates. In fact the SCS may even interact with a number of different types of networks, all performing the same kind of function (e.g. call control) but each using a different interface or protocol to achieve it. A Client Application using such an SCS would be able to control a mobile call, PSTN call or VoIP call using the very same Parlay operations.

At its simplest though, an SCS will translate to only one underlying network type, as is shown in Figure 5.3. This still provides significant advantages as this means that in principle a Client Application written to use a call control SCS offered by one network will be able to use the call control SCS offered by another network.

A few paragraphs ago we said that the SCS's job is essentially one of translation. This is by no means a simple job, however. The underlying network will consist of a number of physical entities and messages take time to move between them. The state machine governing the behavior of the network entity will be different to that defined by Parlay for the SCS. It may be that a single operation over the Parlay interface translates to more than one operation in the network and this requires the SCS to be able to correlate these operations and deal with what happens if one of these operations fails.

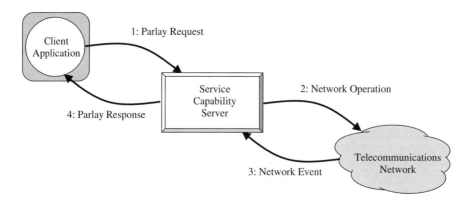

Figure 5.3   Service Capability Server interacting with the network, on the Client Application's behalf

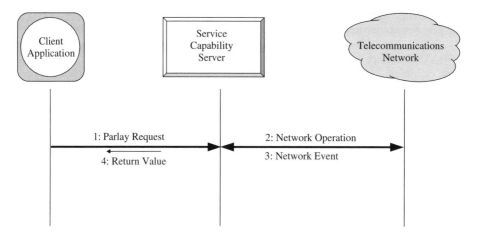

Figure 5.4   Synchronous method invocation

For these reasons the Parlay interfaces offer a mixture of synchronous and asynchronous communication modes. If it can be guaranteed that a request (method invocation) from a Client Application results in an immediate response from the underlying network, then the response can be synchronous. That is, the response can be delivered in the return value of the method. A sequence diagram showing synchronous method invocation behavior is depicted in Figure 5.4.

More generally though, any operation in a network has a finite duration. As a result, a request from a Client Application will result in a response delivered to the Client Application separately. For example, if an application requests that a phone call is initiated between two users in the network the allocation of resources and routing of messages will take a significant amount of time (100 s of ms). A sequence diagram showing asynchronous method invocation behavior is depicted in Figure 5.5.

The SCS must keep track of requests by the Application and correlate any responses to them. In some circumstances, as shown in Figure 5.6, multiple responses may have to be aggregated into a single reply to the Client Application.

There is also a class of requests that a Client Application can make that result in an indeterminate number of responses. These can be thought of as the setting of triggers in the underlying network that fire whenever a particular event or situation occurs. This behavior is outlined in Figure 5.7.

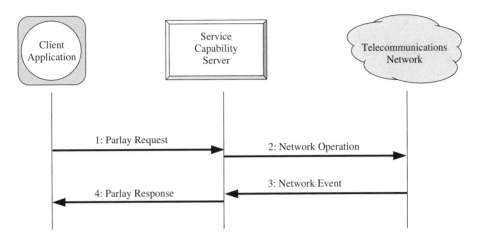

Figure 5.5 Asynchronous method invocation

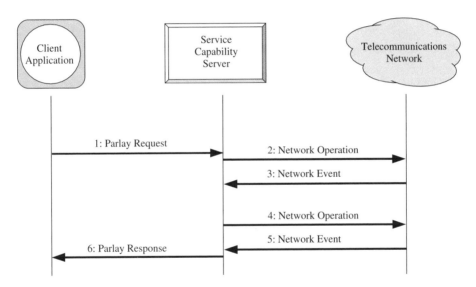

Figure 5.6 Asynchronous method invocation with aggregated response

An example of this is part of the Location SCS. The Client Application is able to request that every time a user moves outside a particular geographical area an event is fired. This will continue to fire until canceled by the Client Application. Of course, this could lead to quite a number of events if the user is sitting on a carousel straddling the area boundary.

5.3.2 Beyond Translation

An extremely important role of the SCS that isn't just translation of messages, is that of policy enforcement. In this context, a policy defines what a Client Application is allowed to do in any particular situation.

Taken at face value, the Parlay definition of the functionality of a particular SCS would seem to suggest that a Client Application could invoke anything that the SCS has been implemented to be capable of doing. The definition of the Call Control SCS includes functionality that allows Client

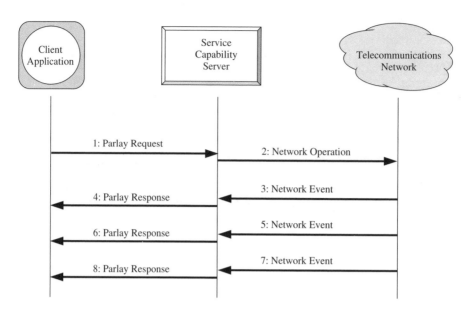

Figure 5.7   Asynchronous method invocation with multiple responses

Applications to intercept phone call attempts in a network (for example, so that the application can forward them to a another number if there is no answer). Nothing in the interface definition itself restricts which phone numbers will be affected by this functionality and so the Client Application could intercept all phone call attempts. Clearly this may not be what is intended and so the SCS may be instructed to reject any attempt to request interception of calls unless a limited range of phone numbers are involved.

Consider a Client Application that supports an insurance company's sales force. One of its roles is to forward unanswered calls to a receptionist in the home office so that the caller always gets through to a real person. Obviously this particular Client Application should only be allowed to work with the set of numbers belonging to that insurance company's employees and not start forwarding calls made to a local pizza delivery service. Or indeed calls that are made to a rival insurance company's sales force.

Perhaps a more vivid illustration of the need to limit Client Application activities relates to calls to emergency numbers such as 999, 112 or 911. It would clearly not be a good thing for a Client Application to be able to accidentally or maliciously intercept such emergency calls.

These simple examples illustrate why an SCS needs to be able to restrict the behavior of Client Applications. There are many other policies that could be defined, and all would need to be agreed before the Client Application starts to use the SCS. Policies may be a set of provisioned rules making up part of the implementation of the SCS, or the SCS may make use of an external entity such as a Policy Management SCS.

The SCS can thus be thought of as the gatekeeper of the network. As well as giving access to the capabilities of the network it must enforce agreements, prevent deliberate or accidental abuse of the network and manage the behavior of the Client Application. In all of this, it is guided in part by the information it receives from the third member of the Parlay triangle and the subject of the next section.

5.4   The Framework

In previous sections we have looked at the Service Capability Server and the Client Application as two elements of the Parlay architecture. We have seen that a Service Capability Server provides

Client Applications with managed access to the functionality previously locked up in telecoms networks. It would be possible for the architecture to be simply a client-server interaction between these two entities. However, we shall see in the coming sections that, to allow the system to be robust and flexible, there are a number of functions that are sensibly separated into another logical entity, and that entity is known as the Parlay Framework. We shall examine why it is sensible to separate out these functions later, but first we need to introduce them.

Clearly there are security implications in providing access to the network functionality that an SCS exposes. Huge sums of money have been invested in the network infrastructure and great effort is dedicated to ensuring high reliability and available capacity. As we have seen previously it is the SCS's job to protect the network by limiting what a Client Application does. There will be certain limits on behavior that will apply to any Client Application so in order to treat each one differently (different functionality or capacity limits for example) its identity must be known. This is where authentication comes in.

5.4.1 Are You Really Who You Say You Are?

The Framework authenticates the Client Application in order to confirm its identity, to confirm that the Client Application is what it says it is. The correct policies can then be applied as it makes use of the SCS. One of the aims of Parlay is that the Framework and Client Applications can all be in different security domains so the Client Application also needs to be sure that it is interacting with a genuine Framework. For this reason authentication is mutual (though always initiated by the Client Application).

It is also the aim of Parlay that an SCS can also be in a security domain different to the Framework and Client Applications. So, of course, mutual authentication can also be applied between the Framework and the SCS (initiated by the SCS).

If mutual authentication has completed successfully, the Client Application or SCS is considered to have established an Access Session with the Framework. For as long as this Access Session is in place the client, whether Client Application or SCS, is able to make use of the services, or interfaces, offered by the Framework. These services are covered in some detail shortly, but some examples are Event Notification, Integrity Management, Service Discovery and Service Agreement Management.

In summary, authentication occurs between the Client Application and the Framework and between the SCS and the Framework. It is the first step in ensuring that the use of network resources is properly managed. Authentication is mutual in each case and is used to establish the identity of each entity involved. Once this mutual authentication is successful, the Client Application or SCS is considered to have established an Access Session with the Framework and can start to use the Framework interfaces.

5.4.2 The Access Session

The Client Application (or SCS) and Framework have mutually authenticated and an Access Session is in progress, but what does this allow the Framework's client to do?

From the Client Application's perspective probably the most important ability that it now has is that it can attempt to start using an SCS (start a Service Session). There are a number of supporting functions that the Client Application makes use of before and during a Service Session. These can help it find the appropriate SCS in the first place (Discovery), and maintain and monitor the session once it is in progress (Fault, Load and Heartbeat Management).

From the SCS's perspective, the Access Session it has with the Framework provides similar capabilities, but its role is a more passive one (the terms of the Service Agreement will determine which functions it needs to make use of). It doesn't need to start a Service Session because this is initiated by the Client Application.

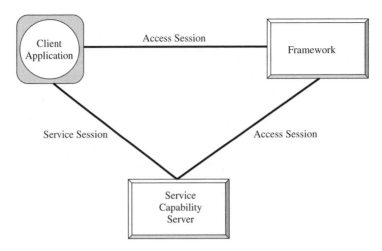

Figure 5.8   Access sessions and service session forming the sides of the triangle

We shall see then that Access Sessions form two sides of a Parlay triangle[1], and can be used to set up and manage the third side (the Service Session), as represented in Figure 5.8.

5.4.2.1 Discovery (and Registration)

All of the proceeding functionality has assumed that the Client Application knows which SCS, or SCSs, it wishes to use. One of the aims of Parlay is that it should be possible to support an 'open market' of SCSs, all providing functionality of use to Client Applications and competing for their business. For a Client Application to make an informed decision in such a market requires an ability to gather information about the nature of the SCSs.

This is where Discovery comes in. The Framework provides an interface to allow a Client Application to specify what kind of SCS it wants and what characteristics are important to it. For example, one of the SCSs defined by Parlay is the Location SCS. Such an SCS allows the location of a particular user (or at least the user's device) to be queried by the Client Application. Since there are different ways that a network can determine the location, this can result in different levels of accuracy. Some Client Applications only need a limited degree of accuracy (for example to determine which town or city a user is in) whereas others (navigation services) may need to locate the user within a few meters. Another characteristic that may be relevant is the rate at which the location information can be updated. Since a frequent update of the information is likely to put a load on the SCS providing it, there may be a higher cost passed on to the Client Application. These characteristics of accuracy and cost are thus ways that one SCS can be distinguished from another.

In order to be able to provide information about an SCS, the Framework needs to have obtained it from somewhere. That somewhere is the owner of the SCS (in Parlay terms the Service Supplier) and the process of obtaining the information is known as Registration. In our example of a Location SCS, the owner of the SCS has a Service Supplier entity that establishes an Access Session (with the Framework) and uses the Registration interfaces of the Framework to provide the relevant information (location accuracy, cost and minimum location refresh interval, for example). This

---

[1] For completeness, it is worth mentioning that there are two further Parlay entities (the Service Supplier and the Enterprise Operator), in addition to the three central entities that we have already described. They too use the Framework and as a result make use of Access Sessions. The Service Supplier is described shortly and the Enterprise Operator will be described later.

information is stored by the Framework and given to Client Applications that are seeking a Location SCS, upon Discovery time.

In conclusion, Registration and Discovery provide the means for the Framework to put Client Applications in touch with SCSs – a lonely hearts club of sorts. Assuming a Client Application has found the SCS of its dreams, they now need to arrange to meet up and talk to each other.

5.4.2.2 Starting a Service Session

The establishment of an Access Session gives the Client Application access to the services (interfaces) offered by the Framework. Ultimately though this can be seen as just a means to an end, since what it really wants is access to an SCS. The Service Agreement Management interface provides the means to get that access.

The procedure is initiated by the Client Application when it decides, perhaps after performing service discovery, on a particular SCS that it wants to use. The Framework and Client Application mutually (and electronically) sign agreement text covering the use of the SCS (the Service Agreement) and the Framework then provides the Client Application with an interface that allows it to start using the SCS. A Service Session has now started.

The Client Application can go on to start Service Sessions with other SCSs available via the Framework, but it will only ever have one session at a time with a particular SCS.

Before we move on, a brief word is necessary to explain what a Service Agreement is. Parlay defines it as a string of text that can be electronically signed using a mutually agreed signing algorithm. No structure or content is defined by the Parlay specification. In the case where a human operator is involved in the Service Agreement process, the text may be read and checked before signing but otherwise it exists only as proof that an agreement has taken place. This may be used at a later point if any dispute between the parties arises.

In summary then, use of the Service Agreement Management interface enables a Client Application to start a Service Session with one or more SCSs and this session allows the Client Application to start using the functionality of the underlying network. The third side of the Parlay triangle is in place.

5.4.2.3 Fault and Load Management

The Fault and Load Management interfaces, along with the Heartbeat Management interfaces covered in the next section, can be used by the Client Application, Framework and SCSs to check and maintain the health of the sessions that have been set up. Collectively they are referred to as the Integrity Management interfaces.

Looking back at the last two sections, we see that during a Service Session (between Client Application and SCS) there are in place three sessions that are intimately related. In addition to the Service Session itself, there is the Access Session set up by the Client Application with the Framework and the Access Session set up by the SCS with the Framework. The latter two can be thought of as the conduits of the Fault and Load Management information. That applies not only to information about the health of the Access Sessions themselves but also to the health of the Service Session. No load or fault information flows directly from Client Application to SCS.

The Framework then acts as a middleman for collection of load and fault information and the information can be thought of as flowing from Client Application to SCS (or vice versa) via the Framework. In addition, load and fault information about the Framework itself can be obtained by the Client Application and SCS (and vice versa), thus checking and maintaining the health of the Access Sessions.

The reality is a little more complex than that as we shall soon see, but first let's return to a question we left hanging earlier: why is a separate Framework entity required? Isn't it possible to incorporate the functions we have just described into the Client Application and Service Capability Service? To answer this, let's consider life without the Framework.

Without the Framework, a Client Application trying to find an SCS (or set of SCSs) would not have a single point of contact. Since it can be assumed that in general there are many more SCSs than 'groups' of SCSs (for example, SCSs owned by a single Network Operator), the problem of finding SCSs is harder without the Framework.

Having found an SCS, a Client Application still doesn't know whether it has the capabilities it wants. With no Framework in place, it has to ask each SCS that it finds (using the discovery interface) what it supports, rejecting those that aren't suitable. This is significantly more time consuming for the Client Application. It is also a burden on the SCS since it potentially has to field enquiries from multiple Client Applications itself, rather than registering once with the Framework and letting the Framework shoulder the burden from then on.

The Framework thus seems to be a good idea as far as discovering SCSs is concerned. We have left a factor out of the above description that makes the benefit of having a Framework even more clear – authentication. If there were no Framework, each Client Application would need to authenticate with each SCS on which it wants to perform Discovery (and subsequently use if suitable). Performing mutual authentication is a significant burden on both parties (especially considering that a match has not been made yet, and some SCSs will be discarded by the Client Application) and the absence of a Framework multiplies this burden. Figure 5.9 illustrates this.

Similar arguments can be applied to the other functions performed by the Framework. This partition of functionality thus seems to be a valid architectural decision. It may also be worth pointing out at this stage that the Parlay architectural entities are of course logical entities. If desired, it would be possible for example for an implementation to include the interfaces and functionality of both an SCS and a Framework, if that made sense in a particular case.

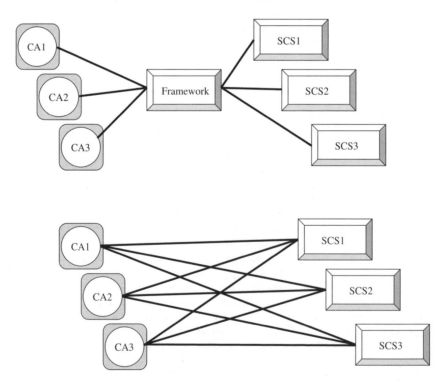

Figure 5.9   The framework as a single point of contact

Having introduced the key entities and procedures of Parlay now is probably the time to bring together the Framework, some Applications and some SCSs to see in detail how they interact. Before leaving this section, however, we outline in broad logical terms, a pictorial view of the capabilities the Framework offers applications and SCSs in Figure 5.10. The specific handshake interaction diagrams with particular methods in each case are represented in [3GPP 2004f], sections six and eight. The interested reader is referred to those sections of the standard for a method level description of the operation of the Framework. The logical sequence diagrams that follow, however, provide an abstracted view of the useful Framework capabilities in each case.

5.5 All Together Now

Let's use a hypothetical scenario to help illustrate how a Parlay ecosystem works. There are three Client Applications (CA1, CA2 and CA3). CA1 and CA2 are owned by one entity and CA3 is owned by another. There are three SCSs (SCSA, SCSB and SCSC, naturally enough) and all are part of a mobile network operator's network. The Framework (FWK) isn't owned by the network operator in this case. It has an agreement with the network operator that it can offer (for a price) the network operator's SCSs and hence is effectively reselling the functionality provided by the network operator.

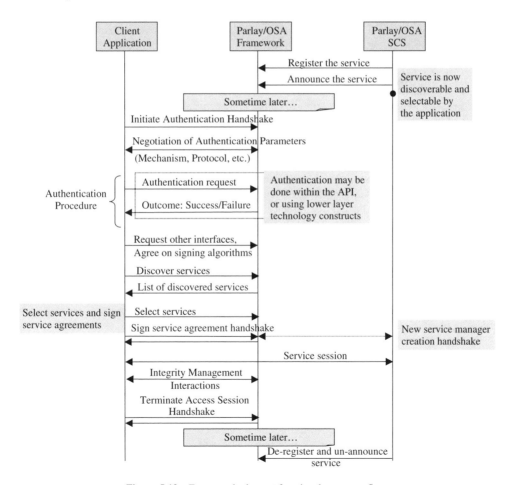

Figure 5.10 Framework abstract functional sequence flows

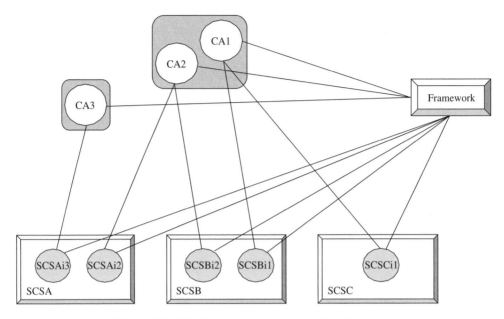

Figure 5.11   The Parlay ecosystem in a hypothetical scenario

We will see that SCSB is a Location SCS whereas SCSA and SCSC are both Call Control SCSs.

Let us have a look at Figure 5.11. How did this picture come about? What were the steps required by each of the entities involved? The previous sections provided a high-level view but we will now dig deeper.

Now if you have been following the discussions closely in the previous sections, something about Figure 5.11 won't look quite right since the three Service Capability Servers (SCSA, SCSB and SCSC) appear to have some inner workings. In order to lay the foundations of an understanding of the Parlay entities, we have made some simplifications. Perhaps the most important of these was to leave unmentioned the Service Instance (there were other simplifications that will become apparent as you read on). We have previously stated that Client Applications set up Service Sessions with SCSs. Though at one level this is correct, the reality is somewhat more complex. When a Client Application signs a Service Agreement to use an SCS, a Service Instance is created in order to handle all operations performed by that Client Application. When other Client Applications subsequently set up Service Sessions, a Service Instance is created for each of them. Thus we see in Figure 5.11 that SCSC has one Service Instance (SCSCi1) and SCSA and SCSB have two Service Instances running (SCSAi1, SCSAi2, and SCSBi1, SCSBi2 respectively), one for each Client Application.

This concept of Service Instances being created for each Service Session between a Client Application and an SCS is a very important one to grasp, since it is essential to understanding the detailed operation of the interfaces between the Parlay entities.

5.5.1 SCS Registration and Announcement

There could be a vast array of SCSs offering different functionality, reliability, cost and capacity to Applications. Faced with such an array of potential suitors, an Application needs to be able to distinguish between them. This leads to the need for a description of each SCS to be available in the Framework for the Client Application to discover. The information is placed in the Framework and then made viewable (and the SCS accessible) in two steps.

The first of these steps (Registration) provides the Framework with information about the SCS (its properties). The second step (Announcement) makes the SCS visible to (discoverable by) Applications.

We have seen previously that it isn't the SCS itself that performs these steps, rather it is an entity acting in the role of supplier, or owner, of the SCS; the Service Supplier. At a time prior to the online Registration and Announcement of the SCS, the Service Supplier will have contacted the owner of the Framework (if they are indeed separately owned) and set up a business relationship. The SCS owner will be armed with a Service Supplier ID and a set of keys – authentication information – to go with it.

Before the Service Supplier interacts with the Framework, it must of course authenticate itself. It makes use of the pre-agreed authentication information in a two-way handshake with the Framework in order that both entities can trust each other. Once authenticated, the Service Supplier may begin an Access Session and is able to choose to do a number of things, but in this case, we are using the Registration and Announcement functions.

Registration, as has been mentioned before, involves providing the Framework with information about the SCS that describes its nature, or characteristics. It will tell its potential suitors – Applications – what it does, and for how much (this is a commercial relationship after all).

There isn't a blank slate on which to write, however. The description must fit a template that the Framework knows about and in turn that the Applications can find out about. This is known as the Service Type and is a way of specifying what the properties of a Service of this type might, or must, have. The reason for defining known types is of course to allow selection algorithms to be implemented in the Client Applications rather than requiring human intervention (which would be required if the description of the SCS was unstructured).

Service Types have been defined for each of the SCFs defined by Parlay. Each has a name and list of properties that can be used to describe any SCS of that type. For example, all Multi-Party Call Control SCSs are of type 'P_MULTI_PARTY_CALL_CONTROL'. There are a number of properties that all Service Types include, such as the Service Name, Service Version and Operation Set. In addition, there are a number of specific properties applicable only to Multi-Party Call Control in this example, such as Maximum Call Legs per Call and Dynamic Event Types. These properties define what is special about the Service Type and the values that the properties can have allow one SCS of that type to be differentiated from another.

Depending on the implementation of the SCS (and perhaps the capabilities of the underlying telecommunications network), different SCSs will register different values for the properties. Our example Call Control SCSs differ in that SCSC can only offer a maximum of two call legs, whereas SCSA can offer up to six (though at the price of being more expensive to use).

It is important to keep in mind that all of the Parlay-defined entities are logical entities. In other words, a physical implementation of them can take many forms and can be as complex or simple as is needed to support the functionality of an entity. For a Service Supplier then, there is the opportunity to take a single physical implementation of an SCS and register it with a Framework as a range of different SCSs. This would allow advertising different functionality and correspondingly different price bands to the Application, while keeping the underlying implementation essentially the same.

Therefore, having chosen a Service Type (template) to work with, the Service Supplier can now register the Service (SCS) by providing the Framework with a set of values for each of the properties. Having received these values and verified that they conform to the Service Type, the Framework allocates an ID that identifies the Service (and returns the ID to the Service Supplier). The Service is now registered.

At this point, the Service is known to the Framework. It has a name (the Service ID) and it has a description. However, it is not yet visible to Applications (it cannot be Discovered) and as a result, Service Sessions cannot be started. Another step is required: Announcement. This can happen immediately after the SCS has been registered, or it can occur at some later point. But either way, the Service is invisible to Applications until the announcement is complete.

Once Announcement is complete, the SCS is discoverable by Client Applications and so the Framework must be in a position to allow Service Sessions to be started. Thus, Announcement involves the Service Supplier providing the Framework with the means to create a Service Instance. The Service Supplier sends the address of an entity known as the Service Factory or Service Instance Lifecycle Manager (SILM) and the sole role of the Service Factory is to create Service Instances when requested to by the Framework. More of this later.

It may not be immediately obvious why the procedure is broken into two steps, since it would seem quite possible to combine Registration and Announcement into a single step. The reason for the split becomes apparent when the reverse procedure is considered: unannouncement and deregistration. Imagine that an SCS needs to be taken out of service (perhaps it is being replaced by a new version, or needs a software upgrade). While an SCS continues to be 'announced', it can be found by Applications and the SCS's Service Factory needs to be ready to create Service Instances. Unannouncing an SCS prevents new Applications from finding it but allows existing sessions to continue (and contracts to be honoured) and allows information about the SCS (and the Service ID) to be retained. Once the necessary changes have been made (a software upgrade or fix to the SCS for example) it can be announced again.

Having registered and announced the Service, the Service Supplier can now end its Access Session with the Framework, its job done (for now). Of course, this Service Supplier may have other Services to register and announce, or other Service Suppliers may have their own Services to offer via the Framework.

In our example ecosystem, all three SCSs are owned by a single Service Supplier so it proceeds to Register and Announce SCSA, SCSB and SCSC. After Registration and Announcement, our Parlay ecosystem now looks something like Figure 5.12.

Not much happening, but there's a lot of potential. Enter three Applications in need of an SCS.

5.5.2 SCS Discovery
We have touched on the need for an Application to find an appropriate SCS, or SCSs. It has a job to do and knows what it needs from a Service (or Services) to get it done. We have seen that Parlay offers a flexible way (Service Discovery) to identify those Services purely from the description held by the Framework. It is always possible for an Application to have obtained the identity (Service ID) of an SCS through offline communication between the Framework operator and Application owner (of course they may be one and the same, in which case this communication is pretty straightforward). This short cut may be desirable in some circumstances and Parlay can support it.

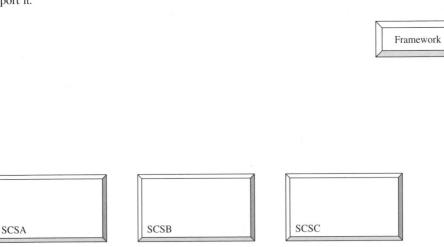

Figure 5.12 Example ecosystem: registration and announcement

In order to start the discovery process, the Client Applications must authenticate with the Framework of course. In our scenario, the owner of CA1 and CA2, and the owner of CA3, will have set up a business relationship of some nature with the Framework owner. This may have involved detailed face-to-face discussions, with extensive agreements being drawn up, policies determined, etc. On the other hand, it could be as simple as the Application owner visiting a web page and providing billing and contact information in return for an Application ID, authentication keys and information on how to contact and authenticate with the Framework.

Whatever the nature of the business relationship, and however that relationship has been arrived at, the owner of the Application is armed with an identifier (the Application ID) and authentication information (keys, etc.). It makes use of authentication information in a two-way handshake with the Framework in order that both entities can trust each other.

Once authenticated an Application may begin an Access Session. As with the Service Supplier, the Access Session opens up a number of possibilities, a number of ways forward. The first of these in our case is SCS Discovery.

As previously described, the SCSs that the Client Application wants to find out about will have been registered against a Service Type. The Client Application can retrieve a list of these Services Types and a list of the properties that define them. A well thought-out set of properties will give the Client Application sufficient information to be able to determine whether the Services registered against this Service Type might just be interesting. The SCSs so far registered and announced with the Framework belong to two types: Location and Call Control. CA1 provides its users with the location of the mobile phones belonging to their children, so it will select the Location Service Type.

Having selected the Location Service Type, CA1 can now plug in some desired property values and ask the Framework to find suitable candidates. There is only one Location SCS (SCSB) in our ecosystem and luckily, it is just what CA1 is after.

It is worth bearing in mind that the process of discovery may be controlled by an algorithm running in a piece of software or may be fully or partially controlled by a human looking at the property values presented to them. The former approach clearly requires that Service Types and the properties that make them up are well defined and machine-parsable.

The Framework may of course have some, one, or no SCSs currently registered and announced that match the requirements of the Client Application. In the latter case, the Client Application may opt to lower its standards somewhat until a match is found, but clearly there may be some requirements it just isn't prepared to relax. If this is the case, the Client Application can either choose to end its access session with this Framework and look into what other Frameworks may be able to offer, or can try again at a later point in time. To help in this, the Framework is able to inform the Client Application of new announcements of SCSs of a chosen Service Type, if the Client Application so wishes.

In our scenario though, CA1 has managed to find a SCS that meets its criteria. All three Client Applications want Call Control SCSs and since two are available, each Client Application will have to make a decision between them based on their needs. As it turns out CA1 only needs a maximum of two call legs on each call (it sets up a call between the parent and their child if they stray outside a 'safe' area) so SCSC is just right. CA2 and CA3 both need to be able to set up conference calls and so SCSA meets their needs since it supports up to six call legs.

CA2 also wants a Location SCS, so once it has also discovered the SCSB, all three of our Client Applications have the Service IDs of the SCSs they want and since they want to get on and start earning money, they begin the process of starting Service Sessions with them[2].

---

[2] Once a Client Application (CA) has the Service ID of a suitable SCS it can choose to start a Service Session immediately, or wait until some later point. The danger with the latter course of action is that the SCS might not still be there to be used so it is generally a good idea to perform Discovery as close to using the SCS as possible.

5.5.3 Service Selection

Service Selection is the process of choosing an SCS and starting a session with it. A straightforward activity in principle, but in Parlay there are a number of steps to be followed, and for very good reasons.

Naturally, the first step is to indicate to the Framework exactly which SCS is of interest. The process of discovery described in the previous section is one way in which the Client Application can obtain the Service ID that uniquely identifies the SCS. There are other ways, outside the confines of Parlay, which can achieve the same result: web page, email or word of mouth perhaps. However obtained, the Service ID is passed to the Framework and in return the Client Application receives another (temporary) identifier known as the Service Token. This is the ID that is used for the rest of the process and can be given a limited lifetime by the Framework. If it expires before being used, the Client Application must perform Service Selection again.

5.5.4 Signing on the Dotted Line

The next step involves an exchange of agreements (Service Agreements) with the Framework. Both parties must electronically sign these agreements, thus providing both parties with a record of the transaction. During this handshake, at the point that the Client Application has signed the agreement text sent to it by the Framework, it is given a reference to an entity called a Service Manager. This is the first piece of the Service Instance seen by the Client Application and implements an interface that can be used to obtain all the functionality promised. In fact, the terms Service Manager and Service Instance are often used interchangeably.

Where did this Service Manager come from? Yes, the Framework gives a reference to it to the Client Application, but the Service Manager is an instantiation of the SCS. It is created by the Service Factory that was provided for the Framework's use when the SCS is Announced. Once the other Client Applications have exchanged signed Service Agreements with the Framework, our ecosystem looks much more complete. Figure 5.13 shows the ecosystem as we have derived it so far.

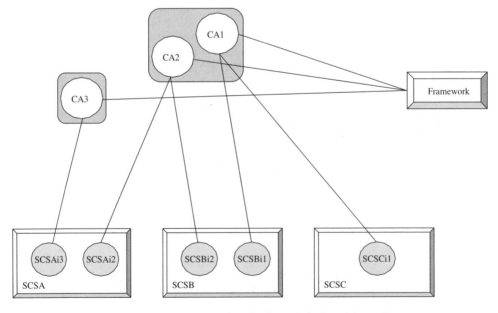

Figure 5.13   Example ecosystem: service selection and signing of the service agreement

The reason why all of the preceding steps exist is to start a Service Session between the Client Application and an SCS. The network functionality that the SCS delivers during this time is what the Client Application needs to deliver in terms of value to its users. At this point, it really is worth money exchanging hands – from the end-user to the Client Application owner and from the Client Application owner to the SCS owner[3].

For this reason, regardless of the nature of the Client Application and the SCS, it may well be worth monitoring the health of the session and providing the means to maintain the quality of service provided. This is one reason why the Parlay Framework provides support for load and fault management of the session, as well as monitoring of both entities involved in the session. We will return to this a little later.

Now let's examine the Service Manager that the Framework has given to the Client Application in a little more detail. No matter what Service Type the SCS is, a Service Manager is the initial instantiation of it and provides the functionality from which all other aspects of the SCS can be obtained. For example, SCSC is a Call Control SCS, providing the ability to control (set up, intercept, or terminate) calls in the underlying network. The Service Manager created for CA1 allows it to create and manage call objects or set up requests for notification that a call to a certain party has begun. All of these operations are part of the Service Manager interface.

5.5.5 The Parlay Triangle Revisited

The essential preliminaries are over. Let's pause at this point and take some time to summarize what is now in place before moving on to examine the life of a Service Session.

There are three basic kinds of entity that are interacting, forming the triangular relationships that are the key to understanding how Parlay works. We have the Client Application, delivering a service to a (hopefully large) group of users. It is interacting with a Service Instance (or Service Manager) of an SCS in order to have access to telecoms network functionality, such as the ability to make a phone call. Enabling, creating and managing this relationship is the Framework.

Reflecting this triangle, forming the sides, are three sessions. The Client Application and Service Instance can each have an Access Session in progress with the Framework. The Application and Service Instance have a Service Session with each other. All of the network functionality is delivered by the Service Session; the directly useful work if you like. The two Access Sessions are there to bring this Session into existence and keep it there for as long as necessary.

5.5.6 Managing the Session

The Client Applications are now using their respective Service Instances. Calls are being made, or locations retrieved perhaps. These activities are as a result of the Client Application's own end-users calling friends, for example, or locating their nearest restaurant. Presumably, they are paying the application owner for the privilege and in return, they probably expect a reliable service. Of course, this expectation of the end-user affects in turn what the Application expects of the Service, and of the Parlay Gateway as a whole.

To help Client Applications and SCSs ensure that the Service Session is there when it is needed, a set of functionality is available. Collectively this is known as Integrity Management. Although use of these interfaces is optional, a robust system is likely to require at least some of the functionality. If on the other hand the Client Application can withstand some downtime, isn't willing to pay for high availability or has some other proprietary means to achieve the same ends, then it is perfectly reasonable to do without.

Earlier we introduced some of the concepts of Integrity Management and that it consists of three main elements: Load Management, Fault Management and Heartbeat Management. We will shortly look at each of these in some detail to see what they provide and why they are useful (or in some

---

[3] Parlay doesn't define how to determine what each party bills each other.

cases, essential), but first, how does the Service Instance of an SCS or a Client Application get access
to them. The Framework is always involved in Integrity Management and so the functionality can
be accessed as part of an Access Session. For the Client Applications in our ecosystem the Access
Sessions are already in place but the Service Instances haven't yet needed one. The process for a
Service Instance to set up an Access Session is exactly the same as for a Client Application. Once
our five Service Instances have done this, our ecosystem is complete. Therefore, in Figure 5.14
we now have gently derived the complete Parlay ecosystem as somewhat abruptly introduced in
Figure 5.11.

5.5.6.1 Load Management

In an ideal world (quite often to be found projected on a screen during presentations) the implementation of the Client Application, SCS and Framework entities in our ecosystem would enable
them to support unlimited demands on their services. Unfortunately, no one has managed to figure
out a way to do this yet. In the meantime the entities need a way to indicate whether the demands
placed on them are going beyond what has been agreed, or even what they are physically capable
of delivering. The Load Management interfaces are used to do this.

The information exchanged allows the entities to modify their behavior in the event of system
overloads. For example by using the Load Management interfaces, CA1 can be informed that
Service Instance SCSBi1 is overloaded. It may then decide to stop creating new calls, or in extreme
circumstances terminate existing calls. It could even decide to start looking for another SCS.

The Framework plays a central role here. We have met before the idea that it acts as a 'middleman' through which all Integrity Management information flows. If CA2 wishes to obtain load
information from Service Instance SCSBi2, it asks the Framework. If Service Instance SCSBi1
wishes to obtain information about CA1, it asks the Framework. And yes, it is sometimes important
that a Service Instance knows whether the Client Application using it is overloaded, for example.
In such circumstances, it could decide that there is little point sending information to the Client
Application until the Client Application is in a position to process it.

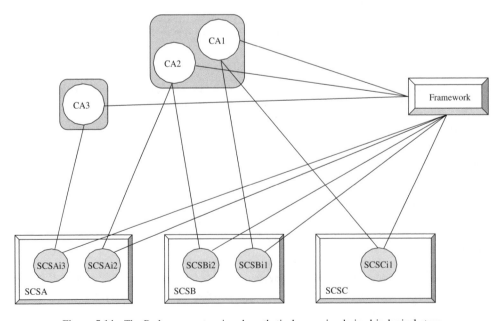

Figure 5.14   The Parlay ecosystem in a hypothetical scenario, derived in logical steps

Let's see how this works in a little more detail. There are two principle types of information that can be obtained via the Load Management interfaces and the Framework's role is somewhat different in each case.

Load Level

The Load Level represents a real-time indication of the load on a particular entity. The Load Level is only reported when it changes, up or down. If an Application wishes to know how loaded a Service Instance is, it requests the Framework to report any changes in Load Level. This request triggers the Framework to ask that Service Instance to report Load Level changes to it. In other words, a change in Load Level is reported by the Service Instance to the Framework and the Framework reports this to the interested Application.

The Load Level can take on one of three values: 0, 1 and 2. 0 indicates that the entity is normally loaded, 1 indicates that the entity is overloaded and 2 that it is severely overloaded. It is on the face of it a very simple and apparently coarse measure of load. It is worth examining the meaning and significance of each of these values in some detail, as it is important to understand that they are not absolute measurements. They may for example be relative to some pre-agreed number of location requests being processed by Service Instance SCSBi2 or an indication that part of the hardware platform that SCSBi1 is running on has crashed and needs some breathing space to recover.

Staying with the above example, a Load Level of 0 (Normal Load) means that a Client Application is using the SCS (or rather, a particular Instance of it) as per prior Service Agreements. For CA1 this may mean that it is setting up no more than five calls per second when using SCSC. For CA2 using SCSA the Normal Load may be up to 20 calls per second. In both cases, the Load Level is 0 and the Client Application can assume that it can continue as it is.

Load Levels 1 and 2 both indicate an abnormal condition. Load Level 1 (Overload) is the Client Application's first indication that something is wrong. A threshold has been reached and a change of behavior may be needed. For example, it may decide not to set up any further calls until the Load Level returns to 0. In general, the Service Agreement between Client Application and SCS will define the expected behavior.

Load Level 2 (Severe Overload) is an indication that despite efforts by the Client Application to reduce the load on the Service Instance (or because of a change in the resources available to the Service Instance) the overload condition has worsened to the point that even existing operations by the Application cannot necessarily be supported. The Application may need to take steps to cancel operations until the Load Level reduces to 1.

Of course, the Client Application too can become overloaded and its load level changes can be supplied to the SCS via the Framework in the same way. One example of an action that an SCS can do to take the heat off an overloaded application is the gapping of or stopping its responses to asynchronous requests.

As you read through the last few paragraphs it may have occurred to you that if load levels are relative to some pre-agreed figure for simultaneous calls or CPU occupancy, then why not just have the load level directly reflect that agreement? For example, if the pre-agreed normal limit for simultaneous calls is 20 then the Service Instance could provide a report of the number of calls and the Client Application could match that against the known limit. There are a number of reasons why this turns out not to be such a good idea.

First, a failure in the Service Instance (loss of hardware, etc) could mean that temporarily it is unable even to support the pre-agreed limit.

Second, having to compare a measurement with a limit requires that the Client Application needs to be provisioned with the limit and implement the comparison.

Lastly, the measurement (or measurements) on which the limit is based could take many forms and thus require the load level to be able to support many kinds of data.

Ultimately what matters is that one entity is able to tell the other either that the situation is normal or that it isn't and something needs to change. Though we have focused on the Client

Applications and SCSs generating Load Level notifications, the Framework is also a critical part of the system and so it too can send them.

Load Statistics

Unlike the Load Level notifications just described, the second type of load information exchanged is non real-time. Load Statistics are historical load measurements that Client Applications and Service Instances are able to request from the Framework. Thus CA1 can make a request that the Framework should ask SCSBi1 for the load statistics for the last 24 hours, for example. SCSBi1 may also decide that it wants statistics about the load experienced by CA1 recently. As with Load Level notifications the Framework may also be asked to provide load statistics about itself.

Since such load statistics may take some time to compile, the operation is asynchronous. In other words, the results aren't returned immediately but will be sent to the requesting entity at some later point. This allows the Service Instance, for example, to prioritize its resources and only calculate the statistics when there is nothing more critical to take care of.

Therefore, what is the information contained in the Load Statistics? The two kinds of load information are the Load Level, which we have already met, and the Load Value, which is a percentage. You might ask: how is this percentage figure arrived at and what does it mean? Parlay doesn't define how it is measured since this would be an almost impossible task. It all depends what the limiting resource is for a particular entity. For example for SCSA it may be disk space, for SCSB it may be CPU occupancy and for SCSC it may be a limitation in the underlying network. Indeed the limiting resource for an entity may change over time. As with the Load Level then, there is no absolute meaning to the Load Value and so it can be seen as providing a more fine-grained measurement of the load on the entity than the Load Level can.

What the requesting entity will see when the Framework replies to its request is a list of Load Levels and Load Values, with an associated timestamp indicating when the load measurement was made. Thus this information shows the history of the Load Level and Load Values changes over a particular time interval. Parlay doesn't define how this information is used.

The Last Resort

There remains one further aspect of Load Management that we should look at. This is the ability for an entity to ask the Framework not to send it any more Load Level notifications for a while. For example, if CA2 was experiencing a critical overload condition, the last thing it wants to have to do is process Load Level notifications from the Framework. It may also need to reduce activity on the Access Session interfaces, of which Load Management is one. Therefore, it can also ask the Framework not to send it any more notifications. Clearly, this is a desperate act as it prevents potentially valuable information from reaching it, but since CA2 may not be able to do anything with it, it might as well ask the Framework to stop.

5.5.6.2 Fault Management

In the same ideal world that we met in the introduction of Load Management section, the implementation of the Client Application, SCS and Framework entities would be such that they would never go wrong. Of course, this ideal world doesn't exist, so there needs to be a way for information about problems to be communicated between these entities so that the appropriate actions can take place. The Fault Management interfaces are used to do this.

The purpose of the fault management interfaces is to allow fault information to be exchanged between entities in the Parlay triangle (Application, Framework and Service Instance). This information allows the entities to modify their behavior in the event of system errors.

Again, the Framework plays a central role. If CA2 wishes to obtain fault information from a Service Instance SCSBi2, it asks the Framework. If Service Instance SCSBi2 wishes to obtain information about CA2, it asks the Framework. In other words, all fault information passes through the Framework.

We now need to dive into this in a little more detail.

The Fault Management interfaces can be thought of as providing three types of functionality. Broadly, they fall into the categories of 'we have a problem', 'are you still working?' and 'tell me about yourself'.

'We have a problem'

For a Client Application, the 'We have a problem' category allows it either to tell the Framework that it has a problem of its own, or to tell the Framework that it can't use the Service Instance with which it has a Service Session. In the first case, the application may have lost communication with an internal resource, for example. This is a controlled (graceful) failure, since the Client Application is still functioning well enough to communicate, and it is a failure that may be recoverable. In the second case, the application will have noticed either that an SCS has stopped responding, or that it cannot continue to use the SCS for internal reasons.

The Framework and the SCS can also use this category of functionality to indicate that there is a problem.

'Are you still working?'

If a Client Application does notice that the Service Instance has stopped responding, one thing it can try to do to check the situation is try an 'are you still working' request. This instructs the Framework to instruct the SCS to carry out a self-test and report the result (if it can). If the SCS is unable to respond to the Framework or the self-test fails, then steps are taken to end the Service Session. The Client Application may also have concerns about the status of the Framework so it can also ask the Framework to carry out a self-test.

As with the 'we have a problem' category, the Framework and SCS can also use this functionality to investigate whether the entities they are communicating with are still working correctly.

'Tell me about yourself'

The final category of requests belonging to the Fault Management interfaces is 'tell me about yourself'. This allows one entity to ask for a record of fault statistics from any of the other entities it is using. Again, the Framework is responsible for gathering the fault statistics and giving them to the Client Application.

There fault statistics contain information about four types of fault. One is a local failure, one is a gateway failure and one is a protocol failure. The fourth type covers faults that aren't defined – in other words faults that don't fall into any of the other three categories.

The Fault Management interfaces thus allow problems to be reported, diagnostic checks to be requested, or fault statistics to be gathered. While it is possible to have Service and Access Sessions running without using these interfaces, it is much harder to design a system that handles problems gracefully unless they are used.

5.5.6.3 Heartbeat Management

The Fault Management interfaces just described make an assumption that the entities have a means to tell whether there are problems with the other entities. Ideally, an entity will detect a problem in itself and report this, but this won't necessarily happen. Alternatively, an entity might decide to request periodically activity tests by the other entities. This has a number of possible drawbacks. A complex activity test puts demands on the requester and any request/response exchange puts additional demands on the requestor. This is likely to mean that activity tests are requested only after a problem is suspected or at infrequent intervals. Heartbeat Management provides an alternative approach.

The Heartbeat Management interfaces allow an entity (the requestor) to ask another entity to start (and also stop) sending it signals (known as pulses, naturally enough) that it is still around.

The time interval between pulses is specified (and can be changed) by the requestor and so if that amount of time has passed since the last pulse, the requestor knows it should start to worry.

There is a very important difference between the Load and Fault Management interfaces and the Heartbeat Management interfaces. Unlike load level reports, activity test requests and so on, requests to send pulses and the pulses themselves are not propagated through the Framework. In other words, the Framework can request the Client Application to send it pulses (and vice versa) and the Framework can request the SCS to send it pulses (and vice versa) but the Client Application can't get the SCS to send it pulses, even indirectly.

For that reason Heartbeat Management might at first sight be thought of as only being useful for monitoring the Access Sessions and not the Service Sessions. This is not quite true as the Framework can use Heartbeat Management to monitor the SCS and report that it is unavailable if the pulses stop arriving. Similarly, the Framework can monitor the Client Application and report a heartbeat failure to the SCS.

5.6   The Enterprise Operator

We have described the three main Parlay entities (Client Application, SCS and Framework) and how they can be thought of as forming the Parlay Triangle. We have also briefly examined the role of a fourth entity (the Service Supplier) in Registering and Announcing SCSs with the Framework. In passing, a fifth entity, the Enterprise Operator was mentioned. We will now describe this entity, and some of the functionality available to it.

In Parlay terms, an Enterprise Operator is an entity, which owns (and/or manages) a set of applications. It can be independent of the Framework owner and therefore must be able to manage its relationship with the Framework and manage its applications' use of services provided by the Framework. A set of interfaces is defined by Parlay to allow it to do just that. Once a business relationship has been set up (offline) between the Enterprise Operator and the Framework (and the Enterprise Operator has been authenticated in the usual way) the Framework's Enterprise Operator interfaces can be used to manage the main entities controlling an application's service use: Contracts, Profiles and Subscription Assignment Groups.

5.6.1 Key Parlay Subscription Model Concepts

By no means all implementations of Parlay will include the Enterprise Operator interfaces, but the underlying data model that they use to manipulate is something that many Gateways will choose to implement at least partially since it is linked to a number of the other interfaces. For that reason we include here details of some of the key aspects of that data model.

5.6.1.1 Service Contracts

In general, the use of an SCS by a Client Application is governed by a contract of some sort. There are contracts and there are Service Contracts. The latter is a Parlay-defined data structure. It contains fields such as the Service Start and End Dates (the period during which an Application may use an SCS), Service ID (the ID of the SCS) and the Billing Contact (a person responsible for billing issues). These, and other fields, are common to all Service Contracts. In addition to these is a set of fields that correspond to the Service Type of the SCS. An implementation of a Parlay Gateway can choose to make use of these Parlay-defined data structures whether the Enterprise Operator interfaces are implemented or not. Naturally, if these interfaces aren't implemented then some other method of populating the data will be provided.

As we have seen before, a Client Application can Discover SCSs via the Framework and decide to use them. For this to happen there must have been an agreement of some nature between the owner of the Client Application and the owner of the Framework in order that Application IDs can be created, security information exchanged and billing arrangements made. This offline agreement is a contract of some description. It typically covers the details of a Client Application's use of SCSs owned by the Framework and may be as simple as a statement that any of the SCSs may be used and the functionality provided by the services will be as described in the service description. This truly gives complete freedom for the Client Application to Discover suitable SCSs with no restriction. On the other hand, owners of Client Applications may prefer to have the means to pre-define what services its Client Applications can use and place additional restrictions on the functionality offered by the services over and above the service description registered with the Framework by the Service Supplier. This is where the Parlay Service Contract comes into play. Rather than Client Applications attempting to Discover what Services are suitable for it to use, the Enterprise Operator decides what Services should be used ahead of time using a Service Contract. As we have seen from the common fields that every Service Contract includes, the Enterprise Operator can also determine the time period during which a Service may be used but it can also determine what service property values apply. A call control Service may have the ability to include up to five call legs but an Enterprise Operator may decide that this should be restricted to just three for a particular Client Application (as a result perhaps limiting the money paid to the Service Supplier). This restriction can be included in a Service Contract.

5.6.1.2 Service Profiles

A Service Profile is, in essence, a Service Contract modifier. A Service Contract can be defined that can be applied to every Client Application using an SCS. If this approach isn't flexible enough for an Enterprise Operator, Profiles are used to further control what functionality is available to each application. In fact, Service Profiles are used even if there is only one application and there are no modifications of available functionality to be applied, but in general, a Service Profile may apply to a number of applications. This is managed through the use of Subscription Assignment Groups, which is the topic of the next section.

5.6.1.3 Subscription Assignment Groups

A Subscription Assignment Group (SAG) is a mechanism for grouping together Client Applications that should be treated in a similar way (from a subscription point of view). Each SAG has a set of Service Profiles associated with it, one per SCS to be used by members of the SAG. Thus, a SAG can be thought of as a convenient way to link a list of Client Applications with a set of Service Profiles that apply to all of them. For example, all of the Client Applications owned by an Enterprise Operator may use the same SCSs but one set of these applications may be considered mission-critical whereas another set are less important. Two SAGs would be created and two sets of Service Profiles associated with them.

Figure 5.15 attempts to illustrate the relationship between Client Applications, SAGs and Service Profiles. It shows two SAGs, one containing three Client Applications and one containing two. SAG 1's Client Applications (the three that have already been assigned, plus any others assigned in the future) have subscriptions to four SCSs. The nature of the subscriptions for each of the SCSs is described in the corresponding Service Profiles. SAG 2's Client Applications have subscriptions to two of the SCSs that SAG1's applications have subscriptions to (SCS2 and SCS3). However, their subscriptions differ since they belong to a different SAG and therefore the profiles are different.

If one of SAG1's Client Applications (CA1 for example) was to be removed and assigned instead to SAG2 its subscriptions to SCS2 and SCS3 would change and it would no longer be subscribed to SCS1 or SCS4. On the other hand, if Service Profile 1a was updated it would change the subscription for the use of SCS1 for all the Client Applications in SAG1.

It is worth noting that in all likelihood the Enterprise Operator that owns all of the Client Applications in these two SAGs would have just one Service Contract for each of the SCSs. For example, Service Profile 2a and 2b would both be derived from a single Service Contract for SCS2.

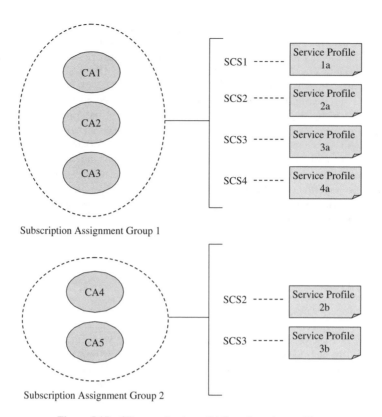

Figure 5.15 Client applications, SAGs and service profiles

5.6.2 The Enterprise Operator Interfaces

The preceding sections have described at a high level the three key entities making up the Parlay Subscription data model: Service Contracts, Service Profiles and Subscription Assignment Groups. Parlay defines a number of interfaces that can be used to manage this subscription data. The Enterprise Operator is able to create, modify and delete Service Contracts and Service Profiles. The Service Profiles can be assigned to Service Assignment Groups (or deassigned) and Client Applications can be added to (or removed from) these SAGs. Naturally, the interfaces also allow the Enterprise Operator to list the details of any previously created Contracts, Profiles and SAGs.

Although we have said previously that it is possible to manage the subscription data using non-Parlay means, the Enterprise Operator interfaces do provide a natural way to do so, and certainly an understanding of these interfaces is necessary for an understanding of the underlying data model.

It may at first sight appear that management of subscription data is only necessary when Client Applications are pre-subscribed to SCSs by their owner. After all, this is the main reason why the interfaces exist. Looked at slightly differently, however, the same interfaces could be used by the owner of a Framework in order to manage Client Applications owned by other entities. As we have seen, use of Service Contracts and Profiles can even apply where a Client Application uses the Discovery interface to find a suitable SCS. This ensures that the application has permission to use the SCSs it has discovered.

This has been a whirlwind tour of the Enterprise Operator entity and the interfaces defined for it. We have seen that there are three key concepts (Service Contracts and Profiles and Subscription Assignment Groups) and associated interfaces that allow the way a SCS is used by a Client Application (or group of Client Applications) to be managed. In principle, these interfaces are

aimed at entities that are independent of the owner of the Framework but they can still be used by
a Framework owner to manage Client Application subscription in general.

5.7 Summary

This chapter has tried to give an overview and an insight into the Parlay architecture. We have
avoided going into detail regarding the exact nature of the interfaces, methods and data types that
make up the architecture (these details are readily available in the next chapter of this book and in
the Parlay specifications themselves) and instead have taken an alternative, narrative approach.

We have seen that there are three key logical entities (Client Application, Service Capability
Server and Framework) that can be considered to be at the core of Parlay. They form a triangle of
relationships that are at the heart of the architecture. Additional logical entities (Service Supplier
and Enterprise Operator) play necessary supporting roles to this 'big three'. In explaining the
nature and role of all of these Parlay entities a number of key concepts (Access Sessions, Service
Sessions, Integrity Management, Registration and Discovery, Subscription) have been introduced
that are fundamental to allowing Parlay systems to be implemented that provide an environment
for the rapid development and deployment of exciting new telecom-based end-user services.

Part II

The Standards in Detail

Standards promote interoperability and seamless inter-working between compatible components developed by different vendors. They enable users of technology products to be able to choose elements from different sources with a reasonable expectation that they will work together without too much bother.

Do Parlay standards measure up to these requirements? Is the standard well enough defined to be worth all this trouble? How does Parlay as a technology really work? What makes it tick? These and related questions are addressed in Part II.

This part of the book is targeted primarily at a technical audience, but a business or marketing oriented non-technical reader can also benefit from some of the examples discussed in Chapters 6 and 7, and the brief technical summaries of standards in Chapter 8.

6

Standards Capabilities and Directions

6.1 Introduction

The previous chapters have introduced Parlay as a technology poised to capitalize on the model of Service Mediation. This took a little working on, as we meticulously followed the yellow brick road that originated with the various network technologies in use today, passed the Parlay pedigree, via the issues faced by the various stakeholders, finally to arrive at the Parlay triangle of Framework, Client Application, and Service Capability Server. Now that we have this under our belt we are fully equipped to start exploring the wide range of Service Capability Features spanning the breadth of the Parlay portfolio. And for that story we must embark on another chapter. We will now shift gears and raise the description of the Parlay technology to the next level.

The Parlay solution is a modular architecture where Service Capabilities are exposed in a secure, controlled and billable manner to application developers. In total, the Parlay solution comprises of twelve Service Capability Features. This section will introduce and discuss each of them, using the specification document structure as outlined in Chapter 4, i.e. the parts of the 3G 29.198 series. For each SCF we will discuss issues relating to how the SCF has evolved, the level of maturity and backwards compatibility, and possible future directions, all from a standards perspective. We will also provide some background information on the various specification versions and the relationship between them.

For the purpose of this book, the 3GPP Release 5 specifications will serve as a basis for the descriptions of the standards capabilities in this chapter. The main reason underpinning this decision is that 3GPP Release 5 is the latest and greatest public release at the time of writing this book. It is not the intention to repeat any of the information contained in the API specifications, and hence this chapter will not contain method signatures with all their parameters and data types, along with their description in full detail. Readers interested in this level of detailed information are referred to the standards specifications themselves.

Table 6.1 lists the exact versions of the Parlay specifications referenced in this chapter.

6.2 Part 1 – Overview

The first part in the series of Parlay API specifications is the Overview [3GPP 2004d]. This specification ties together all parts of the entire suite and provides additional information applicable to all parts. For instance, the abbreviations used and definitions applicable in all parts are only contained in the Overview, and referred to from all the other parts.

Parlay/OSA: From Standards to Reality   Musa Unmehopa, Kumar Vemuri, Andy Bennett
Copyright © 2006 Lucent Technologies Inc. All Rights Reserved

Table 6.1 Parlay specification versions used in Chapter 6

Part name	Specification version
Part 1 – Overview	29.198-1 V5.7.0 (2004–09)
Part 2 – Common Data	29.198-2 V5.8.0 (2004–09)
Part 3 – Framework	29.198-3 V5.8.0 (2004–09)
Part 4-1 – Call Control Common Definitions	29.198-4-1 V5.7.0 (2004–09)
Part 4-2 – Generic Call Control	29.198-4-2 V5.8.0 (2004–09)
Part 4-3 – Multiparty Call Control	29.198-4-3 V5.8.0 (2004–09)
Part 4-4 – Multimedia Call Control	29.198-4-4 V5.8.0 (2004–09)
Part 5 – User Interaction	29.198-5 V5.8.0 (2004–09)
Part 6 – Mobility Management	29.198-6 V5.6.0 (2004–09)
Part 7 – Terminal Capabilities	29.198-7 V5.7.0 (2004–09)
Part 8 – Data Session Control	29.198-8 V5.7.0 (2004–09)
Part 11 – Account Management	29.198-11 V5.6.0 (2004–09)
Part 12 – Content Based Charging	29.198-12 V5.7.0 (2004–09)
Part 13 – Policy Management	29.198-13 V5.6.0 (2004–09)
Part 14 – Presence and Availability Management	29.198-14 V5.7.0 (2004–09)

Some of the more important pieces of overall introductory information that serve as a lead-in to the remaining parts covering the various Service Capability Features are summarized below.

6.2.1 Versions and Releases

In Chapter 4 we have already alluded to the relationship and correspondence between the various release schedules and versioning schemes in use by the participating organizations in the Joint Working Group. A much more detailed discussion of these relationships is included in the Overview specification, capturing the information at the granularity of point releases, and per 3GPP plenary publication cycle. The table for 3GPP Release 5 is reproduced in Table 6.2.

6.2.2 Methodology

The Parlay APIs are modeled, designed and specified using the UML methodology. In order to aid in this process, and to ensure consistency in look and feel across all the various SCFs, a number of agreements have been laid down and recorded in the Overview.

The namespace root for Parlay is 'org.csapi'. This namespace is used to scope constants and data types, and is used as root for the hierarchical package of all SCF interface and data type definitions. For instance, org.csapi.mm.idl contains the definitions for all the Mobility Management

Table 6.2 Specification versions for OSA Release 5

ETSI OSA Specification Set	Parlay Phase	3GPP TS 29.198 version
–	–	Release 5, March 2002 Plenary
ES 202 915 v.1.1.1 (complete release)	Parlay 4.0	Release 5, September 2002 Plenary
ES 202 915 v.1.2.1 (not parts 9, 13, 14)	Parlay 4.1	Release 5, March 2003 Plenary
–	–	Release 5, June 2003 Plenary
–	–	Release 5, September 2003 Plenary
–	–	Release 5, December 2003 Plenary
–	–	Release 5, March 2004 Plenary
–	–	Release 5, June 2004 Plenary
ES 202 915 v.1.3.1 (complete release)	Parlay 4.2	Release 5, September 2004 Plenary

SCFs, whereas org.csapi.SP_MY_CONSTANT scopes a proprietary constant definition for use within Parlay implementations.

Naming conventions are defined for interfaces, method names, exceptions, parameters, data types such as sets and structures, etc. For instance, you will see that all Parlay types have the prefix 'Tp', whereas all interface names start with 'Ip'. Such naming conventions improve the readability of the detailed API specifications and introduce conformity across all interface definitions within the Parlay suite.

Whilst uniformity and structure are two reasons for such conventions as introduced above, other conventions are in place to allow for as many technology realizations as possible and avoid any unnecessary restrictions in the use of specific languages. One example of such a principle is for methods to use return values rather than out parameters, as out parameters may not be supported in some of the languages to which one may wish to map the API definitions.

6.2.3 Interface Design Principles

Apart from conventions, the Overview document also summarizes a number of the design principles applied across the definition of the Parlay APIs. A varied collection of principles is amassed in the Overview specification, some examples of which include the use of NULL values and the use of the service factory pattern.

In this section we shall discuss in detail one of the more noteworthy principles which deals with notification handling and the use of callbacks for event notifications. Two classifications for notifications are introduced. One classification distinguishes notifications based on who enables them. Notifications can be created upon specific request by the Client Application, using the createNotification method. Notifications can also be provisioned by the network operator as a management or service provisioning operation. Examples where this latter mechanism may be preferable include for instance when user data for target addresses (that is, those end-users for which notifications need to be generated) is stored in a network user data store and the Client Application has no access to this store. Or consider for instance the case where bulk provisioning of the triggers is much more efficient using some offline means, rather than using a Parlay method invocation by the Client Application. Once the notifications are provisioned by the network operator, the Client Application invokes the enableNotification method to arm the triggers.

The second means of distinguishing notifications is by their monitor mode. Notifications can either be used to merely inform an interested Client Application of the occurrence of an event of interest (MONITOR mode), or processing in the network can be suspended and the notification doubles as a request to the Client Application on how to continue SCS operation (INTERRUPT mode). We have seen examples in Chapter 5 where an insurance Client Application may be notified of a call attempt and decide to forward the call to an alternative destination.

6.2.4 Shapes and Forms

We have familiarized ourselves with the three technology realizations for the Parlay API interfaces in Chapter 4. The Overview specification contains three annexes each describing patterns, principles, conventions, and mapping rules for each of the technology realizations.

The OMG IDL annex describes for instance how certain straightforward Parlay data types map to CORBA primitive types (e.g. *TpString* to *string*). Some less clear-cut mappings are outlined as well, where some misalignment may exist between OMG IDL and CORBA data types. For instance an OMG IDL sequence maps to a CORBA struct.

The W3C WSDL annex describes all Parlay WSDL namespaces, and the root Parlay WSDL namespace (http://www.csapi.org). Furthermore, the UML to WSDL mapping rules are defined, which we will discuss in further detail in Chapter 16.

The Java Realization annex describes how a local J2SE API and a Java RMI J2EE API are generated from the technology independent UML model. Most of the work is performed in the area of adapting the Parlay API patterns into usage patterns that are more common in the Java developer community, as well as converting towards naming conventions and best practices commonplace in a Java environment. Especially for the J2SE specific conventions, with the creation of the local API, the transformations and conversions are rather involved. For more detail, the reader is referred to Annex C Java Realization API [3GPP 2004d].

6.3   Part 2 – Common Data Types

Part 2 of the specification set [3GPP 2004e] again does not define an SCF, but rather defines a number of data types that are common across two or more SCF specifications.

Data type definitions may be common for several reasons.

1. There are quite a number of base data types that are either used a lot in most all SCFs directly, or used to create more complex compound data types. Examples include data definitions for Boolean, string, and integer.
2. Other data types are defined explicitly in anticipation of frequent use, in order to ensure conformity across the board. For instance the data types for user addresses, time and date, and price and amount related types belong to this category.
3. Yet other data types start out as SCF specific, but are reused in subsequent SCF definitions. This is for instance the case with QoS parameters that were introduced for Data Session Control, and then later reused in Call Control.
4. Last of all there are data types that are intrinsic to the definition of an interface specification, and hence common by definition. These include the address of a generic interface instance, and the reference to such an address.

The language definition files for the common data types may be included in or referenced from each of the definition files of the specific interfaces.

Reuse of common data types not only ensures a consistency in look-and-feel across the entire set of APIs. It also facilitates composite applications that make use of multiple Parlay service interfaces to offer a more feature rich and integrated service experience. Service attributes obtained through one Parlay service interface may be provided as input to a second Parlay interface in order to acquire the attribute of interest. Or multiple attributes of interest, each obtained through individual Parlay service interfaces, may be presented to the end-user in a consolidated fashion. Using the same data type definitions and encodings for the attributes spurs such composite Parlay applications, as well as facilitates interoperability between Parlay applications.

6.4   Part 3 – Framework (FWK)

Part 3 is the first part that actually specifies a Parlay SCF [3GPP 2004f]. Specifically, it specifies the Framework SCF. The functionality that the Framework performs has already been covered in quite some detail in Chapter 5 and therefore will not be repeated here. In this section, we will suffice by enumerating the interface groupings that together make up the Framework SCF, in Figure 6.1.

Some of the interfaces are replicated between the Application and the Framework, and between the Framework and the Service Capability Servers. In order to distinguish between these two groupings, the interfaces between Application and Framework are prefixed with IpXXX and IpAppXXX, while the interfaces between Application and SCS are prefixed with IpSvcXXX and IpFwXXX.

6.5   Part 4 – Call Control (CC)

The Parlay specifications include a suite of SCFs providing the application with control of connection-oriented calls in the network, the so-called Call Control SCFs. The Call Control SCFs can be divided in two classes.

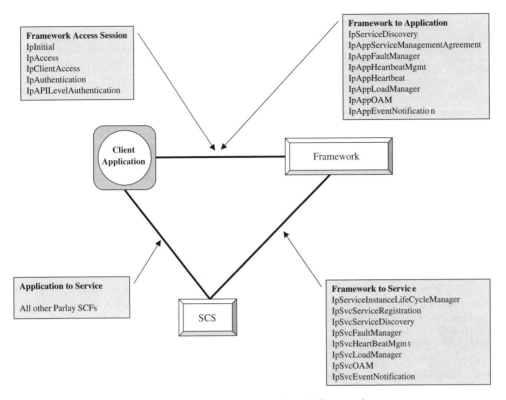

Figure 6.1   Interfaces supported by the Framework

1. *Generic Call Control.*
   The Generic Call Control SCF (GCC) offers a simple means of controlling a call in the network, without the ability to manipulate the call legs to the individual parties in the call [3GPP 2004h]. GCC includes an event notification mechanism, which enables simple call related services like call forwarding to voice mail, or the barring of calls to certain destinations.
2. *Enhanced Call Control.*
   In order to offer more capabilities and richer functionality to the application developer, above and beyond GCC, the Multi-Party Call Control SCF (MPCC) supports a more advanced call model [3GPP 2004i]. The call legs to individual call parties can now be controlled separately, allowing for instance to apply distinct charges to all parties involved, play voice recordings to a specific party in the call, etc.
   The Multi-Media Call Control SCF (MMCC) extends MPCC, by adding the ability to attach media streams to a call [3GPP 2004j]. For instance a video stream can be added to an already existing voice call. The charging capabilities are expanded by the introduction of volume based charging, in addition to time based charging.
   A further extension to MPCC and MMCC is provided by the Conference Call Control SCF (CCC), which adds the ability to set up conferences, reserve conference resources such as a bridge, splitting off sub conferences, and some basic floor control functionality [ETSI 2005a].

   Both GCC and MPCC have been under development within the Parlay standards community for a number of years now, and hence have reached functional maturity as well as operational stability.
   The derivatives of MPCC, i.e. MMCC and CCC, are more recent additions to the suite of Parlay Call Control SCFs, and have therefore been among the last Parlay capabilities to reach a state of

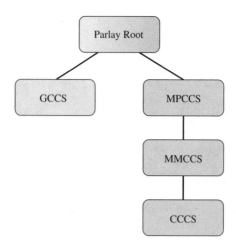

Figure 6.2 Parlay Call Control inheritance structure

maturity. We may expect that they are currently in the process of undergoing the detailed level of scrutinized peer review and developer feedback to make them sufficiently stable for commercial deployment. Minor modifications and corrections may still be anticipated.

All SCFs in the Call Control suite were initially designed as specializations of one another, starting at GCCS all the way through to CCCS. However, being the first call control API to be defined for Parlay, GCCS was very much tailored towards the capabilities of the CAMEL phase 2 specifications defined in 3GPP Release 99. In order to allow derivative Call Control SCFs to be defined with more flexibility and protocol independence, a decision was made to break the direct inheritance structure between these SCFs. This is depicted in Figure 6.2. This decision then allowed the Parlay community to be very conscious of the growing number of GCCS applications in the market. The availability of deployed application puts in place additional restrictions on the amount and severity of functional additions and error corrections. For this reason, only essential error corrections are applied to GCCS, whilst functional additions, feature enhancements and new requirements are considered for the parallel inheritance path only. Despite the two inheritance tracks for call control APIs, there remains a common root, exemplified by a number of common data types and definitions used by both tracks [3GPP 2004g].

6.5.1 GCCS

As explained in Chapter 1, phone calls in a network are represented in terms of a call model or state machine. After 125 years of evolving the telephony systems, with the profusion of bells and whistles, abundance of value added services, and myriad regulatory required features, the telephony state machines can make your head explode when you try to model and document them. And this is only PSTN. Imagine having to familiarize yourself intimately with the call models of a multitude of popular communication technologies in order to ensure as large an addressable market for your call control application as possible. You will quickly come to the realization that there is enormous value to be gotten in a simple abstraction that effortlessly hooks into all of these models.

A gentle introduction to Call Models can be found in Appendix A [Parlay@Wiley]. In this section we shall demonstrate how abstraction of complex call models may aid in the design of call control applications.

The diagram in Figure 6.3 shows the state machines for the Originating Basic Call State Model, or O-BCSM, in IN (on the left) and the state transition diagram for the Originating Call Model of the MultiParty Call Control SCF in Parlay (on the right). It is not the intention here to provide a detailed

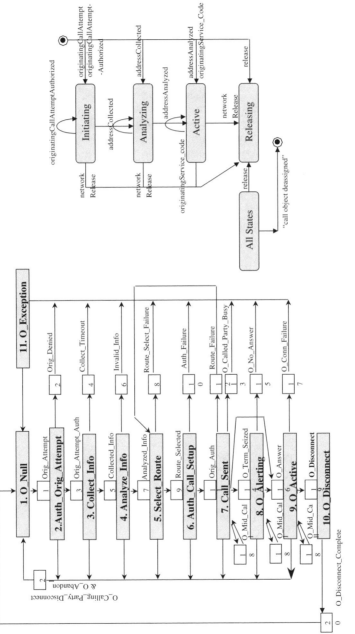

Figure 6.3 Originating call models compared: IN and Parlay

description of the IN O-BCSM; interested readers are referred to [Faynberg 1996]. The diagram is shown to demonstrate what we mean by saying that the Parlay Call Control SCF provides an abstracted view of the call control models in the network, in this case IN. Without explaining the O-BCSM, it is quite evident, if you hold this page of the book at arms length and sort of squint your eyes, that the view of call control behavior as abstracted by Parlay is clearly less complex.

Yet another means of demonstrating the implication of abstraction is by looking at the type of specific network events that can be conveyed through the abstraction to the Client Application. Parlay has opted for both the case to abandon certain protocol specific event types that have no significance in other service control protocols, as well as the case to include certain protocol specific event types that have proven to be extremely useful in one protocol, though not supported in others. In Table 6.3 we provide examples for the IN and SIP protocols.

It is interesting to note how different protocol characteristics are hidden because of the abstraction. For instance, Parlay events are the result of a trigger firing in a CAP scenario, whereas in the case of an IMS network, these are the result of SIP protocol messages being forwarded by the S-CSCF based on so-called filter criteria[1]. Towards the application this is all transparent. Whether there was a CAP Release or a SIP BYE because of the destination party hanging up the phone, the application will observe this as the Parlay event P_CALL_ATTEMPT_TERMINATING_RELEASE. This transparency caused by abstraction is depicted in Figure 6.4. For an interesting approach on how to combine the service creation capabilities of Parlay with the session control facilities of SIP, the reader is referred to [Kozik 2003,Unmehopa 2002a].

The GCCS SCF adheres to the principle of 'single point of control'. This means that although several applications may have registered interest in a single call in the network, only one of them can actually apply control through the Parlay Call Control API. Effectively, all interested applications receive notification of the occurrence of the event (through the method invocation callEventNotify), but only one of them receives the event in INTERRUPT mode. The remaining applications get notified in MONITOR mode. The single point of control principle ensures that no unpredictable behavior will take place. Consider for example two applications that have registered their interest in any subscriber who is roaming into a given foreign city. Application A is an application deployed by the city government, welcoming the visitor to their city, whilst Application B is an application operated on behalf of the local grocery chain drawing the visitor's attention to special discounts for

Table 6.3 Abstracted event types in Parlay

Parlay Events	IN Events	SIP Events
P_CALL_ATTEMPT_ORIGINATING_CALL_ATTEMPT	OriginatingCallAttempt	INVITE
P_CALL_ATTEMPT_TERMINATING_CALL_ATTEMPT	TerminatingCallAttempt	INVITE
P_CALL_ATTEMPT_ADDRESS_COLLECTED	AddressCollected	INVITE
P_CALL_ATTEMPT_ORIGINATING_CALL_ATTEMPT_AUTHORIZED	Originating-CallAttemptAuthorized	INVITE
P_CALL_ATTEMPT_TERMINATING_CALL_ATTEMPT_AUTHORIZED	Terminating-CallAttemptAuthorized	INVITE
P_CALL_ATTEMPT_ADDRESS_ANALYZED	AddressAnalyzed	INVITE
P_CALL_ATTEMPT_ALERTING	Alerting	180 Ringing
P_CALL_ATTEMPT_ANSWER	Answer	200 OK
P_CALL_ATTEMPT_ORIGINATING_SERICE_CODE	Midcall	N/A
P_CALL_ATTEMPT_TERMINATING_SERICE_CODE		
P_CALL_ATTEMPT_ORIGINATING_RELEASE	Release, NetworkRelease	BYE, CANCEL
P_CALL_ATTEMPT_TERMINATING_RELEASE	Release, NetworkRelease	BYE, 3xx, 4xx, 5xx
P_CALL_ATTEMPT_REDIRECTED	N/A	3xx
P_CALL_ATTEMPT_QUEUED	N/A	182 Queued

[1] The service filtering mechanism in IMS networks is briefly explained in Chapter 1.

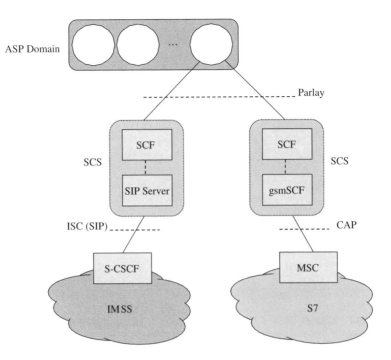

Figure 6.4 Parlay abstraction of IMS and SS7 network events

tourists or business travelers. If both these applications wish to push their content to the unwary visitor through a voice announcement that is being played as soon as the visitor connects for her first call in the new city, things will go awry. Which application will receive control over the call object representing the call in the network may depend on service provider policies, service level agreements, etc.

The single point of control principle is enforced by the SMG by checking for overlapping trigger criteria, as depicted in Figure 6.5. The criteria may overlap if both originating and terminating address ranges overlap, and if the same numbering plan is used and applies to the same notification type (i.e. applies to the originating or terminating call model). But this alone is not enough. Some events are mutually exclusive in that they can never both occur during the same call, and hence will never lead to applications interested in these events to compete for control over the call. For instance, a person cannot at the same time answer and not answer a call, and therefore the events for 'busy' and 'answer' will never both occur during the same call and therefore are said not to overlap.

The Generic Call Control service has a number of limitations in terms of the functionality and control it provides. For example, only two call legs can be modeled (prohibiting for instance the modeling of conference calls) and these legs cannot be controlled individually (prohibiting for instance the playing of an advice-of-charge announcement to one part in the call only). Application initiated calls (sometimes referred to as third party calls) cannot be supported either. The reason for this is historic and is a consequence of modeling GCCS closely after the CAMEL service in 3GPP networks. Since the InitiateCallAttempt operation from the IN-CS2 model is not supported in the CAP protocol, there is no standardized mapping towards the createCall API method supported in GCCS. Invoking createCall results in the creation of an IpCall object in the Call Control SCS, and not to the creation of an actual call in the network

The IpCallControlManager interface provides the manager interface for the GCCS service. Trigger Detection Points can be armed using enableCallNotification and, if fired, are reported up to

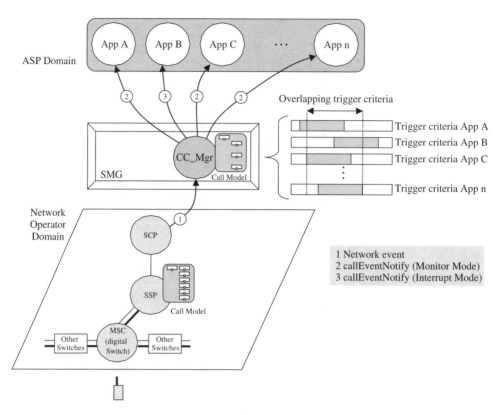

Figure 6.5   Single point of control

the application using callEventNotify. For an introduction of Trigger Detection Points and Even
Detection Points, the reader is referred to Appendix A on Call Models [Parlay@Wiley]. In addition
to the management and reporting of static triggers, the IpCallControlManager interface can be used
to perform load control on calls in the network controlled using the GCCS APIs.

Event Detection Points are armed in the network using the routeReq method of the IpCall interface of GCCS, and hence are reported back up to the application using the routeRes method. Since
multiple Event Detection Points can be armed for the same call, the routeRes method may be
invoked multiple times for the same call. Time-based call supervision, for the purpose of prepaid
charging, is supported through the superviseCallReq method. Further charging capabilities are supported through setAdviceOfCharge, to allow the sending of call charging information to the caller,
and setCallChargePlan, to set a specific charge plan for the call. Furthermore, additional digits can
be collected, using getMoreDialledDigitsReq.

6.5.2 MPCCS

There are many similarities between GCCS and MPCCS. For instance, we see the overload control
capability in the manager interface, and the call supervision and charging capabilities in the call
interface recurring. We see them recurring, since as you recall the inheritance relation between
GCCS and MPCCS was broken.

In addition to similarities, however, there are also significant differences. For instance, the pattern
for notifications in the manager interface is different. In addition to the mechanism to arm Trigger
Detection Points and report them back to the application (createNotification and reportNotification),

there is also support for enabling of triggers that were bulk provisioned via some offline OA&M mechanism (enableNotifications and disableNotifications). We see this new pattern for the management of notifications return throughout the Parlay Service Capability Features.

As in GCCS, call supervision, advice of charge, and charge plan functionality is supported on the IpMultiPartyCall object. Routing to individual parties in the call is performed by invoking the createAndRouteCallLegReq method. There is no support for a createAndRouteCallLegRes method, as the arming of Event Detection Points occurs per individual leg, and hence is supported on the IpCallLeg interface.

The createAndRouteCallLegReq method is a so-called convenience function, as it combines the functionality of creating an IpCallLeg object (createCallLeg) with routing it to a specific destination.

As the MultiParty Call Control Service provides leg manipulation capabilities, in addition to the IpMultiPartyCall interface, there is the IpCallLeg interface to control individual parties in the call. The routing function now occurs on the level of the CallLeg object, rather than the Call object, as was the case with GCCS. There is no routeRes method supported, as the arming and reporting of Event Detection Points occurs through the eventReportReq and eventReportRes methods.

In order to build multiparty calls incrementally call leg object can be attached (attachMediaReq) and detached (detachMediaReq) from their associated call object.

6.5.3 MMCCS

The MultiMedia Call Control Service inherits from the MultiParty Call Control Service, and hence includes all the leg manipulation functionality. The IpMultiMediaCallControlManager interface, in addition, supports notification specific to media streams. This includes for instance the direction of the stream, certain audio codecs (e.g. G-711) in the case of an audio stream and video codecs (e.g. MPEG-1) in the case of a video stream.

The call object for MMCCS (IpMultiMediaCall) introduces call supervision based on volume, rather than based on time.

6.5.4 CCCS

The Conference Call Control service allows the client application to setup multimedia conferences (multimedia, as this interface inherits from MMCCS). Conference resources are managed via the IpConfCallControlManager interface.

The CCC SCF provides management capabilities for the entire conference, such as basic floor control, chair selection and speaker appointment, as well as management of sub-conferences, including the ability to create, merge, and split sub-conferences and move parties from one sub-conference to another.

Sub-conferences may be used for instance to build up a large conference out of smaller conferences, or can be used for a subset of participants to have a private consultation. Conferences can be created at a specific time determined in advance, based on a reservation, or can be created directly, through the invocation of the createConference method. Several conferencing policies can be supported through the changeConferencePolicy method. For example, participation in the conference may be upon invitation only. In another example policies may determine whether the video stream is assigned by the conference chair, or automatically assigned to the conference participant who has the floor at the time.

6.6   Part 5 – User Interaction (UI)

One straightforward way of making applications more personalized and adapted to a subscriber's specific wishes or preference profiles, and thereby enhancing the end-user experience, is by directly interacting with the subscriber when deploying and delivering the application. User interactions can be used to guide subscribers through a decision process or menu structure (press '1' for hot,

'2' for cold), as a kind of heartbeat or pacing instrument (press 'OK' to continue), explicitly to solicit input (please enter your 4-digit PIN), or to provide feedback on the ongoing session ('you are being redirected, please hold', or 'you have 8 dollars and 20 cents left').

Network capabilities used to realize user interactions may range from recorded voice announcements and playing DTMF tones in band and SMS or USSD out of band from the traditional telecommunications world, to pop-windows, or menu-driven and browser-based dialogues in the packetized data communications world. Another way to express this distinction in possibilities is to recognize two categories of user interactions: those that occur within the context of a call and those that do not. The first category is referred to as Call Related User Interaction, whereas the second category is termed Non-Call Related User Interaction.

In a Parlay context, the User Interaction SCF [3GPP 2004k] provides an important capability that may be leveraged in the context of a charging call flow, to obtain user approval for a transaction that impacts her account balance. It is also useful that users be notified of events important to them, such as when certain location-based applications are trying to obtain their location in order to provide location-specific content to them. In either case, a means has to exist for the SCS to notify the user of a certain occurrence during application request processing, and to obtain her approval, feedback, or other type of input if needed. Another example of where this may be used is to notify users of changes in the presence or availability status of other users, etc.

Such a capability presents the Service Mediation Gateway with a convenient mechanism to push alerts to a handset and receive responses to these notifications and take these contents into considerations during further request processing.

A specific example may be a WAP Push alert to a user's handset, sent by the UI SCF through communication with a WAP Push Proxy [WAP Push]. The user may generate responses by loading the URL reference from the alert, and interacting with the server that hosted this page via HTML forms, while the server in turn extracts these data using a CGI-script, and making the message contents available to the UI SCF in question. This innovative user interface enables users to personalize features and services on their phone very easily.

The User Interaction SCF is the link between Parlay applications and their users. The User Interaction SCF allows client applications to send notifications to an end-user and to play announcements and collect information from an end-user. Users can access an application from different devices (PCs, mobile phones) using different presentation protocols (HTML, VoiceXML, SMPP) and using other network resources, e.g. Intelligent Peripherals. The purpose of the User Interaction SCF is to provide client applications with a generic interface to handle these different access types.

Figure 6.6 shows pictorially an example of how Call Related User Interaction can be used to play an application-initiated announcement to Alice, the end-user. The panel in Figure 6.6 shows the sequence of steps involved in the example scenario.

An example scenario for an application using the Non-Call Related User Interaction capabilities to push content to end-user Alice is illustrated in Figure 6.7. Here, the User Interaction SCF interacts with the GMSC in the network to govern the delivery of SMS stock quotes.

There are striking similarities in design and use of patterns between the call-based User Interaction service and the Call Control APIs, which makes the UI SCS easy to understand. One exception that is immediately obvious is the different naming convention for the method in the manager interface to report on the firing of trigger. The IpAppUIManager interface supports the reportEventNotification, whereas, according to naming conventions, we would have expected reportNotification. The reason for this is backwards compatibility and not supporting method overloading. In case an error is found in the definition of a specific method, this is corrected in the standards specifications. However, because of backwards compatibility considerations, existing methods cannot be updated thereby creating two methods of the same name with slightly different behavior. Hence, a new and correct method is introduced, with a new but similar name. The existing and incorrect method will be deprecated.

When looking at sample User Interaction scenarios, one quickly realizes there is a close relationship between the Call Control service and the User Interaction service. For instance, the UI

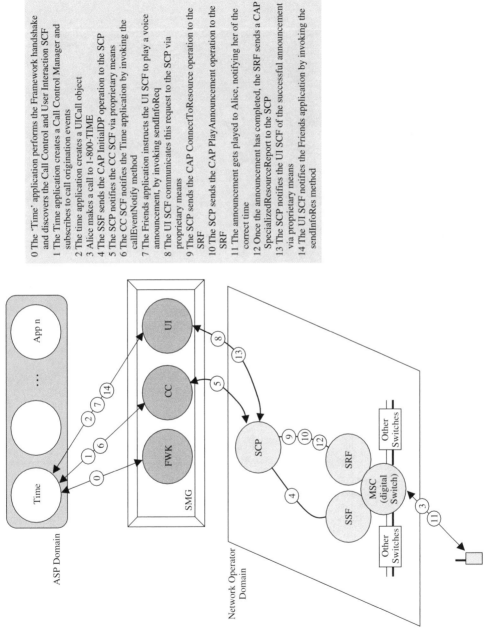

0 The 'Time' application performs the Framework handshake and discovers the Call Control and User Interaction SCF

1 The Time application creates a Call Control Manager and subscribes to call origination events

2 The time application creates a UICall object

3 Alice makes a call to 1-800-TIME

4 The SSF sends the CAP InitialDP operation to the SCP

5 The SCP notifies the CC SCF via proprietary means

6 The CC SCF notifies the Time application by invoking the callEventNotify method

7 The Friends application instructs the UI SCF to play a voice announcement, by invoking sendInfoReq

8 The UI SCF communicates this request to the SCP via proprietary means

9 The SCP sends the CAP ConnectToResource operation to the SRF

10 The SCP sends the CAP PlayAnnouncement operation to the SRF

11 The announcement gets played to Alice, notifying her of the correct time

12 Once the announcement has completed, the SRF sends a CAP SpecializedResourceReport to the SCP

13 The SCP notifies the UI SCF of the successful announcement via proprietary means

14 The UI SCF notifies the Friends application by invoking the sendInfoRes method

Figure 6.6   Call-related user interaction: announcement example

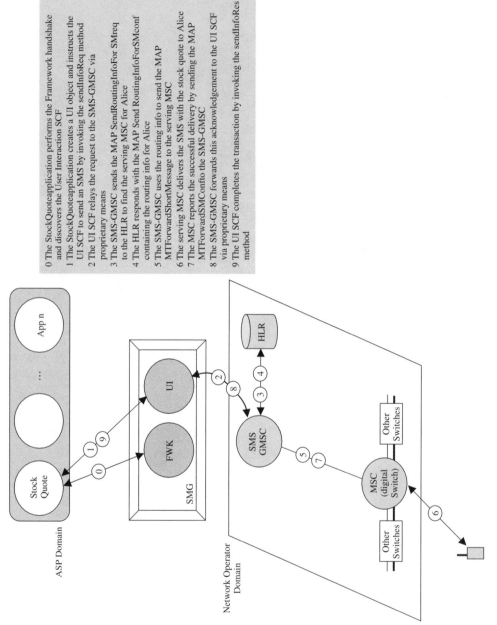

0 The StockQuoteapplication performs the Framework handshake and discovers the User Interaction SCF

1 The StockQuoteapplication creates a UI object and instructs the UI SCF to send an SMS by invoking the sendInfoReq method

2 The UI SCF relays the request to the SMS-GMSC via proprietary means

3 The SMS-GMSC sends the MAP SendRoutingInfoFor SMreq to the HLR to find the serving MSC for Alice

4 The HLR responds with the MAP Send RoutingInfoForSMconf containing the routing info for Alice

5 The SMS-GMSC uses the routing info to send the MAP MTForwardShortMessage to the serving MSC

6 The serving MSC delivers the SMS with the stock quote to Alice

7 The MSC reports the successful delivery by sending the MAP MTForwardSMConfto the SMS-GMSC

8 The SMS-GMSC forwards this acknowledgement to the UI SCF via proprietary means

9 The UI SCF completes the transaction by invoking the sendInfoRes method

Figure 6.7 Non-call-related user interaction: SMS example

SCF can be used to play a recorded voice announcement to parties in a call setup using the CC SCF. Upon closer inspection, we see the relationship is even tighter. The createUICall method is used to create a new user interaction object for call related purposes. It takes as input the identifier of the call or call leg object representing the call with which the user interaction is associated. So we see the call or call leg object's reference is passed to the UI Call object to provide means to correlate those two in the Parlay gateway. Note that no standardized API exists for sending the object reference from one SCS to another, and proprietary means would be required. If both SCSs are supported on the same physical node, this may be easy to realize. But if both are supported on different physical nodes, or provided by different manufacturers, this may not be possible. When deploying a Parlay solution in your network, or when designing your Parlay Gateway platform, these considerations are important to bear in mind

The IpUICall interface provides the application with the ability to prompt the end-user for input and subsequently record a message (recordMessageReq). The sendInfoReq method on the IpUI interface can be used to play back the recorded message. Previously recorded messages can be deleted using deleteMessageReq.

6.7   Part 6 – Mobility Management (MM)

If one feature can be said to characterize mobile networks, it is of course the fact that the users are mobile. That is rather obvious, but it does introduce the complex task of mobility management to the already involved procedures of basic call processing in telecommunication networks. Hand-over of voice channels needs to occur when a user moves between coverage areas, users have to be paged before a call attempt can be terminated, roaming agreements are required to be in place when changing networks, subscriber data are copied from the home location to whichever switch is serving the user at a particular moment in time, etc. However, the result of these added complexities is that it is possible to know where a mobile user is and this knowledge can serve as a value-add to applications. Indeed this feature is often presented as the hallmark of mobile applications. Information related to the position of a user can be taken advantage of and be embedded in the service logic of location aware applications. Both behavior of an application as well as the content may be adapted depending on the user's whereabouts. The Mobility Management SCF of Parlay covers both the user location as well as the user status [3GPP 2004l]. User Location, as the name suggests, pertains to the location of a user, be it geographical location or network location, whereas User Status concerns the network availability of a user.

Before we take a more detailed look at the various mobility management interfaces, we first introduce a number of communication patterns used across these interfaces. The asynchronous request pattern, constructed by the triplet of Request, Response, and Error, is a common pattern across the entire Parlay suite as we have seen, and is supported on each of the interfaces within the Mobility Management SCF. This pattern can be used for one-off or irregular location or status requests. The triggered request pattern allows the arming of a trigger for the generation of reports, based on for example entering or leaving a given location area, or a change in status. The triggered request pattern supports a start and a stop request on the server interface, an error message and the actual triggered report on the application interface. The periodic request pattern is a variant of the triggered request pattern where the trigger is generated by the time out of a periodic timer. Like the triggered request pattern, the periodic request pattern supports a start and a stop request on the server interface, an error message and the report itself on the application interface.

6.7.1 User Location

The User Location SCF allows the application to obtain the geographical location of both mobile as well as fixed subscribers, although perhaps periodic location reports may reveal a certain lack of moving about regarding these latter types of users. Two types of asynchronous location requests are supported, i.e. a simple request for the location of one or more users and a more advanced location request. The simple request merely takes one or more user addresses as input, and delivers the location. The location is provided in terms of geographical location coding as defined in [WGS84][2], which defines a location as a certain shape or surface (an ellipsoid[3]) at a given latitude-longitude pair. The advanced location request, called extended location request, allows the application to be more specific when requesting the location. For instance the application has the ability to specify a certain response time for the report or accuracy of the returned location data. As a result of such an extended request, the returned report is also more involved, i.e. in addition to the geographical position as returned with the simple request, the additional information indicated in the extended request is returned as well.

The current whereabouts of a given subscriber is privacy sensitive information. In most cases, unlawful, unsolicited, or inappropriate access to privacy information is safeguarded by local, national or regional authorities or regulatory organizations. Access to location information by third party applications, through the User Location SCF, is bounded by the privacy restrictions implemented in the network due to such regulations. For instance in 3GPP networks, a distinction is made between various service types seeking to access location information. Access must be granted to emergency services (as we will see later in this chapter), the network operator may access location data for the routing of calls, but access may be prohibited for tracking services or location based information services. In addition to regulatory provisions in the network regarding access rights to location data, the Service Level Agreement between the Parlay application and the network operator, for use of the User Location SCF, may add auxiliary restrictions to safeguard further the privacy of user data.

The User Location SCF supports both the triggered as well as the periodic request pattern. Both these requests return the extended location reports, rather than the simple location reports. The supported criteria for triggered requests, which are contained in a separate interface IpTriggeredUserLocation, are either the entering or leaving of a certain location area, again defined using geodetic location coding [WGS84].

Figure 6.8 provides an example of the use of the User Location SCF, where the Parlay API invocations are mapped towards the LIF MLP operations towards the GMLC. The network architecture and protocol aspects of Location Based Services were introduced in Chapter 1.

6.7.2 User Location Camel

As opposed to the User Location SCF, which retrieves geographical information for the application, the User Location Camel SCF provides the ability to attain network-related location information. This network location information, such as a Cell Identifier, has no direct relation to geographical location. The User Location Camel SCF supports all three patterns of direct asynchronous, triggered and periodic location requests.

The trigger criteria differ from those of the User Location SCF, as geographical areas have no significance here. Rather, network location reports can be generated as a result of a location update occurring within a given VLR area, or when a user moves from one VLR area to another.

The returned location reports differ as well. Network location information is returned to the application, as opposed to geographical information. The CAMEL location report may contain a VLR number, which gives a very crude and far-flung indication of location. The CAMEL location report may also contain a slightly more accurate location indication, in the form of a Cell Global Identification (CGI) or a Location Area Identification (LAI). A LAI consists of a mobile country code, a mobile network code, and a location area code. A CGI consists of the same, and in addition a cell identification. Network location information provided by the CGI or LAI can be mapped onto geographical coordinates, although the granularity of latitude-longitude can never be

---

[2] WGS84 is a location positioning reference system used for GPS satellite navigation and air traffic control systems. It is useful as it provides a way to express any given location in a universal manner.

[3] An ellipsoid is a smooth mathematical surface that best fits the irregular shape of the earth's surface.

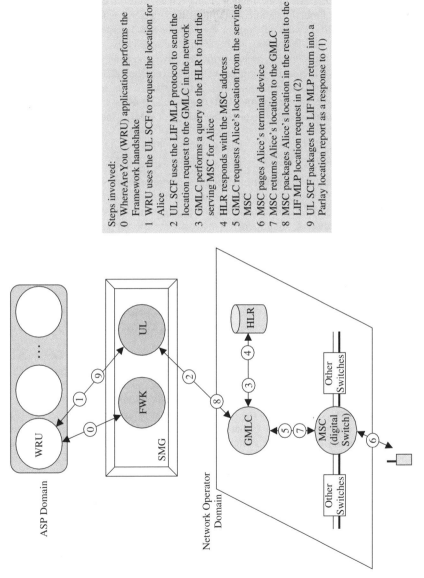

Figure 6.8 User location: location based services example

Steps involved:

0 WhereAreYou (WRU) application performs the Framework handshake

1 WRU uses the UL SCF to request the location for Alice

2 UL SCF uses the LIF MLP protocol to send the location request to the GMLC in the network

3 GMLC performs a query to the HLR to find the serving MSC for Alice

4 HLR responds with the MSC address

5 GMLC requests Alice's location from the serving MSC

6 MSC pages Alice's terminal device

7 MSC returns Alice's location to the GMLC

8 MSC packages Alice's location in the result to the LIF MLP location request in (2)

9 UL SCF packages the LIF MLP return into a Parlay location report as a response to (1)

achieved. However, depending on for instance the size of the particular cell site a certain precision can be reached. More specific information on CGI or LAI encoding can be found in [3GPP 2004c].

The astute reader will have noticed that the User Location Camel SCF runs aground in the shallow waters of network dependence. It is fair to comment that the User Location Camel SCF violates one of the basic design principles of Parlay. The reason for this is historic, if you will. The User Location Camel SCF was designed in the days when 3GPP and Parlay were not yet working together in unison, as they are now in the Joint Working Group. 3GPP required a location API that could expose the somewhat limited location capabilities supported by the CAP and MAP protocols. There was not yet any 3GPP network support for the collection and dissemination of geographical data. For practical purposes and for reasons of backwards compatibility, the User Location Camel SCF continues to be supported. Finally it is worth mentioning that the User Location SCF can be implemented on a CAMEL based system as well, however, the accuracy of the geographical information returned is restricted by the granularity of the network user location information.

Figure 6.9 shows the same WhereAreYou (WRU) application as in Figure 6.8, though in this example scenario the network capability is provided through CAMEL functionality.

6.7.3 User Location Emergency

Emergency calls in mobile networks are treated as a special type of call. For instance they use a specific numbering plan (e.g. '911' in North America does not adhere to the scheme of local and long distance numbers) and often have a higher priority in the network than regular calls (or may even pre-empt ongoing calls), to ensure guaranteed completion of the emergency call in busy hours. Often there are regional regulatory requirements that mandate the support for retrieving location information regarding the user placing the emergency call, such as the FCC's E911 requirements. Such additional information accompanying an emergency call originated from a mobile phone will aid in timely dispatching emergency medical support or law enforcement staff to respond to the emergency call, in case urgency is required or in case a victim is incapacitated and unable to provide the location.

Location requests for emergency calls are obviously only applicable to wireless users. For a fixed line phone call to an emergency number, the location is known in advance as any fixed line is registered at a given address ('just follow the wire').

The User Location Emergency SCF allows an application to obtain the location of a mobile user who initiated an emergency call, using the information that accompanied the original emergency call. This information may include generally available data such as the user's phone number and mobile phone equipment identifier. The information may also include very specific emergency service related parameters. An example is the telephone number of the emergency service provider (such as a dispatcher) and its associated Location Services (LCS) client, or the telephone number used to route the emergency call from the switch where the call originated to the emergency service provider. The actual location request itself is the same as used with the extended location request in the User Location SCF, that is, the very accurate geographical position.

The application also has the possibility to subscribe to triggered emergency location reports, based on whether emergency calls are originated or released. Such a subscription does not relate to an individual emergency call, but to all emergency calls by all possible mobile users placed during the time of the subscription.

Here again we see that Parlay seems to have strayed from the network and protocol independence design principle, however arguably for better reasons than is the case with User Location Camel. A conscious decision was made to depart from the principle in order to fulfill very specific requirements, and mandated regulatory requirements at that.

6.7.4 User Status

A telephony switch maintains data on the status of a user for the purposes of both basic call processing as well as switch-based or IN services. Such network user status may for instance indicate

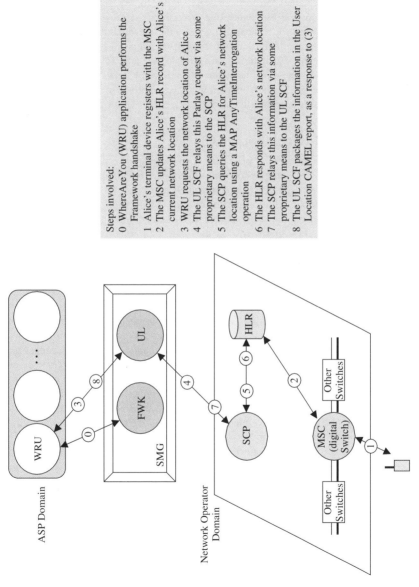

Steps involved:

0 WhereAreYou (WRU) application performs the Framework handshake
1 Alice's terminal device registers with the MSC
2 The MSC updates Alice's HLR record with Alice's current network location
3 WRU requests the network location of Alice
4 The UL SCF relays this Parlay request via some proprietary means to the SCP
5 The SCP queries the HLR for Alice's network location using a MAP AnyTimeInterrogation operation
6 The HLR responds with Alice's network location
7 The SCP relays this information via some proprietary means to the UL SCF
8 The UL SCF packages the information in the User Location CAMEL report, as a response to (3)

Figure 6.9 User location Camel: CAMEL example

that a given user is already engaged in a telephone call ('busy' status), and hence completing the voice path for a call attempt to that user would be wasteful of processing and scarce resources. Instead, the switch could send a call waiting indication to the user. Similarly, a network user status of 'not reachable' may be used to redirect a call attempt to that user's voice mailbox.

Since such network user status is maintained and available in the network, it may be considered a service capability. Applications may employ the User Status SCF to obtain the network user status of a given user. Three possible values for user status are supported, i.e. 'reachable', 'not reachable', and 'busy'. Note that this is network user status, that is to say the status of a user as it is perceived by the network. A user that shows as 'reachable' may still be unavailable to take a call. Use cases that make use of the status and availability as experienced or indicated by the user are dealt with when discussing the Presence and Availability Management SCF later on in this chapter.

The User Status SCF supports the direct asynchronous request pattern, and the triggered request pattern. No trigger criteria are set by the applications, as a simple change in network user status would result in a triggered status report to the application.

6.8   Part 7 – Terminal Capabilities (TC)

Just as context sensitive applications may make use of knowledge of a user's whereabouts, availability or personal preferences to enhance the user experience, information about the equipment currently in use by a subscriber can enrich an application and boost its appeal. Advances in the capabilities of, for instance, mobile phones, sporting ever-increasing high-resolution color displays, allow application providers to exploit these device features and offer their end-users attractive applications and compelling user experiences. At the same time, older, less capable devices are deployed in the market in their millions and are expected to remain in wide use for some time to come. It is therefore equally important to adjust the delivery of an application, for instance, to suit the presentation on a small footprint, monochrome device. In fixed wireline environments the diversity in devices is even more abundant, ranging from simple POTS phones to sophisticated and highly capable VOIP clients on powerful high-end desktop PCs.

The Terminal Capabilities SCF [3GPP 2004m] provides the application with the ability to attain such capabilities of the end-user equipment. The application has two means to retrieve the terminal capabilities, either via a direct synchronous request, or through enabling triggered terminal capability reports. Reports can be triggered by a mixture of changes, including hardware changes or even complete handset replacements (i.e. placing the SIM in a new model) and software updates, ranging from firmware retrofits to alterations in user preferences or terminal configurations.

In addition to polling for characteristics, the Terminal Capabilities SCF offers the application the ability to be notified of changes in the terminal capabilities, using the methods supported on the IpExtendedTerminalCapabilities interface.

The capacity to support the functionality offered through the Terminal Capability SCF is itself bound by the qualifications of the terminal as well. That is, the end-user device has to be capable to notify its capabilities to an entity requesting it. If the device does not support the dissemination of its features and functional attributes, it does not make sense for an application to ask for them. There is of course the possibility to store such features and attributes in a static database and have the application consult the database, however such an approach would never be able to cater for dynamic changes and would have to deal with issues revolving around keeping such data up to date. But perhaps more importantly, this scenario relies on a one-to-one relation between an end-user and her terminal device, whereas a typical end-user can be connected to the network using any number of devices.

The requested terminal capabilities are returned to the application in the form of a Composite Capabilities/Preference Profile header (CC/PP). CC/PP headers can be used to express presentation and input capabilities of the device such as the dimensions of the display, the number of colors, or the input device (a touch pad, a keyboard, or just the numeric phone key pad), as well as communication capabilities of the device, such as the supported video codecs (e.g. MPEG-1, H.261) or the supported bearers (e.g. SMS, USSD, GPRS). In addition CC/PP can be used to articulate end-user preferences. For instance, the browser on a device may be capable of displaying rich marked-up text, but the end-user may prefer to read text messages in black and white plain text. Or the device may support a Java virtual machine and be capable of software download, but for security reasons the user does not wish to accept the download of Java scripts.

Note should be taken that the terminal capabilities are returned to the application in a format defined as a string. As CC/PP headers conform to a specified and published schema, the application can unambiguously interpret the string data element and consequently no a priori arrangements have to take place between network operator and application developer to avoid interoperability problems.

Summarized, the Terminal Capabilities SCF provides the application with the ability to retrieve the capabilities of the terminal, in order to be able to adapt both content as well as delivery mechanism to best address the terminal in use by the subscriber and offer a personalized, context aware, user experience.

6.9   Part 8 – Data Session Control (DSC)

The Parlay specifications support the notion of a 'data session' and provide APIs to facilitate application control of such sessions. This Data Session Control SCF [3GPP 2004n] enables third-party applications to provide value added interception, intermediation and content-based billing related capabilities in data session contexts. In this section we study the API itself, the functions it supports, and then present a simple example to illustrate how it may be used.

The Parlay Data Session Control APIs enable an application to initiate and control a data session on behalf of an endpoint. In other words, the end-user can request the establishment of a data session. When the application receives this request, it utilizes the Data Session Control APIs to request the setup and control of the associated session characteristics. The API supports two logical objects for Data Session Control – a data session manager (that manages data sessions), and a data session (that indicates characteristics of particular sessions and associated methods). A given data session manager can control multiple data sessions, but any given data session is controlled by only one manager.

The DSC SCF predominantly builds off of GPRS PDP Context establishment, defined in 2.5 G GSM networks. The functionality supported includes connection setup and tear down, volume based supervision (for Prepaid charging), notification of quality of service (re-)negotiation, and event notifications (session setup, session established, and QoS changed). Other core network technologies that may provide support for DSC include Internet Content Adaptation Protocol (ICAP) [RFC 3507] and Open Pluggable End-Services (OPES) [RFC 3835,RFC 3897] capabilities being defined in the IETF.

Services developed using the DSC API may vary from one network to another based on the capabilities afforded by the supporting infrastructure components. Generally speaking however examples include:

a) generic content-based billing;
b) re-writing URLs and redirecting HTTP requests (e.g. block inappropriate content to minors);
c) supporting Credit-Earning scenarios (browse some web pages, get credit for calls);
d) interjecting advertisements, coupons and other content into HTTP responses.

DSC thus enables service providers to leverage more profitably their partnerships with content providers and charge for content requests on their behalf in a totally seamless manner supporting capabilities such as content based charging and billing based on time of day, data volume, flat rate with ceiling, web site accessed, quality of service (QoS), per event, per percentage of retail purchases, etc. Switching from flat-rate to content based billing greatly improves profitability.

Data sessions are distinguished from Call sessions, as supported by the family of Call Control interfaces, in that data sessions are connection-less whereas call sessions are connection oriented. In other words, data sessions provide a logical connection between two endpoints without a dedicated bearer path being setup along which the packets are sent. Although this pulls against the cardinal virtue of underlying network technology independence, the attributes and features of data sessions and call sessions differ significantly such that distinct control mechanisms are required and hence justified.

Examples of these differences include:

- Call sessions are typically charged for the duration of the connection whereas data sessions are by and large charged for the volume of data exchanged.
- The state machines for data sessions and call sessions are different and hence the events associated with state transitions are different as well. This implies that the events reported through the manager interface differ. For instance, when it comes to data sessions there is no equivalent of the 'off-hook' event, which is so common and familiar in call sessions.
- The connectionless nature of data sessions means that endpoints are connected through setting up packet links between various hops in the networks, whilst connection oriented denotes that calls are routed, in traditional telephony networks, through a fixed path from origination to destination.

Despite these differences, in terms of interface patterns there are striking similarities between the DSC SCF and for instance the GCCS SCF. Both these SCFs sport a manager interface; and both feature a session interface (be the session a Call session or a Data session) that supports setting up connections, charging for those connections (supervision), resuming sessions after interrupts, tearing down sessions and cleaning up resources.

An example scenario of where the Data Session Control SCF can be used is pointed up in Figure 6.10. A ParentalGuideApplication (PGA) regulates access to certain content in external data networks by Beth, Alice's daughter. Here, DSC uses the network's capabilities to control GPRS connections to either restrict or allow access to certain content from being completed.

6.10   Part 11 – Account Management (AM)

The Account Management SCF [3GPP 2004o] is neither about Customer Relationship Management (CRM) and the management of sales accounts, nor about managing user accounts for network access, as the name may suggest. Rather, the Account Management SCF pertains to the account governing the user's payment for both communication charges as well as imbursements for goods and value added services acquired through electronic transactions using the network. Whereas the Content Based Charging SCF deals with the usage of an end-user account (such as credit and debit operations), the Account Management SCF provides the application with management capabilities regarding the end-user account. The Account Management SCF allows the application to view and monitor account data and view credit availability for a given user. Both SCFs could have been combined into a single SCF, as both present the application with access to the same object, the end-user account. However, as the nature of the operations differs, and hence the target applications are expected to differ as well, the functionality is allotted across separate SCFs.

Such separation allows for distinct Service Level Agreements and differentiation in for instance authorization stringency. For example this separation allows a network operator to provide an application with the means to debit an end-user account for a service rendered, without revealing information on the transaction history for the account. Transaction history could be used to predict which customers are most likely to respond to a particular product or service promotion. Including the ability to retrieve the transaction history for a specific end-user account in the Account Management SCF, allows only properly authenticated and authorized applications to access account information.

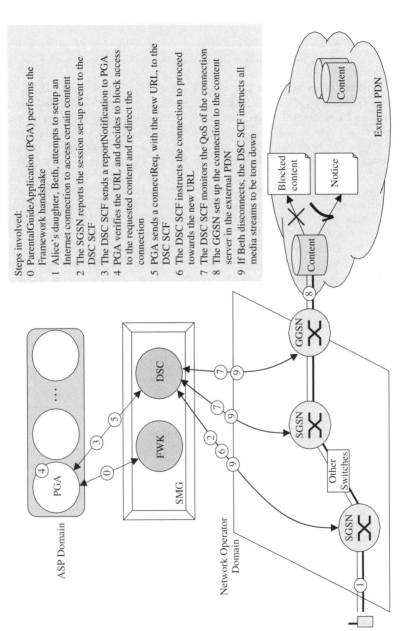

Figure 6.10 Data session control: content access restriction example

Steps involved:

0 ParentalGuideApplication (PGA) performs the Framework handshake

1 Alice's daughter, Beth, attempts to setup an Internet connection to access certain content

2 The SGSN reports the session set-up event to the DSC SCF

3 The DSC SCF sends a reportNotification to PGA

4 PGA verifies the URL and decides to block access to the requested content and re-direct the connection

5 PGA sends a connectReq, with the new URL, to the DSC SCF

6 The DSC SCF instructs the connection to proceed towards the new URL

7 The DSC SCF monitors the QoS of the connection

8 The GGSN sets up the connection to the content server in the external PDN

9 If Beth disconnects, the DSC SCF instructs all media streams to be torn down

The transaction retrieval request is an asynchronous operation, as some slow filing systems may need to be consulted for backed up records, and the history may have to be composed by pulling records from different stores. As such a transaction history may be quite lengthy, depending on the age of the account, or the frequency of transactions, the application has the ability to specify a time interval for which the history is to be supplied.

Apart from retrieving the transaction history, an application is offered the means to view the account balance for a given end-user. This is again an asynchronous operation, as the execution may involve a database query to some offline financial system or account repository. Only monetary accounts can be queried[4].

The Account Management SCF also supports a notification mechanism that allows the application to receive notifications of events regarding the end-user account. The account management application can ask to be notified when another application is charging (debiting) the end-user account, or whether the end-user has topped up the account. Furthermore the account management application may subscribe to be alerted when a given end-user account dips below a certain threshold or is exhausted (empty). Finally, a notification can be triggered when the account of an end-user is being disabled.

The pattern used for the notification mechanism is the same as the one used in other SCFs that implement a manager interface. However, despite the use of this pattern, and notwithstanding the fact that the name of the interface (IpAccountManager) adheres to the naming convention of manager interfaces, Account Management SCF does not employ a true manager interface. We have seen that manager interfaces serve as a factory to create on demand, session related objects that have a limited life span. A better name thus should have been IpAccountManagement, and perhaps a different pattern for the notification mechanism may have avoided this potential cause of confusion.

Figure 6.11 shows the example scenario where, depending on the domain hosting the application, access to certain network capabilities and the information stored in the network is either permitted or barred.

6.11   Part 12 – Content Based Charging (CBC)

Rolling out innovative, value added applications to attract new customers and to reduce subscriber churn is of paramount importance to service providers in the highly competitive and de-regulated telecommunications market. But this objective misses the overall target of any commercial outfit if that subscriber base does not generate any revenue. The service provider would require a means to charge for those applications; it would require an instrument to charge for the value-add, presented through the application. Not only transport or communication costs (either duration or volume-based) are of interest, the service provider would also want to recover charges for the provided value-add, the content. And also, for the purpose of retaining and growing the subscriber-base, the ability to charge and bill your customers accurately, timely, and transparently is imperative. In addition, charging for applications is used to recover financial investments in networking infrastructure, application hosting facilities, and the like.

The Content Based Charging API [3GPP 2004p] would interface to billing systems that collect, rate, and calculate charges for use of applications and the content delivered through them. Of course, as for many other programmatic interfaces, the API towards the billing systems of a service provider could be proprietary. However, even more so than for other SCFs, an open and standard interface for charging is a crucial enabler in order for the third party access paradigm to take off in any significant way. The reason is that it is unlikely to see a proliferation of applications, many of them developed by small and independent software vendors, if they all need to undergo the tedious, complex, and hence costly integration process with arcane and proprietary back-end billing systems, for each network where the application gets deployed.

---

[4] In the Content Based Charging SCF later on, we will see the support for both monetary (amount) as well as non-monetary (unit) accounts. Non-monetary charging may include such systems as loyalty points or air miles.

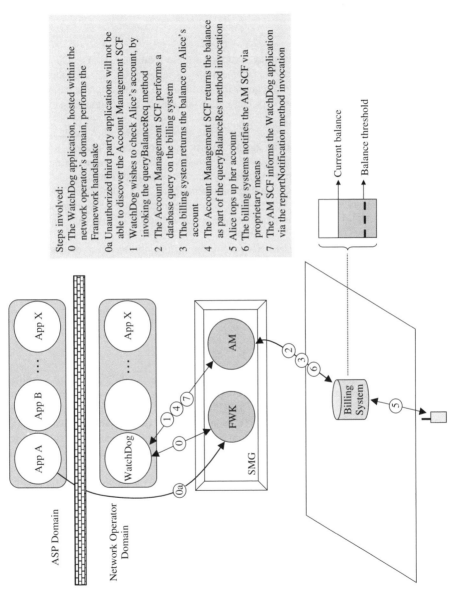

Steps involved:

0   The WatchDog application, hosted within the network operator's domain, performs the Framework handshake

0a Unauthorized third party applications will not be able to discover the Account Management SCF

1   WatchDog wishes to check Alice's account, by invoking the queryBalanceReq method

2   The Account Management SCF performs a database query on the billing system

3   The billing system returns the balance on Alice's account

4   The Account Management SCF returns the balance as part of the queryBalanceRes method invocation

5   Alice tops up her account

6   The billing systems notifies the AM SCF via proprietary means

7   The AM SCF informs the WatchDog application via the reportNotification method invocation

Figure 6.11   Account management: account information access example

In order to plug revenue leakages (caused by careless or unrefined management of account balance thresholds) and to increase revenue generation capabilities, the Parlay Content Based Charging solution needs to offer a multitude of pricing, rating and billing options. Such options include one time fees, subscription fees, BOBO (billing on behalf of), free trials, discounts, bundling, time based, volume based, event based, flat fee, reverse charging (coupons, rebates), monetary charging versus unit charging, prepaid versus post-paid, direct charging versus reservation-based charging, etc.

6.11.1 Service Considerations

We will now have a closer look at the Parlay solution to these requirements. The Content Based Charging interfaces support an IpChargingManager and IpAppChargingManager to enable the application to create a management object and request the beginning of a charging session. Once such a session is created, the associated objects IpChargingSession and IpAppChargingSession are used to invoke charging related events.

Content Based Charging allows the service provider to monitor a subscriber's credit level (e.g. getAmountLeft) and take appropriate action against a particular subscriber if needed. Rating capabilities (rateReq) exist that allow an application to request an item to be rated by the charging service, to find out how much a subscriber should pay for an application transaction.

Content Based Charging supports the notion of reservations on a subscribers account (e.g. reserveAmountReq). Reservations can be extended both in terms of time, as well as in terms of the amount reserved. Additional credits may be requested or unused balances returned, according to subscriber activity.

Based on usage patterns and the subscriber's purchasing history, the service provider can anticipate and inform subscribers, e.g. through the User Interaction SCF, when they are about to reach their minimum balance thresholds. As an example, prepaid user access to the customer care center may continue for a limited duration after expiration of the account, whereas premium revenue generating services such as video downloads are suspended and only resumed after the account is topped up.

6.11.2 Reliability Considerations

In addition to all the required functionality for charging and the flexibility in terms of billing options, there are very stringent requirements on reliability of transactions as well. There is a greater perceived need for the Content Based Charging SCS to address failure and recovery scenarios more carefully and completely than other SCSs mainly because Content Based Charging deals with monetary transactions. It is also rather more important for Content Based Charging to recover from failures in a more timely fashion and restore user balances to their pristine, uncorrupted, accurate state in recovering from as many errors as possible. Charging is a critical application where transactions either need to be *idempotent*, or there needs to be a clean way of handling rollback and commit operations. Callback functions are used to indicate to the application whether each charging request was successful or failed for some reason (e.g. directDebitAmountRes). If the callback fails due to a network error, the server knows to rollback the charge, since the application will assume that the charge request did not go through.

Transaction idempotence refers to the concept of enforcing the appropriate kind of retry semantics that holds in all cases, even in case of failures. There are several different kinds of retry semantics in general, such as 'at most once', 'at least once' and 'exactly once'. Since we need to ensure that a given user is billed only once for each kind of transaction, we need to support the 'exactly once' retry semantics for the Content Based Charging SCS for balance affecting transactions.

This is achieved by adding the notion of a requestNumber to every transaction. The CBC SCF tracks this requestNumber for the duration of the transaction and for some residual period thereafter. The Content Based Charging APIs only permit one outstanding operation per charging session in progress. This is implicit in how the APIs are defined, but will not be obvious to the unwary reader. The API requires that the requestNumber be used when a request is made, but the request number for the next request is only made available with the 'res' or 'err' method, which implies that no new requests can be made until a final response to the original request has been issued to the Client Application by the Content Based Charging SCS (this is so because the request number for the next request is needed for the next request that the Client Application may make).

Figure 6.12 demonstrates an example for the use of the Charging SCF. The Charging SCF can be used by the MoviesForYou (M4U) application to bill Alice's account for content delivered to her terminal device.

6.12   Part 13 – Policy Management (PM)

Policy Management in Parlay [3GPP 2004q] pertains to the use of policies at the services layer. In particular, it pertains to the role of policies in creating and managing high-level communications services in networks that allow access to third party applications. [Hull 2004] provides a good discussion on policy enabling the services layer. The origins of Policy Management are deeply rooted in the area of Quality of Services (QoS) management of differentiated networks and resource and configuration management for network equipment such as routers. In Parlay the concepts of Policy Management are now applied to the area of services and applications.

The Parlay model, as we have seen in previous chapters, represents tremendous revenue potential, but also the risk that accompanies relatively open access to network resources by third party applications. The use of policies to express and enforce service criteria is a valuable addition in the set of tools that are being used to define and design next generation services architectures and associated products. At the same time, however, the model exposes a number of challenges especially those that relate to network integrity, security and performance. The use of policies entails the need to manage these challenges. The scope of policy management increases considerably when we allow third party applications to access network services.

Policies are formalisms that are used to express business, engineering or process criteria. We are familiar with their use in specifying routing or triggering criteria, e.g., as in routing tables for network routers or as expressions for call triggers in network switches. To ensure that policies are well defined, a policy information model is needed. Similarly an architecture that supports the creation, storage and execution of policies is needed.

This section will introduce some service scenarios and use cases to illustrate how Policy Management can add value to Application Service Providers deploying third party services to end-users subscribed to a communications network. A distinction is made between the use of policies in Parlay service specific business logic and the use of policies in the operations and deployment of a Parlay Gateway.

6.12.1 Service Scenarios

To some extent, policies can already implicitly be implemented and used by an application without the explicit support of a Policy Management API. Certain policies can be implemented and enforced as an embedded part of the business logic of an application. A frequently cited example is the check for sufficient credit of a subscriber for an e-commerce application. As part of the business logic, this application may verify the prepaid credit of a subscriber given the current charge of the transaction, before allowing the actual transaction to take place. Not only does this check have to be performed for each subscriber for every transaction, similar e-commerce applications involving chargeable transactions would in all likelihood have to perform analogous checks. Rather than embed this credit check in the business logic of each and every application in its service offering, a third party Application Service Provider (ASP) may decide to facilitate a Policy Rule to be hosted in the Policy Engine of the Network Operator. Each time a transaction is requested, the Network operator will

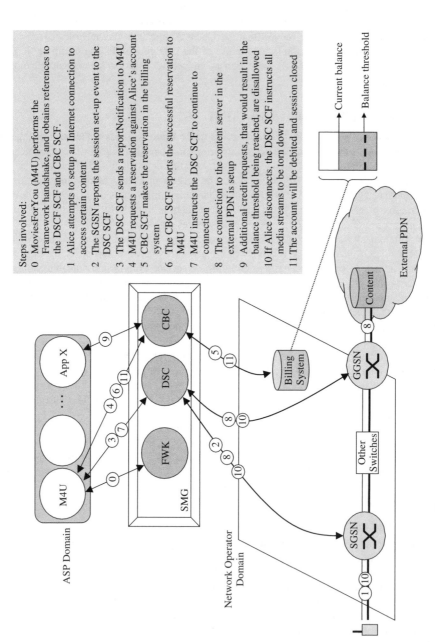

Steps involved:
0 MoviesForYou (M4U) performs the Framework handshake, and obtains references to the DSCF SCF and CBC SCF.
1 Alice attempts to setup an Internet connection to access certain content
2 The SGSN reports the session set-up event to the DSC SCF
3 The DSC SCF sends a reportNotification to M4U
4 M4U requests a reservation against Alice's account
5 CBC SCF makes the reservation in the billing system
6 The CBC SCF reports the successful reservation to M4U
7 M4U instructs the DSC SCF to continue to connection
8 The connection to the content server in the external PDN is setup
9 Additional credit requests, that would result in the balance threshold being reached, are disallowed
10 If Alice disconnects, the DSC SCF instructs all media streams to be torn down
11 The account will be debited and session closed

Figure 6.12 Content based charging: third party charging for content example

perform the credit check on behalf of the Application Service Provider. So commonality and reuse may be a reason for farming out policy enforcement to the Network Operator. Another motive could be efficiency. The verification of a subscriber's prepaid account for sufficient credit may be executed far more efficiently directly on the prepaid system, i.e. the policy enforcement takes place on the prepaid system, the application is notified of the result, and the policy decision subsequently takes place in the application.

A second example shows the use of policies in a Messaging application. A familiar feature of messaging applications is the ability to filter incoming messages according to certain filtering criteria. This capability can be realized using policies, where the filter criteria, i.e. the policy rules, are enforced in the network on behalf of the messaging application, rather than in the messaging application itself. Personal filters, specific to individual subscribers, are most likely to be enforced in the application itself, e.g. using preference settings. An example is a filter to separate business related email from private emails in different message folders. Policies common to the entire messaging application, for instance for performance reasons, or policies common to the Application Service Provider's business policies, for instance for legal reasons or based on the SLA between the ASP and Network Operator, are candidates to be enforced in the network. An example is the screening of incoming messages for adult content. This message filter criterion may need to be applied (enforced) on every incoming message for every subscriber based on legal agreements recorded in the Service Level Agreement (SLA). An example of a policy enforced for performance reasons is an upper limit on the size of attachments to incoming messages.

6.12.2 Operations Scenarios

The previous section provided some example deployment scenarios for the use of policies by third party applications. This section presents the use of policies for operational and business aspects of the Service Mediation Gateway itself.

Parlay introduces the concept of providing third party applications with open, secure, and regulated access to core network service capabilities, while maintaining the integrity of the network. Policy Management can provide the mechanisms and infrastructure for ensuring regulated access and network integrity preservation. Consider the following example. During the discovery process, the application is polling the Framework on whether it supports certain Service Capability Servers (SCSs), based on a set of service property values provided by the application. The Framework will try and match these service property values with those of registered Service Capability Servers. In normal operation, the Framework will conclude the discovery process by returning the service identifiers of those SCSs with matching property values. One could think through a number of motivations why a Framework operator does not wish to disclose all SCSs that fit the profile of the request by the application. For instance a particular SCS could be a high performance, fault tolerant SCS only available to a limited number of mission critical applications, hosted within the Network Operator's domain. Or a particular SCS could be a system under test only to be accessed by a trusted group of system testers. Another rationale could be privacy constraints. It is imaginable that not all third party applications are allowed to discover the User Location SCS, as a subscriber's exact geographic whereabouts may be subject to privacy regulations.

Using policies to provide a more mature, user-friendly, stable and feature rich Service Mediation Gateway product can be regarded as another example of non-service related employment of Policy Management. Consider for instance the situation where the Framework is unable to create Service Managers for a specific Service Capability Feature on a Service Instance Lifecycle Manager (SILM). Normal operation specifies that an exception must be raised towards the Framework. However the behavior towards the application is not clearly defined. Should the application try again after receiving such an exception? Even more, the behavior towards other applications that wish to use the same SCS is unspecified as well. A policy can be introduced to ensure that a failure to create a Service Manager on a Service Instance Lifecycle Manager, results in the SCS being 'unannounced' by the Framework so that the SCS is no longer discoverable by third party applications.

6.12.3 Service Properties versus Policies

Before the introduction of Policy Management in Parlay, the only way to configure and tune an SCS according to certain behavior definitions, technical limitations or business criteria was to use service properties. The predominant difference between service properties and policies is the level of flexibility to express these criteria. Service properties are static and are assigned at the time when SCSs register themselves with the Framework. Policies can be installed at registration time, application provisioning time, but also dynamically at application run time. Let's have a look at a Call Control service property defining the maximum number of parties that can be involved in a single call. As service property, this upper bound on the number of involved parties is instantiated when the SCS registers with the Framework. The application becomes aware of this upper bound at service discovery time. The implementation of the application business logic needs to be constrained by this upper bound. Defined in terms of a policy rule, however, the upper bound can vary dynamically. For example, the upper bound may be further limited based on changing load statistics in the network, previous behavior of the application, credit history of the ASP, etc.

6.12.4 Business Opportunities

The service and operations scenarios outlined above showed use cases where Policy Management can be used to manage policy rules, either on behalf of the application, or on behalf of the Framework. As such, Policy Management can be used to further enhance the existing roles and relationships defined in the Parlay architecture. However, a whole new interesting employment of Policy Management looks beyond just managing policy rules, and extends to managing the Parlay client application itself. This provides new business opportunities for the Service Mediation Gateway operator. For instance, the feature of enabling and disabling policy rules can be incorporated and exploited in a Service Creation Environment. Client applications can be tailored to address specific end-user audiences or deployment scenarios, using network hosted policy rules. As another example of new business opportunities, consider the case where policies are used to enforce certain QoS related contracts defined in the SLA, e.g. where the Application Service Provider (ASP) has agreed to pay for a specific throughput. Based on collected traffic statistics and usage profiling, the Service Mediation Gateway operator may discover that the client application is running at its transaction limits, and additional traffic may result in increased data loss and delay. The Service Mediation Gateway operator may wish to notify the ASP of this fact and offer the possibility to upgrade dynamically to a higher QoS class and change the SLA accordingly. With the functionality provided by Policy Management, such scenarios can be supported in a flexible, non-service interrupting way.

6.12.5 The Policy Management Interfaces

The interface classes supported in the Policy Management SCF are quite extensive, with some classes consisting of fifteen to twenty methods. The interested reader is referred to [3GPP 2004q].

Policies, or policy rules, capture operational or business criteria that can be managed, defined, and executed by the Policy Management SCF. Within the scope of Parlay PM, policy rules are defined by the Policy Rule interface. The semantic used by the Policy Rule interface to represent these policies is of the form of productions, i.e. '*If* **Condition** *then* **Action**'. Hence, interface classes are introduced for policy conditions and policy actions. Creating a condition and an action, and associating the condition and action with the rule, create policy rules. The functionality to associate the condition and action with a rule is provided by methods defined in the Policy Rule interface. The creation of conditions is done through the use of the Policy Condition interface classes, whereas the creation of actions is achieved via the Policy Action interface classes.

Three types of conditions are supported, each defined by their own interface:

1. *Expression Condition* – The Expression Condition interface is used to define a condition in terms of an expression that needs to be evaluated.

2. *Event Condition* – The Event Condition interface is used to define events that can trigger the evaluation of a policy rule.
3. *Time Period Condition* – The Time Period Condition interface offers a means of representing time periods for which a policy rule is valid.

Complex conditions can be built from single conditions, either using Disjunctive Normal Form (i.e. an ORed set of ANDed conditions) or Conjunctive Normal Form (i.e. an ANDed set of ORed conditions), offering maximal flexibility in the definition of conditions. A BNF (Backus Naur Form) grammar is supported, which allows an expression to be defined in terms of variables of certain types (e.g. integers or strings) and some arithmetic and comparison operators (e.g. '+' and '>='). An example of an expression condition could be 'call_legs.allowed > call_legs.requested'.

Two types of actions are supported, again each with their own interface, i.e. Expression Action and Event Action. An action list can be constructed using several individual actions, each possibly assigned with a priority indicating the relative order in which the actions need to be executed. The Event Action can be defined to generate events, e.g. to report back to the application the result of a policy rule evaluation. The Expression Action interface is used to define an action in terms of an expression that needs to be evaluated. The possible expressions are somewhat more limited than the expression conditions, i.e. a more limited BNF is supported. Also, no comparison operators are supported, but instead the assignment operator is supported. An example of an expression action could be 'call_legs.assigned = call_legs.allowed'.

An example of a policy rule that could be constructed using these interfaces could be [IF 'call_legs.allowed > call_legs.requested' THEN 'call_legs.assigned = call_legs.allowed']. In this example the Parlay application attempts to request a call to be set up to more call parties than is allowed according to the Service Level Agreement. In that case, the number of call legs assigned to the call is reduced to equal the upper bound agreed in the SLA.

Policies can be grouped and managed at certain levels of granularity. The highest level of grouping is a Policy Domain (IpPolicyDomain interface). Within a domain, certain Policy Groups may be created (IpPolicyGroup interface). The Policy Rules themselves (IpPolicyRule) exist within a Policy Group. Rules then, as we have seen, consist of Policy Actions and Policy Conditions.

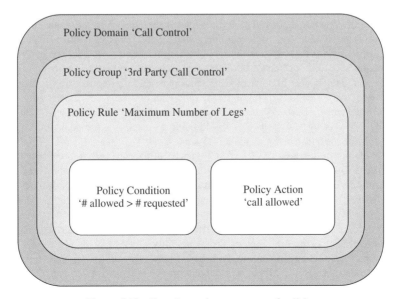

Figure 6.13   Grouping and management of policies

Several administrative type of methods are supported in all these interfaces, for instance iterators to search through groups, the ability to create and delete groups and domains, and to find the domain for a specific group, etc. The grouping relationships are depicted in Figure 6.13.

Figure 6.14 illustrates how the Policy Management SCF can be used to policy-enable the Framework and the User Location SCF. In this particular example, full access to location information of all subscribers is restricted to 'Gold' applications, whereas 'Bronze' applications are granted somewhat reduced access.

6.13   Part 14 – Presence and Availability Management (PAM)

Everyone has played a game of phone tag at some point ('Hi, I got your message about my message, please call me back') or has been involved in a high-speed chase on the electronic highway ('I sent you an email and left messages on your mobile and desk phone about the SMS I sent you, where are you?'). Your communication experience could be greatly enhanced if you could simply 'see' when your colleagues or family and friends are online and available, and how they prefer to be reached. There is tremendous value in real-time information about the ability and willingness of contacts to communicate and about the best means for that communication. Such information is often available in the network for various basic communications processes (e.g. handset registration, call processing) but often dispersed across various repositories and locked in through the use of dedicated protocols. The goal of Presence and Availability Management [3GPP 2004r] is to establish a standard for collecting and mining information about identities throughout the network, and maintaining and publishing the information about these identifies towards applications.

The benefits for end-users clearly involve an enhanced user experience, as it allows you more accurately and predictably to contact other people, as well as provide you with means to manage and filter inbound communications more flexibly. For network operators there are gains to be had as well. The capability can make communication more efficient owing to more intelligent call routing based on preferences expressed in terms of presence and availability information. For example there is no need to page a mobile using scarce radio resources if that subscriber has indicated not to be available for communication. Moreover, providing a caller with alternative means to reach a certain subscriber may result in increased successful and hence billable sessions in the network, compared to just a single incomplete attempt to a destination where the subscriber happens to be unavailable.

The concepts of presence and availability are related though subtly different. Presence denotes whether a subscriber is able (e.g. online/offline) and capable (e.g. busy/idle) of communicating. Availability further refines that information by indicating the subscriber's ability and willingness to share information about him or herself or to communicate with another subscriber. Availability can be established through aspects like the mode of communication ('Redirect to voice mail if I have roamed out of coverage'), who is trying to reach you ('Out of all my colleagues, only my boss can call me in the weekend'), or subscriber preferences put in place by the addressee ('During my daily commute I do not wish to be disturbed'). So although presence is a prerequisite to availability, a subscriber might be present but not available (to some).

Figure 6.15 demonstrates how the PAM SCF can be utilized to realize the Friends application that allows Alice to manage her availability towards her buddies, based on her user preferences. After working hours, Alice is available to friends only. Any call attempts by colleagues for instance during the weekend are notified by the network towards the Friends application, and subsequently barred.

The PAM Access SCF consists of four interface classes. There are interfaces to get and set presence (IpPAMIdentityPresence), get or set availability (IpPAMAvailability), and to compute availability in a given context (IpAppPAMPreferenceCheck). In addition, there is a manager interface (IpPAMPresenceAvailabilityManager) to obtain access to the PAM service.

The PAM Event SCF deals with registration to presence events. This SCF supports an event handler (IpPAMEventHandler) and a manager interface (IpPAMEventManager).

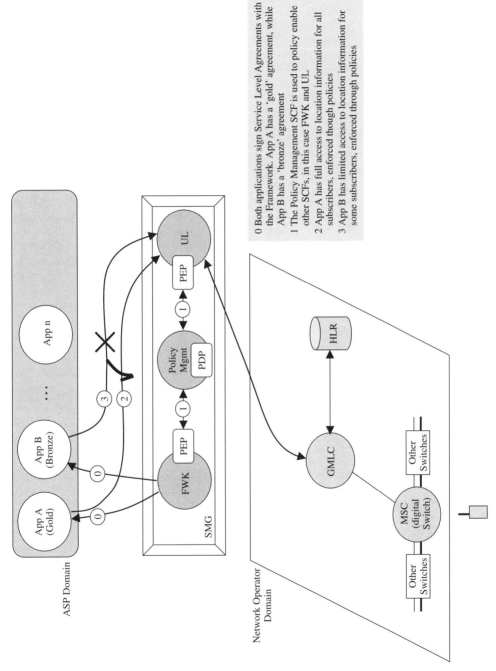

0 Both applications sign Service Level Agreements with the Framework. App A has a 'gold' agreement, while App B has a 'bronze' agreement
1 The Policy Management SCF is used to policy enable other SCFs, in this case FWK and UL
2 App A has full access to location information for all subscribers, enforced though policies
3 App B has limited access to location information for some subscribers, enforced through policies

Figure 6.14 Policy management: policy-enabling example

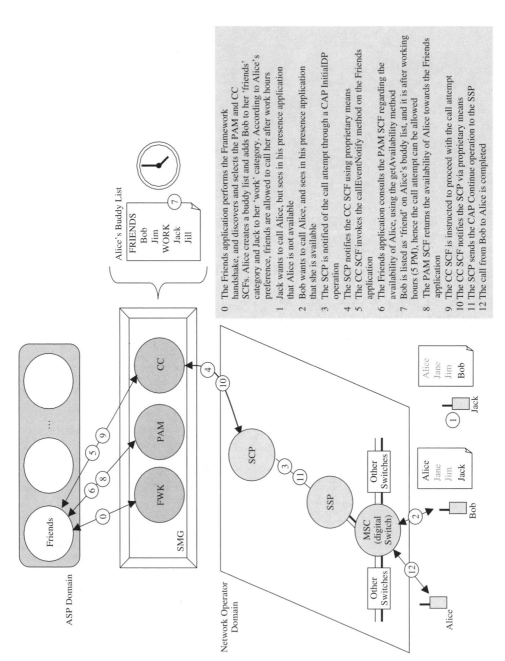

Figure 6.15 Presence and availability management: buddy list example

0   The Friends application performs the Framework handshake, and discovers and selects the PAM and CC SCFs. Alice creates a buddy list and adds Bob to her 'friends' category and Jack to her 'work' category. According to Alice's preference, friends are allowed to call her after work hours

1   Jack wants to call Alice, but sees in his presence application that Alice is not available

2   Bob wants to call Alice, and sees in his presence application that she is available

3   The SCP is notified of the call attempt through a CAP InitialDP operation

4   The SCP notifies the CC SCF using proprietary means

5   The CC SCF invokes the callEventNotify method on the Friends application

6   The Friends application consults the PAM SCF regarding the availability of Alice, using the getAvailability method

7   Bob is listed as 'friend' on Alice's buddy list, and it is after working hours (5 PM), hence the call attempt can be allowed

8   The PAM SCF returns the availability of Alice towards the Friends application

9   The CC SCF is instructed to proceed with the call attempt

10 The CC SCF notifies the SCP via proprietary means

11 The SCP sends the CAP Continue operation to the SSP

12 The call from Bob to Alice is completed

6.14   Other Standards-defined SCFs

In addition to the ones covered thus far in this chapter, there are more SCFs that are part of the Parlay standards specification suite. They are not covered in dedicated sections on their own in any amount of detail, for one of two reasons: the SCF is considered dormant or the SCF is only a recent addition to the entire suite and hence deemed to be still maturing. In the subsequent sub-sections, these SCFs are briefly summarized, for the purpose of completeness.

6.14.1 The Generic Messaging Service (GMS)

The Generic Messaging Service [ETSI 2005b] has existed outside of the 3GPP subset of Parlay specifications, as the messaging requirements within the 3GPP domain could be satisfactorily ful-filled using the User Interaction SCF. Partly as a result of this circumstance, no standards attention in the form of detailed peer review, input contributions or developer feedback has been paid to the GMS SCF for a number of years. As such, the GMS SCF is considered to be dormant. This is despite the fact that it is generally recognized and accepted that a number of faults, ranging from minor to significant, are contained in the API definitions. In addition to the errors, the GMS API is considered complex for simple messaging use. A good example to illustrate this point is the necessity to create a mailbox object, followed by a folder object, followed by a message object in order to send even a simple message. It is quite evident this layered design provides a processing overhead for Instant Messaging applications.

Given the popularity of messaging applications, new requirements for messaging were defined for Parlay. Rather than performing a major design overhaul of GMS, the Parlay standards commu-nity has decided to design a messaging API from scratch, reflecting the improved understanding of messaging technologies and fulfilling newly evolved and more advanced requirements. This MultiMedia Messaging Service (MMM) SCF is described in Section 6.14.3.

6.14.2 The Connectivity Manager (CM)

The Connectivity Manager SCF [ETSI 2005c], like the Generic Messaging Service SCF, is not part of the 3GPP subset of Parlay specifications. And like GMS, it has been a dormant specification since its initial publication. For the purposes of this book, we shall suffice with a brief description of its functionality.

The Connectivity Manager SCF provides an enterprise operator with the ability to manage its network, which is being physically realized by resources and assets in the network of the Network Operator. For instance, functionality is provided to setup and manage a Virtual Private Network (VPN) consisting of virtual leased lines. The VPN can be made up of multiple sites, and the Connectivity Manager SCF allows the enterprise operator to manage the VPN's quality of service (QoS) characteristics, and to manage individual pipes in the VPN.

In addition, the Connectivity Manager SCF can be used by the Network Operator to offer the enterprise operator various templates for a virtual pipe in the VPN. Templates may be defined for instance for high quality video conferencing or voice quality audio traffic, using QoS attributes specified by both the enterprise operator and the network operator. The templates are then used to create the virtual pipes in the network, on behalf of the enterprise operator.

6.14.3 The MultiMedia Messaging Service (MMM)

The MultiMedia Messaging service [3GPP 2005c] is the most recent addition to the Parlay speci-fication set[5]. It has been designed in honor of the newly evolved and more advanced requirements for messaging functionality and in recognition of the shortcomings and dormant state of the GMS

---

[5] Note that the MultiMedia Messaging SCF is added in 3GPP Release 6, whereas for the purposes of this book the discussion of all other SCFs is based on their 3GPP Release 5 version.

SCF. Since the MultiMedia Messaging service can still be considered to be maturing as a standards specification, for the purpose of this book we will not include a detailed description of its functionality. Rather, we will suffice with presenting an overview of its main design characteristics and a brief summary of the functionality it provides.

The MMM SCF provides the client application with the ability to send, receive, and store messages. Various messaging mechanisms are supported, including multipart messages (where the header and the body parts can be accessed independently), text messages, mail messages (including voicemail and email), and multimedia messages. Additional address parameters typically used for messaging, like a cc-address and a reply-to-address, are defined for the MMM SCF.

For the mailbox messaging paradigm, mailbox management in the form of creating, opening and closing of mailboxes, and the management of various folders within mailboxes, is supported. Use of the mailbox interface however is optional so that non-mailbox messaging paradigms are efficiently supported as well.

6.15 Support for Non-Standard SCSs and Value-Added Extensions

As we have seen in the previous sections of this chapter, the OSA standards define a set of 13 SCFs, each of which provides programming interfaces to a given functional domain. Examples of these include User Location, User Status, Presence and Availability Management, Charging, etc. However, Service Providers might want to pre-package functional capabilities tied to other network element assets (from non-OSA functional domains) and make them more easily accessible to application developers with a view to leveraging their investments more efficiently. Alternatively, they might want to access additional non-OSA-defined capabilities from even the standards-defined SCSs. These requirements may be easily met through the development of non-OSA SCSs, and through value added extensions to OSA-compliant SCSs. These topics are discussed in what follows.

6.15.1 Standards-defined and Proprietary SCSs

Service Providers that have certain network elements already deployed may decide they would like the capabilities supported by these elements to be exposed to third party application developers to enable them to build applications that can utilize this functionality, thereby leveraging their deployed assets more efficiently, and using them more effectively in increasing subscribers' reliance on their networks.

Wherever these functional capability sets fall within the context of the standards-defined Service Capability Function (SCF) specifications, it is indeed possible for a Service Capability Server (SCS) vendor to build an SCS with targeted device specific mappings over suitable protocols to use the network element in question.

When the functions provided by the network element do not relate directly to the standards-defined SCFs, however, Service Provider requests for an API abstraction to the underlying functional set through interfaces similar to those defined by Parlay may still be met. Service Providers and Network Operators are typically more interested in these kinds of scenarios when they find there are no standards-defined SCFs for the specific functional capabilities they want to make immediately accessible as they start deploying Parlay/OSA-based solutions.

This is achieved by building a programmable interface atop the protocol primitives supported by the underlying network element in a manner that 'appears' similar to (i.e. has the same overall structure as) how the Parlay/OSA SCFs are defined. These SCSs are commonly referred to as proprietary, non-standard, or non-Parlay SCSs.

6.15.2 Standards Directions

Again, as we have stated in previous sections, the Parlay specifications define a modular and extensible architecture, whereby service capabilities can register themselves with the Parlay Framework in order to make themselves discoverable by client applications. Registration occurs based on a name (e.g. P_USER_LOCATION) and a set of service properties. Properties can be generic (e.g. 'service version' and 'supported interface operations') as well as service type specific (e.g. 'supported address plans' and 'support of GPS positioning'). A client application obtains the interface definition for a given service capability (e.g. the IDL definitions, or the WSDL files), and then contacts the Parlay Framework to discover whether such a service capability is available, and what the specific service properties are.

The Parlay specifications have standardized this mechanism for a dedicated set of service capabilities (i.e. the Parlay SCFs), by defining a set of service type names and the service properties, for each of the service capabilities. However, the mechanism supports proprietary service capabilities as well. Service Providers can extend the set of service capability names, by adding the name of the proprietary service capability to the set of standardized service type names, and by defining a specific set of service properties. In addition, the service interface definition needs to be published to the client application.

Proprietary Service Capability Features that are registered with the Framework through the aforementioned standardized mechanism can be discovered by client applications like any other standardized SCF. Furthermore, the Framework will support all the Integrity Management functionality such as Load and Fault Management for the proprietary SCF, in exactly the same manner as it would for any standardized SCF.

Service Providers and Network Operators now have a means to provide client applications with access to their existing value added service capabilities as well, in a controlled, secure, manageable, and billable fashion, through the provisions of the Parlay standardized architecture. This can open up new revenue streams for existing assets. Furthermore, new proprietary value added service capabilities can be integrated horizontally into the Parlay service layer, reusing the common infrastructure and offering the same look and feel towards client applications.

6.15.3 Example Proprietary SCFs

This section provides three examples and possible rationales for a Service Provider to include proprietary SCFs into their overall Parlay infrastructure. The reader should note that these examples are indicative only, as there may indeed be several reasons to deploy proprietary SCFs.

1. The standards process typically takes time (e.g. involving multiple design iterations, scrutinized reviews, etc.), sometimes creating a tension between being first to market and waiting for a standards solution that enables a larger total addressable market. A Service Provider may opt, while waiting for the standards process to take its due course and publish a standardized API specification, to deploy a proprietary version of the SCF being standardized. The proprietary deployment may serve to capture initial market share as well as to supply proof of concept. Initial feedback may be submitted to the standards process, thereby adding rigor to the SCF design and definition. Once the standards specification is published, the Service Provider may decide to register the newly available SCF and replace the proprietary one. Needless to say, updates are required to the client application, possibly resulting in a temporary disruption of service.
2. One of the design principles of Parlay is to be agnostic of specific communication protocols and to design abstract interfaces. One of the benefits is the applicability of your client application to a multitude of underlying network protocols and architectures. If however a Service Provider will ever only deploy their highly customized client application using a given specific protocol in their network, an abstracted API may result in loss of certain parameters and hence reduced level of control and functionality. A proprietary SCF, expressly designed to take advantage of certain specific protocol characteristics most effectively, may be the best solution in such a case. In all likelihood, only niche applications will be able to benefit from such a proprietary SCF.
3. Parlay APIs expose network capabilities towards third party applications. For application delivery, however, also non-network capabilities may provide useful functionality. For example, a

User Interaction application may use the sendInfoReq method to send binary data to the end-user's terminal. The binary data may contain an MPEG encoded audio file. A particular Service Provider may decide to offer a proprietary SCF, exposing an MPEG audio encoding API to the client application, as part of its overall service delivery platform. The client application may use this API to encode the binary data element to be sent using sendInfoReq, interacting with a single integrated Parlay infrastructure, incorporating a proprietary SCF.

6.16 Summary

It's been a long time coming, but this chapter finally introduced the Parlay APIs. We have chosen not to provide a laundry list of all Service Capability Features in terms of the interface classes and their methods. For this, the interested reader is referred to the standards specifications. Rather, a general description of the functionality the APIs offer to applications is discussed, along with examples and explanations.

7

Standards Capabilities and Directions II – Scenarios and Details

7.1 Introduction

In the first chapter of this book, the reader was exposed to various scenarios that discussed the problems, uncertainties, frustrations and doubts faced by service providers, application developers and end-users of the different networks. In the chapters that followed, we have looked at various factors that contribute towards costs of building and integrating new services and applications into today's networks, the value these changes and new additions offer (revenue acceleration is typically the main business driver behind this constant evolution), and how enhanced end-user experiences result from this. Subsequently, we derived a set of requirements that would enable a quicker, more cost-effective integration of new services and applications into networks, and demonstrated how Parlay and OSA technologies try to meet these goals.

In this chapter, we first study the Parlay ecosystem and value chain, tying this to the discussion of value-chains and service models from Chapter 3. We will build upon the technical content of the discussions from these previous chapters to drive home these points, through a simple (but powerful) and illustrate just what a difference standards-based programmatic APIs can make in terms of end-user experience – this is our primary focus here. There are other lower level aspects of application deployment relating to application stability, certification, high availability, reliability, performance, and so on, and these are relegated to a later chapter dedicated to these issues (Chapter 13). Similarly, deployment considerations for Service Mediation Gateways themselves merit separate discussion on their own, and these are addressed in Chapter 11.

7.2 The Parlay Ecosystem and Value-Chain

In Chapter 3, we discussed various service models, and how the one that defined a set of service enablers hosted within a services layer, is the one finding more widespread use today. We also talked about how reuse of components from this services layer enabled the transition from the paradigm that required re-integration of applications with individual lower level protocol elements ('the smoke stacks' model) to one where the integration was done once and then leveraged multiple times from within the context of a standards-defined API.

Parlay/OSA: From Standards to Reality   Musa Unmehopa, Kumar Vemuri, Andy Bennett
Copyright © 2006 Lucent Technologies Inc. All Rights Reserved

Subsequent chapters, Chapter 5 and Chapter 6 in particular, have provided more background around this, illustrating how a given application might leverage multiple SCSs as it processes a single transaction from an application or end-user context.

The Parlay value chain may be viewed as the set of 'tentacles' that the application has into the services layer – one to each SCS. The more tentacles, the more value is provided by the gateway. Applications grow like mold on bread – the service mediation gateway is the substrate, and the richness of the services layer (the number, type and usability of services) contributes directly to the value provided to the applications. Admittedly, there is some symbiosis here as well, since the greater the number of applications that 'grow on' this substrate, the greater the value the service provider realizes for the SMG investment. A later section in this chapter will discuss the idea of 'mixed mode applications', which will serve to clarify further how both the applications and the gateway can co-evolve to provide best value end-to-end.

Let us take one more look at the Parlay ecosystem. So far, we have concluded that two components exist in this – the Parlay-compliant application, and the Service Mediation Gateway. Obviously, the end-user is part of the equation somewhere (he/she connects to the application), but is not directly involved in the ecosystem per se.

If we were to take a closer look, we see that from a business standpoint, each component in the Parlay ecosystem individuates into separate elements – both the SMG and the Parlay application software need to be implemented atop some platform. Telecom networks normally require 'carrier-grade' (or high availability, high reliability equipment. The term 'carrier-grade' is explained in more detail in Chapter 10) platforms upon which to host these components.

Thus, we have four parties that are involved in any service layer build-up – the providers of the carrier grade platforms for services and applications (which may be the same), the provider of the SMG software, and the provider of application software. The platform supplier may provide value added capabilities such as application server containers, application hosting capabilities, OAM&P[1] tools and features, etc., to entice more application and service vendors to use its, instead of competitors,' platform. This extended Parlay value chain is depicted in Figure 7.1.

It is important for the reader to note that there are several companies that specialize in one or more of these areas today, and some areas are getting more crowded than others as Parlay technology takes off in a big way, while others have seen some form of consolidation over time already. For instance, the application servers have been in use for some time, and that technology was not developed to be Parlay specific. Thus, there has been some consolidation in that area already with two or three clear winners emerging. A similar process is in the offing for Parlay gateway and application vendors, with the more traditional Telecom Equipment Vendors[2] and Application Integrator firms well positioned to take the lead. Of course, as with all other classifications, there are exceptions, and some small companies have been successful in making inroads with customers outside their more traditional bases with their differentiating capabilities. Since this book is not a report on the state of the industry today, nor a competitive intelligence report, we do not endorse any particular products in any of these areas.

---

[1] The term OAM&P is commonly used in telecom parlance to refer to Operations, Administration, Management and Provisioning. These capabilities are critical to the development of any telecom-grade service and often Telecom Equipment Vendors will spend large sums of money building these capabilities into their products. In many cases, Service Providers take these capabilities and features for granted, and assume that no telecom-grade service will ever be delivered without them.

[2] We posit that these companies are better poised because of their deeper understanding of network integration issues and due to their wealth of experience in handling the arcane and involved aspects of the details of the underlying telecommunications protocols, many of which require very specialized knowledge. However, at the same time we concede that given the new free market economy, and the post Internet bubble talent dispersal, such knowledge is no longer limited to Telecom Equipment Vendors, though it is still likely to be concentrated there. The Telecom industry has been through some rather turbulent times in the past few years.

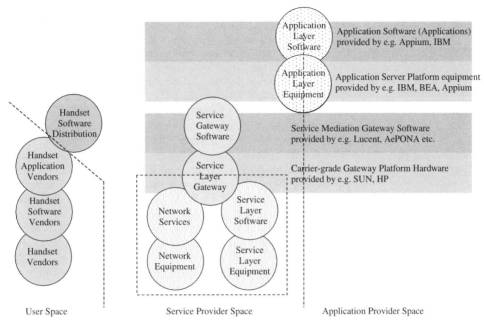

Figure 7.1   The expanded Parlay value chain

7.3   Example Scenario

Scenario: Alfie goes to town!

Alfie was in Chicago on business and was staying at an expensive downtown hotel. After a long day of meetings, he decided to take a walk and see some of the sights. Being a frequent traveler, he had subscribed to the 'Tourism Genie' application offered by Utopia! Wireless. He had noticed how Utopia!, not really a leader in services just a few years back, was now able to offer so many new applications to their subscribers so quickly. Come to think of it, this just started happening since their press release a few months back stating they had integrated a Parlay/OSA service mediation gateway into their network.

Alfie did not know much about Parlay technologies, and quite frankly, he didn't care. What he did appreciate though, were the services offered to him. He was spending quite a few pounds more on communications services these days, he smiled to himself, well, that was the price of convenience. Tourism Genie came bundled along with a number of other services, and now he was finding out that it alone was worth the price of the full bundle.

Alfie exited his hotel, and started walking down the street. His mobile was out and on, and he activated the Tourism Genie (TG) program as he stepped out. Nice weather, he thought, as he clipped his blue-tooth earpiece and microphone on. TG asked him if he wanted to take a walking tour of the neighborhood. He said he'd very much like to. The application took a moment to initialize factoring in its new surroundings, and was then ready to act as his tour guide.

Alfie told TG he did not want to be disturbed during the walking tour by phone calls from business associates, though he'd still like to take calls from his wife and son if they were to call him. TG assented. TG interacted with him in multiple modes – some cues were visual and sent to his mobile's screen, while others were auditory. Many were synchronized. TG guided him along Jackson Street to the river, and waited as he stood there on the bridge, looking down at the ferries. After a minute or two, TG started telling him about the river, history, other details, just as a real tour-guide would. Alfie knew he had the option to shut down the commentary at any time, by simply staying 'Stop!'. He could similarly use a vocal cue to restart the narrative at any point in their journey as he wished.

Next, he asked TG for a list of points of interest in the vicinity. TG splashed a map onto his screen with these clearly marked out with icons. He asked for a brief summary description of each, and chose to walk by the Sears Tower, the Art Institute of Chicago, the Buckingham Fountain, Lake Michigan, and the Field Museum, as TG read through the summaries. In response to his request, TG dutifully plotted the optimal course past these landmarks indicating his approximate tour duration, and also indicated other landmarks that may be visible along with photographs on the mobile screen so Alfie could identify them.

As he passed by the Sears Tower, TG asked Alfie if he wanted to go in to the Skydeck, and observe the City from there. If so, he could buy the ticket through the program, and save some time (no waiting in line to purchase tickets). Alfie thought this was a good idea, and made the purchase – this would appear on his phone bill. Utopia! TG had deals with many tourism hubs, and he even got a discount on the purchase. A voucher receipt and code were presented to Alfie, so agents there could verify payment prior to his being permitted onto the Skydeck.

As he walked around on the Skydeck, TG continued to present him with facts, pictures, etc. relating to the Sears Tower, thereby making his experience more enjoyable. Returning to street level again, Alfie was exhilarated with the quality of the experience. He saw the jealous stares he received from several other tourists as he walked out.

TG guided him next along Adam's Street past the Art Institute. He was asked if he wanted to visit that too, but declined. He wanted to take his time and visit it properly later, rather than rush through the exhibits. It was getting late. He told TG he wanted to drop the Field Museum from his trip now, just see the lake and return to his hotel. TG agreed and re-computed his route.

Alfie walked past the flowerbeds at Buckingham Fountain, and spent some time sitting on a bench overlooking Lake Michigan absorbing some of the serenity from the atmosphere there. TG asked him to look to his right, and pointed out the Field Museum, the Adler Planetarium and the Shedd Aquarium to him along with pictures on his mobile so he identified the landmarks correctly.

Finally, he got up. It was getting close to dinnertime, and he wanted to grab a bite to eat at a nearby restaurant before returning to his hotel room. TG was only too glad to oblige with a listing of nearby eateries once Alfie had provided preference information. TG connected Alfie to the restaurant where he made reservations, and then guided Alfie towards the place. Nice to have technology you can rely on, thought Alfie. First, GPS receivers in cars so you didn't get lost driving around, and now helpful interactive agents for tourists in cities new to them. I wonder what they will think of next.

7.4   Under the Covers – How it Actually Works

The TG application from the above scenario looks very complicated, but in reality, it is rather simple to implement with Parlay technology (or to be more fair in our assessment, it is rather simpler to implement with Parlay technology than without).

The entire user experience from that scenario can be implemented with support for just four Parlay service SCSs – User Location (UL, part of the Mobility SCS), Presence and Availability Management (PAM), Charging (CH), and Call Control (CC). As the reader will see, simpler versions of the TG service could be implemented with subsets of these SCSs, though the user experience would be modulated somewhat by the constraining choices of such an implementation.

The TG application essentially uses network information for the following functions:

1. to locate Alfie and then correlate his progress on a map that indicates his walking tour circuit;
2. push relevant context-sensitive information to the mobile regarding points of interest, photographs, restaurants, maps etc.;
3. to charge Alfie's prepaid phone account for any purchases he may make along the way (or permit him to recharge his account with more money as he deems appropriate);
4. block or allow calls from the many people Alfie knew, and who were listed in his address book under folders labeled family, friends, business associates etc.;
5. connect Alfie to various businesses such as hotels and restaurants from within the context of his current experience.

Of these, (1), (3), (4) and (5) could be easily accomplished through SCSs – (1) by UL (Mobility), (3) through CH, (4) through a careful usage of PAM profiles and preferences, and (5) by CC. (2) could just as easily be implemented through usage of the non-call-related User Interaction SCS – through suitable mechanisms such as WAP or SMPP to the handset, but it is expected that applications as sophisticated as TG would want to control their own access channels to the handset for more efficient user interactions.

Let's look more specifically at the usage of the various SCFs in the application call flow (please refer to Figures 7.2 and 7.3). From the previous chapter, we know that the User Location SCF offers applications the ability to request periodic reports on the location of a particular cell-phone. This method could be used by TG to locate Alfie as the tour progresses (1).

The CH SCF provides a method for the direct debit of prepaid or postpaid accounts via the DirectDebitAmountReq() method call once a charging session is established for the user with the SCF by the application. TG can use this technique to bill Alfie for purchases made along the way (3). Other methods in the API could be used to support credits for refunds (say he bought a ticket to the Skydeck but later changed his mind), or recharges if Alfie asked that more money be added to his account by billing it to his credit card.

The PAM SCF supports a number of different methods that deal with application access to provisioned preference information. These are included in the PAM Access SCF and enable applications to set or retrieve the presence and availability parameters of individual users. When Alfie indicated he did not want to be disturbed, TG could simply mark him as 'unavailable' to his friends and business associates, but mark him still as accessible to his family. The SetIdentityAttributes() and SetAgentAttributes() method calls, for example, could be used to achieve this in (4).

Last, but not least, the CC SCF gives applications the capability to set up calls between two parties for third-party call control (see, for example, the createCall() API call). This enables TG to connect Alfie with other businesses as needed for (5).

Figures 7.2 and 7.3 show 'call flow blocks' on a functional basis per SCS, though the actual application implementation will most likely interleave method invocations to the various SCSs to enable access to the network hosted capabilities on an as needed basis as user requests to the application are processed.

7.5 Mapping APIs to Protocols

The above is all well and good from an API perspective for programmability, but a full appreciation of what truly happens is only possible if we also explore how network capabilities are leveraged as the scenario progresses.

The standards documents in 3GPP, in particular – recall that in Chapter 4 we studied how 3GPP provides standards documents for the OSA APIs – provide two views of the API. The first view, in the documents in the 3GPP TS 29.198 series, is a description of each SCF and its API in detail,

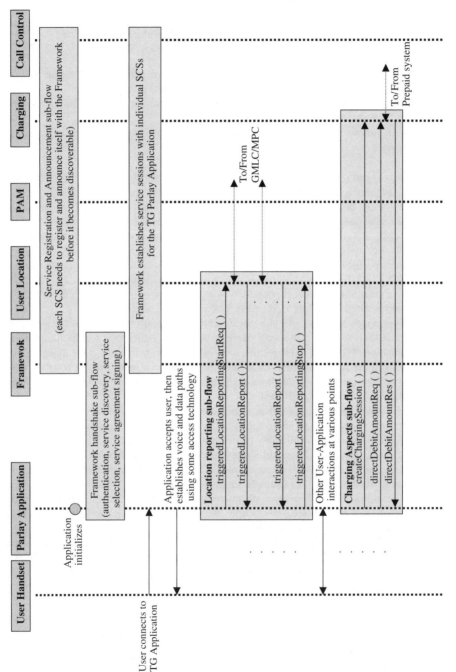

Figure 7.2 TG Application scenario call flow sample with sub-flows per SCS

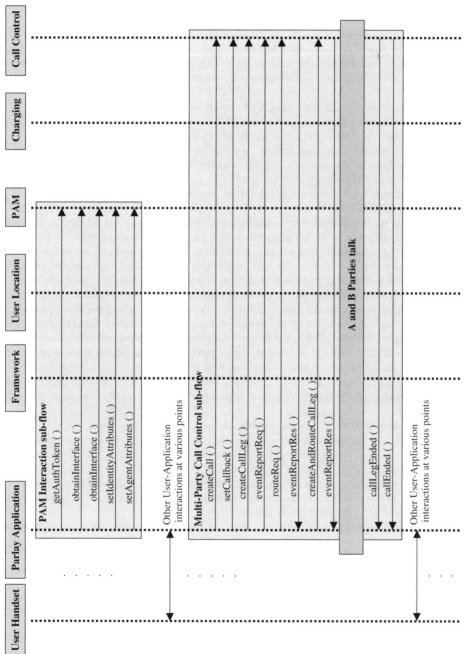

Figure 7.3 TG Application scenario call flow sample with sub-flows per SCS (continued)

including method parameters and related details, which is very useful to application developers that build these Parlay/OSA-compliant applications. This series of documents is normative and binding on implementations from a standard interface perspective. A second view is also available – this latter one in the 3GPP 29.998 series of documents, provided merely as informative guidance to implementers of service mediation gateway elements (such as telecom equipment vendors), is a set of API to protocol mappings for the most frequently used protocols in today's networks, that can enable service providers to leverage deployed infrastructure for Parlay services. Let us now look at our example from the network point of view. This theme is also touched upon in the next chapter, which explains how one could navigate most effectively through the large set of standards documents.

When the application requests the UL SCS to provide periodic updates to Alfie's location, this SCS interacts with network elements such as the GMLC (in GSM networks) or MPC (in CDMA networks), the HLR, the MSC, etc., (with the HLR and MSC either directly or indirectly – the GMLC/MPC may interact with these instead) and gets location reports including details such as the X/Y/Z co-ordinates of the mobile's location, the cell ID and sector, and other information. Various location-finding algorithms such as AFLT, EFLT, Network Assisted GPS etc., are used by the underlying network elements to achieve this. The reader will recall that this was briefly discussed in Chapter 1.

Once a location fix is obtained, it is propagated back in the reverse direction by the network element, eventually to arrive at the TG application. TG can now use this location and factor it into any context sensitive information it wants to provide Alfie. For instance, it may do an application database dip, and generate a new set of web pages with points of interest, photographs, new summary narratives for these etc., and then feed them to Alfie via the interfaces available to it.

Similarly, when Alfie makes purchases, the TG application would invoke debit methods on the CH SCF interface as previously described, and these would translate within the service provider network to directives to be executed by the prepaid application hosted within. Standard or prepaid vendor proprietary interfaces could be used for this (depending on deployment), so long as a decent mapping between the CH API and the underlying protocol is possible (although there may be a wide degree of variance in the capabilities supported by different vendors for a prepaid solution, there are some common aspects to these). When TG learns a debit was successfully made, it can then make the purchase on behalf of Alfie and send him the voucher to his handset as proof of payment.

As far as blocking calls from unwanted parties during the tour is concerned, simply marking Alfie as unavailable to buddies such as friends and business associates would merely discourage these parties from making phone calls, but not prevent them from doing so. Calls may be blocked if suitable policies were set to influence the operation of the CC SCS, or through the support for other, more advanced, code based routing schemes[3] through the use of the user interaction SCS, during call setup. Blocked calls could be courteously transferred to voice mail. For simplicity, these aspects are not covered in this example in any detail.

As far as TG application initiated calls are concerned, the CC SCS could issue the appropriate directives (acting in the role of an SCP from the traditional IN model) over the service control protocol to have the switch initiate two call legs, one to Alfie's phone, another to the target business, and then bridge them. These kinds of operations typically require the use of a network element called a Service Node (SN) to perform these operations. The SN implements the SRF and CCF functional elements from the traditional IN call model (please see Chapter 1 and Chapter 4 for more details) and bridges the call legs and tracks call progress.

---

[3] Code-based routing schemes are gaining in popularity. Alfie's wife, Joanna, calls him, say, from a payphone, but the system offers her the opportunity to enter a code to talk to Alfie. There may be different codes Alfie gives out – one for family, one for friends, one for business associates and so on. Once the code is entered, depending on the policies that are set for that user group, Joanna's call may or may not go through. In our example above, Alfie would receive her call while on tour, because he has indicated he wishes to be able to take calls from family. Bob, his boss, would not be able to get through.

Application developers and service providers alike would be wise to pay close attention to the protocol to API mapping considerations, however. Sometimes, the underlying protocols do not support a clean mapping for some capabilities required by certain applications, which are available by the standards-defined API. For example, the ANSI-826 protocol used in CDMA networks, does not provide for third-party call control capabilities, so the techniques described in the previous paragraph cannot be used seamlessly to achieve the desired goals. In such cases, other workarounds may be used to provide the same end-user experience, or the application itself may be configured or dynamically configures itself to operate in degraded mode, thereby providing the best user experience possible given the current network circumstances. These, and other related issues, are addressed in Chapter 13.

7.6 Toolkits for Application Development

As can be seen from the above description, there are several tasks that different applications perform that are common, need to be done repeatedly, and can be re-factored and reused from a coding standpoint. If this is done, then common procedures could simply be re-linked, re-compiled into, or re-invoked, thereby further reducing development and testing costs through a sharing of usable software assets across multiple applications. This is so because the re-factored libraries (also called Software Development Kits or SDKs) that contain code that is commonly used across many or most applications need only be developed and tested once, and then reused seamlessly over and over again.

This leads to a schism or split in the way application logic may be perceived. Application software now consists of two parts – one, the common component that performs the required generic functions (available from the SDK), procedures such as (for example) application authentication, service discovery, selection and service agreement signing with the Framework, and two, the business logic that intelligently drives the methods available from the common component as well as provides any additional value add and competitive differentiators when compared with similar application business logic from other competing offerings.

These Application SDKs for Parlay/OSA APIs may be made available by Parlay Gateway vendors themselves[4]. They may do this to promote their own products with application development shops thereby leveraging these re-factored technological components also as an effective marketing tool for their gateways. However, if the gateway element is truly standards-compliant, it is rather likely that the Application SDK components provided by one vendor will be transparently and seamlessly reusable even as the developed applications interact with other vendors' gateways – though indeed some peculiarities in particular vendors' implementations of the gateway and SDK components will be observed as these kinds of interoperability tests are performed. This is also largely due to the fact that the standards themselves specify merely the interfaces, not the behavioral aspects of the interactions between the client application and gateway (server) components of the architecture. This is as it should be, since standards are meant to drive interoperability, push through wider adoption and deployment of new technologies and promote innovation, not stifle creativity. It is this very degree of freedom that provides more impetus for new products and new ways of leveraging technology to build still newer and more exciting applications.

We must also note that as this pattern of re-factoring and multi-use of software is becoming more prevalent in the Parlay/OSA domain, a number of gateway vendors are now applying similar techniques to the server components as well – some productize these capabilities and make them

---

[4] It is relatively easy for application developers to incorporate standards-defined informative IDL mappings into their code. An SDK may provide convenience classes to simplify further application development, as well as provide a complete environment including an IDE (Integrated Development Environment) or the ability to integrate seamlessly provided classes into a general purpose IDE. These are additional advantages. In fact, some rather complete SDKs are provided with simulators of actual SCSs developed by the vendor. Such SDK packs are all the more useful since they can serve not just as development but also as test environments for new applications that developers might wish to build and deploy in said vendors' environments.

available to their service provider customers while others treat this as an internal capability that gives them a competitive edge in being able to more quickly, effectively and cheaply build and deploy new services or SCSs than their competitors.

7.7   Mixed Mode Applications

As Parlay/OSA applications are developed, capabilities defined within the various SCF APIs are leveraged advantageously to obtain network and other contextual data for use within the transaction-processing logic. In certain deployments, additional capabilities may be available in terms of either additional proprietary or specialized, not-too-widely-deployed SCSs that can be used to perform the required operations even more efficiently, or (on the flip side) some capabilities expected by the application when it was designed, may not be available due to a certain type of SCS not being deployed in that network.

This opens up a number of interesting issues relating to the design, development and testing of applications in different network contexts, and some of the more advanced topics in that area are discussed in Chapter 13. Here, we study a few design patterns in common use today to accommodate either 'still in progress', developing or not-so-mature SCF APIs, more flexible deployment options (these permit the application vendor to have a greater 'total addressable market' (TAM) with the single offering, though at a greater investment), and ease of accounting for some network or protocol peculiarities in particular situations.

Let us revisit the same example from earlier on in this chapter. We have seen how the TG Application may invoke method calls on the UL SCS to obtain Alfie's (the user's) location information and factor this into its transaction processing. An application developer may decide to either implement this functionality completely independent of the UL SCS by letting his application talk directly to underlying network elements such as the GMLC or MPC using some protocol such as LIF MLP[5] to obtain this information from the network, while continuing to use Parlay APIs for all other capabilities offered by the application. Alternatively, he may choose to implement both modules – the one that leverages the Parlay UL SCS and the one that utilizes the LIF MLP interface – with some kind of provisioned or configurable flag or parameter that dictates which mechanism would be enabled in each particular network deployment. The latter option offers greater flexibility and TAM (thereby contributing to access to larger sources of potential revenue) but also at greater development cost.

The key idea from this section is that Parlay/OSA-compliant applications that employ additional alternative mechanisms to get at network context information also available through the Parlay APIs may be called 'mixed mode' applications. They may do so for flexibility, to attract wider deployment, or to overcome some obscure protocol issues in cases where the Parlay API defined functional set does not map well with the underlying protocol. This does happen, though rarely, and where it does occur, it is generally because someone wants to use a protocol to perform functions it was not specifically designed to do, thereby force-fitting it to the API.

Mixed mode implementations are quite prevalent today, since some SCF interface definitions are much more mature and well developed from a technical standpoint than others from the Parlay family. Those better defined are more widely used, and functional capabilities from those still in development are incorporated into logic that will gradually evolve to utilize the standards-defined method calls as that API matures. This enables Parlay to be successful and gather momentum as a technology as it 'crosses the chasm' into wider acceptance, while other capabilities can continue to be developed to meet as yet unforeseen needs, thereby permitting greater penetration and even more widespread usage in the time to come.

---

[5] The reader will recall that this protocol was explained in Chapter 1 in Section 1.7.2 on 'Location-based Services'. LIF MLP is an XML-based protocol – MLP is Mobile Location Protocol – developed by the Location Interoperability Forum or LIF, a consortium defining standards, since subsumed by the OMA or the Open Mobile Alliance.

7.8 Summary

In this chapter, we have discussed a simple example to illustrate the value of Parlay services, and their impact on the end-user experience. From the description here it should be clear to the reader that as more Parlay-compliant applications are built, and the services infrastructure gets used more and more, new application development becomes easier, cheaper and faster and the gateway pays for itself as more revenue is collected through these new and exciting applications.

Now that we understand at the highest level of abstraction how Parlay applications work, and how they may be built to leverage network capabilities, in the next section of the book, we shall discuss service mediation gateways, implementation considerations for these from both the standards and development and deployment perspectives, and then similar issues for client applications before moving on to more advanced topics such as the Parlay Proxy Manager, Feature Interaction issues, Web Services etc., in the rest of the book. But first, the next chapter reinforces some of the standards learnt in this section of the book while also helping the reader make sense of the large number of standards documents by summarizing their contents and explaining their relationship with each other.

8

Standards Capabilities and Directions III – The Lay of the Land

8.1 Introduction

In Part II 'The standards in Detail' so far we have focused mainly on the Parlay API specifications themselves and the service mediation architecture and interfaces they define. This chapter will expand that view somewhat. We aim to summarize other related standards documents and try to demonstrate how they contribute towards a deeper understanding and appreciation of Parlay and the application of the Parlay APIs. The chapter is not intended however to cover any complete network architecture and every specification describing it, as provided in great breadth and depth by various standards development organizations, like for instance 3GPP. But rather we aim to show how Parlay fits in with its environment, and interacts with it.

8.2 Navigation

Standards specifications form a landscape that is difficult to navigate without some basic directions and a few landmarks. This section will provide some insight into the numbering scheme in use and the categorization of specifications into stages, which will hopefully help the reader take the helm and pilot her way through the sometimes rough terrain of standards documentation.

Three stages are used to categorize standards specifications, each targeted at a different audience and produced by another part of the standards community. A 'Stage 1' specification contains the high-level service requirements for a given service, from a service-user's point of view. A 'Stage 2' specification is the architecture document for a given service, and provides an analysis of the problem into functional elements and the information flows between these elements. A 'Stage 3' specification defines the actual definition of the protocols or interfaces between the physical elements onto which the functional elements, identified in Stage 2, have been mapped.

The 29.198 series, with parts for each of the Parlay APIs, collectively forms the Stage 3 for Parlay. The concept of specification stages is used by 3GPP[1] and 3GPP2[2], but not the Parlay Group.

---

[1] 3GPP specifications can be found at http://www.3gpp.org/specs/numbering.htm, see also Appendix B [Parlay@Wiley]

[2] 3GPP2 specifications can be found at http://www.3gpp2.org/Public_html/specs/index.cfm, see also Appendix B [Parlay@Wiley]

Parlay/OSA: From Standards to Reality   Musa Unmehopa, Kumar Vemuri, Andy Bennett

Table 8.1 3GPP specification numbering
scheme

Stage	Series
Service aspects ('stage 1')	22 series
Technical realization ('stage 2')	23 series
Signaling protocols ('stage 3')	29 series

The Parlay Group, through ETSI, publishes the API specifications, in conjunction with a number of informative white papers, whereas 3GPP and 3GPP2 introduce the Parlay APIs as part and parcel of the suite of normative technical specifications defining a comprehensive service architecture in a third generation wireless network environment.

3GPP uses a numbering scheme to support the specification categorization, where the first two digits in any specification number, also called the series, signify the particular stage. The remaining three digits are merely a sequence number within the series. Table 8.1 shows this numbering scheme.

8.3   Parlay in 3GPP Environments

None of the sections that follow do the specifications they describe any justice by even the flimsiest of criteria. However, the intention is to provide the backdrop against which to tell the tale of Parlay. In isolation, Parlay may be just another interface. In its natural habitat, however, we can fully appreciate how this technology is entirely comfortable in and perfectly adapted to a multitude of network environments. References are provided for the interested reader wishing to explore the specifications in all their gory detail.

Figure 8.1 shows the 3GPP environment for Parlay.

8.3.1 The Service Concepts

3G TS 22.101 'Service Principles'
This specification [3GPP 2004s] describes the service philosophy for networks specified by 3GPP, which is defined as the ability to provide subscribers of 3GPP networks with the ability to use

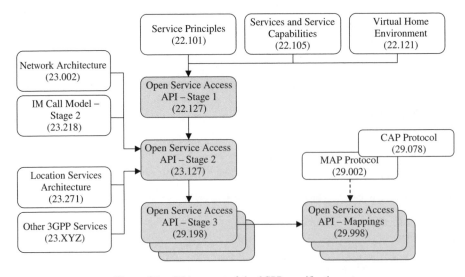

Figure 8.1   OSA as part of the 3GPP specification set

personalized communication services based on the multimedia capabilities offered by the network. These services should allow for efficient use of network resources, be compatible with global standards, support service evolution from second to third generation wireless networks, and support roaming users, while complying to regulator requirements regarding for instance emergency calls and national numbering plans.

3G TS 22.105 'Services and Service Capabilities'
This specification [3GPP 2002b] describes the principle of standardizing service capabilities, rather than fully specifying and normalizing individual services, in order to allow for increased service differentiation and product innovation. A service capability is defined by any means on resource available in the network that can be utilized to realize a service. These include core network elements and their protocols, certain bearers and their parameters, and network repositories. When offering a standardized application interface to access its capabilities, a service capability is referred to as a Service Capability Feature, or SCF. An SCF then forms a building block, which, either stand-alone or in composition with other SCFs, can be used to compose end-user services.

3G TS 22.121 'The Virtual Home Environment'
This specification [3GPP 2002c] introduces the concept of Virtual Home Environment (VHE) as one particular application of the service concepts outlined in the two specifications above. The VHE concept is defined as presenting the end-user with a consistent service experience, in terms of both behavior as well as appearance, irrespective of which network the end-user has currently roamed to, and independent of the device currently in use (though within the capabilities of the device and the network). Parlay is one of the service toolkits identified as means to realize the VHE concept.

8.3.2 The Overall 3GPP Architecture

3G TS 23.002 'Network Architecture'
This specification [3GPP 2003] describes how the 3GPP PLMN is configured of various architectures and sub-architectures, comprising the functional entities and their interfaces. Interfaces are defined between PLMN entities themselves, and between PLMN entities and entities in domains external to the PLMN, including user equipment, access networks, and different service platforms.

A high-level view of the 3GPP network architecture has been provided in Chapter 1. In Chapter 4 we introduced the protocol mapping recommendations and explained the informative nature of these mappings. Given that there is typically more than one way to deploy a Service Mediation Gateway in any network, using a variety of standardized signaling protocols or proprietary interfaces, the basic 3GPP PLMN configuration in 3G TS 23.002 does not contain any Parlay entities or interfaces.

However, certain sub-architectures and configurations include Parlay explicitly, specifically in those cases where Parlay is an unequivocal part of the architecture, or where the Parlay API is a defined interface. Such instances include the Parlay API as interface between the OSA SCS and an external LCS client (Location Services), and the functional architecture for service provision in the IP Multimedia Sub-system (IMS), where the IMS Service Control (ISC) interface is defined between the S-CSCF and the OSA SCS, and where a reference point is defined between the HSS and the OSA SCS.

3G TS 23.218 'IP Multimedia (IM) Session Handling; IM Call Model; Stage 2'
This specification [3GPP 2005r] further elaborates on the functional architecture for service provision in the IMS, as part of the IMS call model. The IMS service architecture supports three mechanisms for service delivery, each aiming to address a different category of applications. First of all, continued, seamless support for legacy CAMEL based services is provided through an Application Server serving as an IM-SSF (IP Multimedia Service Switching Function). SSFs have been introduced in Chapter 1 as the Intelligent Network component that interacts with the SCP for the delivery of IN-based services to the end-user. Incorporating an IM-SSF in the IMS architecture allows the reuse of SCP based IN services to any end-user connected the IMS network. Second, lightweight SIP based applications are intended to be supported through the inclusion of a SIP

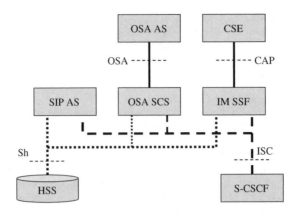

Figure 8.2   OSA Service Control as part of the IM Session Handling

Application Server to the IMS architecture. The third option pertains to an OSA Service Capability Server (SCS) for the support of more feature-rich, value added applications, possibly in the third party domain. These three options are depicted in Figure 8.2.

8.3.3 Services Making Use of OSA

3G TS 23.271 'Functional Stage 2 Description of Location Services (LCS)'
This specification [3GPP 2004t] provides the architecture to support location services in 3GPP networks. Even though there is no standardized protocol mapping available between the OSA SCS and the GMLC, OSA is explicitly included in the LCS architecture as one of the two possible interfaces towards an external LCS client application, as shown in Figure 8.3.

8.3.4 The Stages of OSA

3G TS 22.127 'Stage 1 Service Requirement for the Open Service Access (OSA)'
This OSA Stage 1 specification [3GPP 2002d] contains the service requirements for OSA, namely it defines functional requirements that are to be fulfilled by the OSA Service Capability Features.

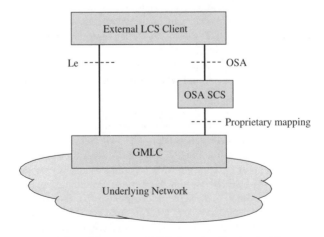

Figure 8.3   OSA as part of 3GPP Location Service Architecture

The service capabilities are defined such that network functionality supported in 3GPP networks can be exposed to application developers in order to build applications.

The OSA Stage 1 does not specify whether certain requirements should be supported by a separate single SCF or distributed over several SCFs. Particular service requirements may naturally exclude specific design options or may be most efficiently realized using specific communication patterns, but no such mandates are included in the Stage 1 and the decision is ultimately left to the Stage 2 or Stage 3 specifications.

In addition to functional requirements for both the Framework as well as the Network SCFs, the Stage 1 includes more general requirements on for instance network, programming language, and operating system independence, harmonization and alignment with other standards bodies (e.g. ETSI and the Parlay Group), and general FCAPS[3] considerations such as the need for secure, scalable and extensible systems.

3G TS 23.127 'Virtual Home Environment/Open Service Access'
This Stage 2 specification [3GPP 2002e] defines the OSA architecture, in support of the service requirements outlined in the Stage 1 specification. The OSA architecture is defined in terms of the three-way model now familiar to us, i.e. OSA Client Application, Services, and the Framework. We will not repeat that discussion here and the reader is encouraged to go back to Chapters 4 and 5, if required. The grouping of functionality into Service Capability Features is described in the Stage 2 as well, and these will correspond to the parts in the Stage 3 29.198 series.

The more advanced architecture deployments outlined in the OSA Stage 2 specification are addressed in more detail in Chapter 9 'Alternative Architectures'.

3G TS 29.198-X 'OSA; Application Programming Interface (API); Part X'
This specification suite consists of multiple parts, one for each network Service Capability Feature, or Service, one for the Framework, one for the common data types, and an overview. These specifications have been introduced in Chapter 4 and described in great detail in Chapter 6.

3G TR 29.998-X 'OSA; Application Programming Interface (API) Mapping for OSA; Part X'
This collection of recommendations encloses the protocol mappings for a number of the Service Capability Features, and a number of protocols. The mapping recommendations have been introduced in Chapter 4. For more detail, the reader is referred to [3GPP 2002f, 3GPP 2002g, 3GPP 2002h, 3GPP 2002i, 3GPP 2002j, 3GPP 2002k, 3GPP 2004u].

8.4   Parlay in 3GPP2 Environments

3GPP2 has been introduced in Chapter 1 as the partnership project for the creation and publication of specifications for CDMA and ANSI based 3G networks. In Chapter 4 we have seen that Parlay is endemic to 3GPP2 networks since the start of 3GPP2 participation in the Joint Working Group. The position of Parlay in the set of 3GPP2 specifications is depicted in Figure 8.4.

8.4.1   The Overall 3GPP2 Architecture

3GPP2 P.S0001-B 'Wireless IP Network Standard'
This specification [3GPP2 2002a] defines the overall requirements on all architecture entities involved in a 3G wireless packet network based on cdma2000 spread spectrum access technology. These overall high level requirements can be considered as the Stage 1 for 3GPP2 networks, addressing capabilities needed to support Quality of Service, Security, Accounting, and Mobility Management. It provides the context and background for all service related specifications for 3GPP2 network, including OSA, as we shall see.

---

[3] Fault Management, Configuration Management, Accounting Management, Performance Management, Security Management

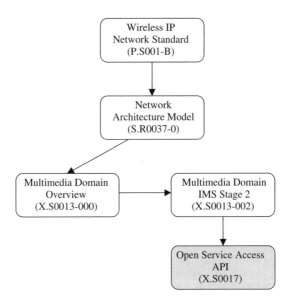

Figure 8.4   OSA as part of the 3GPP2 specification set

3GPP2 S.R0037-0 'IP Network Architecture Model for cdma2000 Spread Spectrum Systems'
This specification [3GPP2 2003k] describes the 3GPP architecture model and hence in scope is
quite similar to 3G TS 23.002 from 3GPP. The 3GPP2 architecture model describes the network
entities and all reference points that exist between them and which collectively comprise the 3GPP2
All-IP network.

The OSA SCS is positioned in the 3GPP2 All-IP network in exactly the same way as in the
3GPP All-IP network. That is, the OSA SCS is supported through the ISC (IMS Service Control)
interface to the S-CSCF. Effectively this means that from the perspective of the OSA SCS, the
underlying network, be it an All-IP network from either 3GPP for UMTS access technology or
3GPP2 for cdma2000 access technology, is transparent. That is, the OSA SCS communicates with
the resources in the network through the ISC interface, irrespective whether these resources are
built to 3GPP or 3GPP2 specifications. This transparency is illustrated in Figure 8.5.

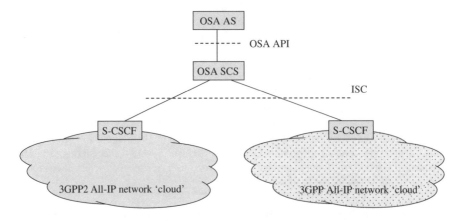

Figure 8.5   3GPP/2 network Independence through ISC interface

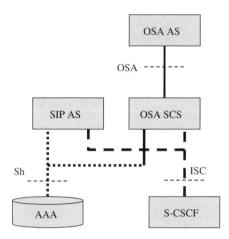

Figure 8.6 OSA Service Control as part of MMD Session Handling

Similar to the All-IP Stage 2 specification for 3GPP, the All-IP Stage 2 specification for 3GPP2 specifically defines an interface between the OSA SCS and the Position Server (compare GMLC in the 3GPP case).

3GPP2 X.S0013-002-0 'All-IP Core Network Multimedia Domain – IP Multimedia Subsystem – Stage 2'
This specification [3GPP2 2003b], which follows from [3GPP2 2003a], defines the elements, protocols, and flows in order to support the IP Multimedia call models in 3GPP2 networks. It can be considered as the equivalent of 3G TS 23.218 [3GPP 2005r], and as such describes a similar role for OSA service control as part of MMD (Multimedia Domain) session handling. This is depicted in Figure 8.6, which bears close resemblance to Figure 8.2 but without the support for legacy IN services.

8.4.2 OSA in 3GPP2

3GPP2 X.S0017-0 'Open Service Access (OSA) – Application Programming Interface (API)'
This specification [3GPP2 2003l] provides the Stage 3 descriptions for Open Service Access APIs in cdma2000© based systems, and as such is intended to aid developers in interpreting the applicability of the 3GPP OSA specifications in a 3GPP2 environment and infrastructure. In the spirit of reuse and to avoid re-inventing the proverbial wheel, the OSA Stage 3 for 3GPP2 is defined as a so-called delta document. This means that the document does not duplicate or redefine any of the API definitions, but rather limits itself to pointing out where there are additions or exclusions. True to the objective for harmonization, the delta document does not define any additions or exclusions from the OSA API definitions. Or in other words, the OSA API for 3GPP and 3GPP2 is one and the same. The only recorded differences apply to the reference sections of each part of the OSA specification set, or to specific informative examples. For instance, a CAMEL Pre-Paid example would not be applicable to 3GPP2 networks.

8.5   Summary

In this chapter we have aimed to summarize standards documents other than the Parlay API interface definitions and tried to demonstrate how they contribute towards a deeper understanding and appreciation of Parlay. A sneak preview, if you will, of various standards specifications from both 3GPP and 3GPP2, with the sole purpose of placing Parlay/OSA in its natural environment. With these navigational pointers, the reader will hopefully get a feel for the standards landscape and may feel comfortable to wander around beyond the Parlay specifications themselves.

Part III

Building a Service Mediation Gateway

This is a small but important part of the book. It deals with some of the more practical issues relating to how Service Mediation Gateway components utilizing Parlay and OSA technologies may be constructed. The emphasis here is on giving the reader a lot of 'food for thought' and helping people think, rather than attempting to give them all the answers (good answers usually depend heavily on the context associated with particular questions). This section is intended for technical readers and designers.

9

Alternative Architectures

9.1 Introduction

In Part I, we showed how the concept of Service Mediation fulfills the objectives we have set for the solution to our problem statement. By means of service abstraction and controlled access we have seen that service capabilities in a communications network can be unlocked by the network operator to a large community of software developers to build converged applications that are both innovative and attractive from an end-user perspective, as well as portable, billable and manageable from a network operator perspective.

The overall system architecture of Parlay, as an embodiment of the Service Mediation concept, has been introduced and explained in Chapter 5. We have outlined its three main components, i.e. Client Applications, Services Capability Servers, and the Framework. For any given architecture it is important that the structure and composition is sound, i.e. the architecture defines the foundations of the system you are going to build and deploy. Mistakes in the architecture can merely be covered up or kludged at a later stage. It is significantly more costly and mostly even impossible actually to solve architectural problems or oversights during the build or deploy phase.

In this chapter we will demonstrate the flexibility of the Parlay system architecture by outlining several architectural alternatives as defined in the Parlay standards. We will then discuss more advanced architecture patterns and explore several detailed and broken down architectures applicable to each of the high-level architecture alternatives. The aim is to validate the Parlay system architecture and establish its applicability in real-time, high performance, and scalable network deployments. Does the architecture hold up and can we reasonably expect commercially attractive, high-performance systems to be built off this blueprint.

9.2 Standard Architectural Alternatives

The high level Parlay system architecture, as defined in its most simple form in the Parlay standards specifications, consists of Client Applications, Service Capability Servers, and the Framework. We shall leave the Enterprise Operator out of the scope of this discussion for the purpose of brevity. This simple architecture makes no statements about deployment models. Applicability to a broad range of existing communication network architectures with a multitude of stakeholders is an important aspect in order for the architecture to achieve wide industry acceptance. No single deployment, or limited subset of possible deployments, should be mandated in any open standard.

The basic architecture model supports full flexibility in terms of potential groupings among the three basic components, or possible integration with existing network elements, without impacting the specified Parlay Application Programming Interfaces. Considerations like performance

Parlay/OSA: From Standards to Reality   Musa Unmehopa, Kumar Vemuri, Andy Bennett

characteristics, especially throughput and reliability, as well as specific characteristics of the network or cost aspects will have a bearing on the chosen deployment architecture and will render a certain architecture alternative more viable or desirable than others. It all depends. The following sections will introduce various approaches for deployment and discuss the respective merits and drawbacks of each of these approaches.

9.2.1 Embedded Approach

You will recall from Chapter 4 that the role of the SCS is to translate service requests from the Parlay Client Application into operations that can be performed by certain elements in the network. One obvious way of deploying Parlay Service Capability Servers then, is to collocate the SCS physically with the network element it interacts with, such as an HLR or an SCP, in order to fulfill a service request from the Parlay Client Application. Or put another way, Parlay Client Applications can be supported by extending the appropriate network entity, or entities, with a Parlay Application Programming Interface. We will term this architectural deployment alternative the embedded approach as the Parlay layer is embedded into the actual network element.

9.2.2 Gateway Approach

Opposite to the embedded approach of fully integrating the SCS with the network entities one could envisage an approach where the SCS is fully separated. The network operator may opt to deploy Parlay by introducing a new node in the network physically separate from any of the supported network entities. This new node then implements the Service Capability Servers and interacts with the network entities required to fulfill a service request from the Parlay Client Application, through standardized communication protocols native to those network entities. The new node acts as a service mediation gateway, and hence we will term this approach the Gateway approach. In this case the gateway supports the Parlay Application Programming Interface, whereas the existing network entities remain Parlay agnostic and continue to communicate through their native protocol(s).

9.2.3 Hybrid Approach

Within the range delimited by the opposite approaches outlined above, several variations can be conceived. Consider for instance a network deployment where several gateway nodes are supported instead of just one monolithic gateway. Or a network deployment where some SCSs are embedded while others are operated on physically separate nodes, essentially combining the embedded approach and the gateway approach in a single deployment. Such architecture deployment variations are collectively referred to as the hybrid approach. There may be several reasons that would yield a hybrid approach. An operator may have implemented a multi-vendor policy where specific vendor solutions impose a mixed architecture deployment. Or for reasons of proof of concept an operator may go for a small scale embedded SCS, before implementing a phased rollout covering the whole network, which evolves into a hybrid architecture deployment.

In each of the above three architecture alternatives, outlined in Figure 9.1, there is one logical Framework for all SCSs. The Framework itself can be embedded into a network element, deployed as part of a Gateway solution, or operate as a stand-alone entity. Similar considerations will apply as outlined above for the SCSs.

9.2.4 Discussing the Merits of Standard Architecture Alternatives

When looking at one of the more evident distinctions between the embedded approach and the gateway approach we can draw a comparison with the Intelligent Network architecture considerations, as discussed in Chapter 1.

In the embedded approach there is an obvious close coupling of the capabilities of the Parlay SCS with the existing network node in which it is embedded. The advantage is that optimizations

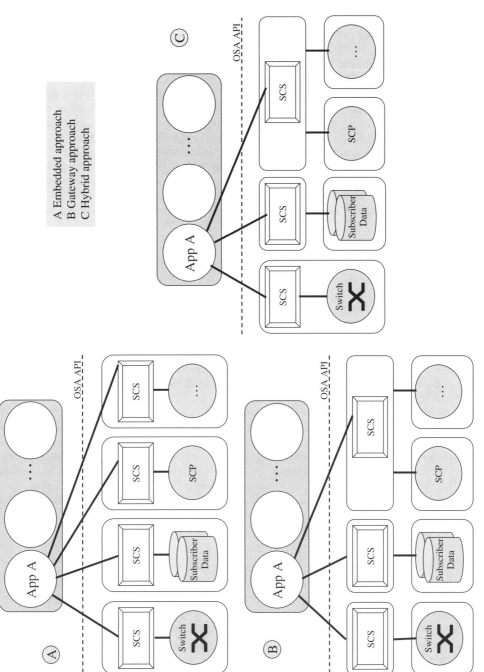

A Embedded approach
B Gateway approach
C Hybrid approach

Figure 9.1 Standard architectural alternatives

specific to the existing network node can be efficiently exploited by the Parlay Client Application accessing the embedded Parlay SCS. The embedded Parlay SCS can make the most of the service features supported on the existing network node. Speed is also an advantage. The benefit of this optimization and tailored SCS deployment however comes at the expense of Client Application portability, as the Client Application is confined by the features and service capabilities of the specific network element.

In the gateway approach, much like the Intelligent Network architecture, a mediation gateway is introduced to decouple the Parlay Client Applications from the network nodes. The Parlay Client Applications can now communicate and interact with a multitude of network nodes in the underlying fabric. The mediation gateway, supporting the Parlay SCS, accesses the network nodes through some southbound communication protocol that is native to the nodes, allowing the nodes in the fabric to remain Parlay agnostic. As with the Intelligent Network architecture, in this model of separation, introducing new Parlay-based services in the network no longer necessitates updating all switching elements in the fabric. Rather, only the physically separated mediation gateway needs retrofitting. Basic service as performed by the existing network elements is not impacted and can continue transparently during the time of the update. There is of course an implied impact on performance, however, this is outweighed by the advantage gained through loose coupling.

9.3   Advanced Architecture Patterns

So far in this chapter we have introduced some high-level architecture alternatives supported in the standards, adhering to the one basic Parlay system architecture of Client Application, Service Capability Server, and Framework. This has given us a simple architecture, which supports various deployment models. So far so good, however, we not only wish to use the architecture to model deployments or design APIs. The goal is to build commercial systems. The further success of this architecture will rely in part on characteristics such as performance and fault tolerance through the support for e.g. redundancy and distribution. This is where the use for multiple service instances comes in. Load can be shared across multiple service instances, whereas redundancy through replicated service instances reduces the vulnerability that would result from single-component failures.

In the remainder of this discussion we shall make use of the concept of service sessions, as introduced in Chapter 5. A brief recap is provided here. You will recall that a service session was defined as the time during which a Parlay Client Application has use of a Service, in order to provide value towards the end-user. Service sessions are set up between the Parlay Client Application and the Service making use of the reference to the service instance, which the application has obtained from the Framework. All service requests pertaining to the execution and delivery of this specific Parlay Client Application take place within the context of this service session. Chapter 5 also explained the Parlay Integrity Management interfaces. We have seen that Integrity Management is performed at the granularity of a service session, and hence of a service instance. So why is this useful? Or asked in another way, why would one support multiple service instances?

9.3.1 Multiple Cloned SCSs

When looking at load sharing, an initial and perhaps trivial solution that springs to mind in order to be able to use multiple service instances, would be to support multiple clones of the same SCS. Here SCS clones are defined as multiple identical SCSs, of the same service type and with the same service properties. Each of the SCS clones is registered with the Framework, and a Client Application discovers each clone individually and engages in a Service Agreement with each and every one of them. As each clone is a full-blown SCS in its own right, it will feature a service factory. Client Applications can engage in service sessions with the SCS clone, which will instruct its service factory to spawn service instances on that SCS clone. Business as usual. However, now the Client Application has the ability to perform load sharing using the multiple clones it has discovered, each featuring their own service instances. When one clone fails, other clones can be contacted by the Client Application and new service sessions can be set up to continue operation. In addition, reliability improvements can be realized by keeping multiple clones simultaneously alive, that is, the Client Application has concurrent service sessions with more than one clone.

Genesis of the term 'Clone'

The term 'SCS instance' is often misused to refer to particular instances of a given type of SCS deployed in an SMG configuration (the term 'instance' here is used in the generic object oriented sense – the SCSs of a given type constitute a 'class', and each SCS of that type, that belonged to that class, is referred to as an 'instance'). For example, if an SMG deployment supported three types of SCSs – for User Location (UL), User Status (US) and Call Control (CC), in addition to the mandatory Framework, with two SCSs of each type, sometimes each of the SCSs is referred to as an instance of that particular type (i.e., we have two SCS instances of UL, two of US, and two of CC, and so on).

Unfortunately, this definition is overloaded, since the Parlay and OSA specifications already use this terminology to refer to instances of service sessions supported by an SCS. In other words, when the specification refers to an 'SCS instance', it refers not to particular SCSs of a given type, but to a service manager spawned by an SCS of a given type.

It is not difficult to see that usage of the term 'instance' without setting the right context could lead to confusion. We therefore, as explained in previous sections, designate particular SCSs of a given type to be clones or copies of each other, and utilize the Parlay standards-defined terminology of instances to refer to particular service session instances – for example, if clients A and B are connected to the same Call Control SCS, then there are two instances of Call Control that are active, one corresponding to the service session and service manager for client A, and the other for B.

A shortcoming of the Multiple Clone SCS pattern, shown in Figure 9.2, is that load sharing and improved reliability is not transparent to the Client Application. The burden of distributing service requests across the multiple clones is placed entirely in the application domain. Integrity Management and safeguarding of the Service Agreement has to be performed on a bilateral basis with each of the clones.

9.3.2 Some Practical Implementation-related Considerations

In this section, we present three distinct but connected problems relating to SCS implementations: SCF Nesting, Dual SCSs, and FSM Synchronization. These issues impact all Parlay/OSA gateway implementations, and, if not carefully addressed, could have adverse impacts on seamless interoperability across different vendors' SCS implementations. It is therefore important clearly to document the behavior of individual SCSs and note precisely how they function with respect to the content already specified in the standards documents. Here, we focus on attempting to capture adequately the issues themselves, and present areas for exploration so a solution may be devised. Particulars of designed solutions may vary, and provide competitive advantages to products.

9.3.2.1 SCF Nesting

As we have seen in earlier chapters of this book, the Parlay specifications present, in each document, the API for a particular SCF or Service Capability Feature. Physical manifestations of these functional components are called SCSs or Service Capability Servers. A single SCS may implement one or more SCFs.

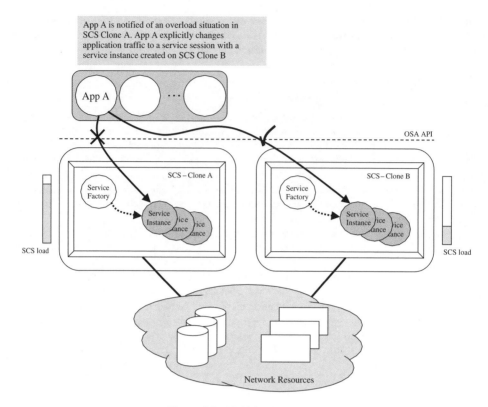

Figure 9.2   Multiple cloned SCSs

Broadly, each document from within the specification describes a single functional entity. Table 9.1 summarizes the kinds of SCSs currently supported by the standard, and the associated service type tags.

Each one of the SCFs defined in individual standards documents supports a single functional area, but supports, in many cases, multiple nested SCFs, each providing particular capabilities. Further, as illustrated in Table 9.1, each of these functional capabilities supports one or more directly accessible interfaces.

When a service registers with the framework component, it is expected to register with a name from the set indicated in column 3 of Table 9.1. Upon successful registration, a Service ID is assigned to that particular service type. Note that some proprietary or customized service provider defined SCFs are also supported. These are identified through tags that begin 'SP_', and could be used to represent either non-standard SCF types or SCF super-types or agglomerations of more than one nested SCF capability where such an implementation is being referenced.

Unless super-types are used, this also means that if we implement a UL SCS that supports all three kinds of User Location capabilities, then the Service Instance Lifecycle Manager (SILM) or Service Factory of the UL SCS must register and announce itself thrice with the framework, once for each of the UL capabilities it supports. Three different Service IDs are then assigned to it, and the Framework makes these three nested SCFs independently discoverable and selectable by third party client applications. Also, when the client applications select a service, and a service manager is to be created, the Framework communicates the service type with the service factory (SILM) and the desired kind of service manager instance is created to process queries over that service session.

Table 9.1 SCSs currently supported by the Parlay standard

OSA Specification	SCS	TpServiceTypeName	Supported 'SCF Manager' Interfaces
29.198-04	Call Control	P_GENERIC_CALL_CONTROL	IpCallControlManager
		P_MULTI_PARTY_CALL_CONTROL	IpMultiPartyCallControlManager
		P_MULTI_MEDIA_CALL_CONTROL	IpMultiMediaCallControlManager
		P_CONFERENCE_CALL_CONTROL	IpConfCallControlManager
29.198-05	User Interaction	P_USER_INTERACTION	IpUIManager
29.198-06	Mobility		IpUserLocation
	User Location	P_USER_LOCATION	IpTriggeredUserLocation (intended to be used as an extension to IpUserLocation)
		P_USER_LOCATION_CAMEL	IpUserLocationCamel
		P_USER_LOCATION_EMERGENCY	IpUserLocationEmergency
	User Status	P_USER_STATUS	IpUserStatus
29.198-07	Terminal Capabilities	P_TERMINAL_CAPABILITIES	IpTerminalCapabilities
29.198-08	Data Session Control	P_DATA_SESSION_CONTROL	IpDataSessionControlManager
29.198-09	Generic Messaging	P_GENERIC_MESSAGING	IpMessagingManager
29.198-10	Connectivity Manager	P_CONNECTIVITY_MANAGER	IpConnectivityManager
29.198-11	Account Management	P_ACCOUNT_MANAGEMENT	IpAccountManager
29.198-12	Charging	P_CHARGING	IpCharingManager
29.198-13	Policy Management	P_POLICY_MANAGEMENT	IpPolicyManager
29.198-14	PAM	P_PAM_PRESENCE_AND_AVAILABILITY	IpPAMPresenceAvailabilityManager
		P_PAM_EVENT_MANAGEMENT	IpPAMEventManager
		P_PAM_PROVISIONING	IpPAMProvisioningManager (excluded from the 3GPP specification)

Notes:

1. The first three parts of the 29.198 series of specifications are dedicated to an Overview, Common Data, and the Framework aspects respectively (this is also explained in Chapter 4).

2. The SCS column merely lists the SCSs that could be built given the standards defined functional areas or SCF APIs. As we have seen in previous chapters, the Parlay standards define functional APIs and implementations can build one or more of these SCF APIs into physical manifestations called SCSs. The SCS column indicates examples of how functional interfaces can be wrapped into physical envelopes.

3. Not all the SCFs listed above are in all the specifications – there remain some minor differences between the 3GPP, ETSI and Parlay APIs in terms of supported SCFs, and these, while highlighted above, are subject to change with time (also explained in Chapter 4 in more detail).

When service managers are created, it is expected that they each implement one of the nested SCFs defined in column 3 of Table 9.1. This means that in the UL SCS example above, an application can request the creation of a service manager for P_USER_LOCATION, another for P_USER_LOCATION_CAMEL, and yet another for P_USER_LOCATION_EMERGENCY, all from the same Mobility SCF implemented in a single UL SCS.

Note also that the inheritance structure of the interface definition needs to be carefully studied as an SCF is being implemented. This has a bearing on the implementation architecture. For example, in CC, a Conference Call Control (CCC) SCS is also a Multiparty Call Control (MPCC) SCS as CCC inherits (via Multi-media CC (MMCC)) from MPCC. There is no need to register a CC SCS for each of MPCC, MMCC and CCC, although you could still register them separately. That is an implementation option.

Likewise, the Triggered UL (ULTr) interface inherits from the User Location interface, and is not considered a different SCF (it's treated simply as an auxiliary capability extension) but the other Mobility-related SCFs each inherit from IpService and are therefore different SCFs.

Where there are nested SCFs whose interfaces may not be independently registered with the Framework, (e.g. MPCC/MMCC/CCC and UL/ULTr, depending upon the particular implementation), the OPERATION_SET property would indicate what is actually supported.

9.3.2.2 The 'Manager' Pattern

The standards have previously utilized the Manager pattern to support streamlined interface access in cases where a single SCF contained multiple subtended interfaces. In such cases, the application would first obtain a reference to a service manager that implemented the 'Manager' interface, and then perform 'obtainInterface() ' operations upon that interface to get access to other SCF 'sub'-interfaces. In fact, some SCFs still support this – for example, the PAM SCF, which now has three nested SCFs, supports managers for each of these three capabilities, whereby further 'sub'-interfaces can be obtained. However, this pattern is not applied consistently throughout the standards specification. This leads to some confusion. This cannot be fixed now in the standards, except for newer SCSs such as PAM, since most of the SCF API definitions have already been baselined. (Note that only two of the three manager interfaces are available from the 3GPP version of the PAM specification – the IpProvisioningManager interface is not made available in the 3GPP specification, though it is supported by ETSI. There are several such minor differences between various standards versions and these need to be carefully studied by anyone attempting to build a standards-compliant implementation or desirous of certification testing, or testing for interoperability.)

If super-types alluded to above are indeed supported by an implementation, one may simply build an SCS that registers with a super-type (say SP_LOCATION), and then returns a 'manager' interface (say 'Location_Mgr') that supports an 'obtainInterface() ' function which the application can then use to obtain either of the three nested UL SCF Ip interfaces. The service manager could be defined to support one or more of these based on the set of properties passed along to the service factory at service manager creation. One issue with this is that although the implementation itself is simpler (in terms of SCS state machine management if nothing else), it is (at least in terms of SCF access) not standards-compliant in how the various nested SCF interfaces are derived.

The problem therefore becomes: how is one to support multiple nested SCFs in the implementation? We have already examined two different ways of doing so. A third (not quite so elegant) option would be to support three different service factories, each of which is capable of returning just one kind of service manager. The benefits of having three different service factories hosted within the same binary are not obvious, and it is clearly not beneficial to be able to support such closely related nested SCF capabilities into separate SCSs.

Generally, it is advantageous if SCFs with nested SCFs are implemented as a single SCS with a single service factory that registers with the Framework once per nested SCF, and can serve multiple types of service managers based on the nested SCF selected by the client application. This, although slightly more involved in the short term, offers some protection from any interoperability issues in the mid- to long-term.

It is perhaps also interesting to note that some implementations segregate the UL and US SCFs (part of the Mobility SCF in the standards) into separate UL and US SCSs. This is an example of a case where a single Parlay defined SCF is implemented as two separate physical SCSs. The opposite problem is of greater interest, however, and forms the basis for the next section.

9.3.2.3 Dual SCSs

The above discussion provides a good lead into the second problem, namely that of dual SCSs. In some cases, the Parlay specifications have defined separate SCFs for closely related capabilities. Implementations could benefit from supporting these SCFs in a single SCS binary since the data models and processing capabilities are very close to each other from a functional perspective.

An example of this is found in implementations of a CHAM (Charging and Account Management) SCS. The Charging and Account Management parts were defined as separate functional entities in the specification primarily due to a distinct set of security requirements, but some implementations may choose to build them into one SCS to save on development and deployment costs, especially since they are so functionally related.

There are five different options here in terms of how the service factories may be implemented. Some implementations may build the Charging and Account Management capabilities as two different SCSs (they each have their own service factory) but contained within the same binary, thus making them a single orderable item. This provides some of the benefits of separate SCSs, but at the same time, tries to leverage a common data model most effectively.

There are several different options for building such SCSs, especially with respect to how many service factories are supported, and what those service factory implementations are capable of. Let us say there are two SCFs that need to be supported – K1 and K2. The following options then become available:

a) Support these two functions in two separate kinds of SCSs, one of which returns service managers of the kind K1 only, and the other of which returns service managers of the kind K2 only; or
b) Support a single SCS with a single service factory that can serve up two different kinds of service managers – K1 and K2, depending on the specific needs of the requesting entity as made known to the service factory at 'createServiceManager()' time; or
c) Support a true 'dual SCS' like the above referenced implementation of the CHAM SCS, with two different service factories, each of which can return only a single kind of service manager – K1 or K2; or
d) Support only a single service manager K that supports both the interfaces for K1 and K2 simultaneously. (K is the union of K1 and K2.) This is similar to the 'Manager' pattern applied to non-standard IDL as discussed above for PAM (K will now support methods that the application can invoke to get access to either or both of K1 and K2 for further use); or
e) Support a single service factory that can support the existing standards-defined Parlay interface (K1) with a single method extension that may be used to request and receive a reference to an interface of type K2, which the client application may then utilize for K2 implemented functions. Note that this will require internal proprietary modifications to the Parlay inheritance hierarchy.

Note that (c) is logically and functionally equivalent to (a), the only difference is the implementation as a single binary from a functional standpoint, though other characteristics such as scaling also differ.

Option (d) merits some additional discussion too. Here, the standard interface is expanded to support proprietary extensions so that clients aware of these extensions could use them while all other clients would simply receive standard Parlay treatment. Thus, this technique could be used to provide some good value-add and differentiators while remaining standards-compliant in implementation.

When we talk of 'dual SCSs' we normally mean implementations that build to the pattern defined in (c). The failure characteristics of such dual SCS implementations are such that the two SCFs are a single process, and they fail and recover as a single element. Note that the issues here, although similar to a single SCF with two nested SCFs implemented in a single SCS, are not the same, at least not from a functional standpoint, though similar implementation constructs could be used to solve both problems.

9.3.2.4 FSM Synchronization

An interesting challenge in the above cases from an implementation perspective is the mapping of state machines. Generally speaking, SCSs support three different kinds of state machines from a Parlay perspective:

a) There is one FSM that governs the operation of the SCS itself – functions such as startup and initialization, registration and announcement of services with the framework, and eventual shutdown (be it graceful or abrupt).

b) There is another FSM that governs the establishment, control, management, and teardown of service sessions on the SCS. This state machine is only operational when the SCS itself is in the operational state, when the SCS has registered and announced itself with the framework, and is implemented by the Service Factory or Service Instance Lifecycle Manager (SILM) component.

The Framework then has the ability to request (on behalf of subtended client applications that have an access session with it), that the SCS create a service manager to host a service session for that particular application, and can pass along one or more parameters that govern the operation of the service session.

Later, after using the service, when the client application wishes to tear down the service session, it simply requests the Framework to tear it down via the established access session, and the Framework instructs the service factory to do the necessary, and this causes the termination of the service manager state machine associated with that particular service session, and the freeing up of the associated objects.

Service session control and management is achieved through support for the integrity management APIs, coupled to the created service managers themselves, which are useful in reporting load, fault conditions and pulsing heartbeats at periodic intervals to convey session health and sanity to the Framework.

c) A third, but no less important set of state machines is implemented by the service manager. These are 'ephemeral' state machines – one per transaction, that support the processing logic involved. These state machines implement the transaction model, are created when a request is received, and die when a final response is sent back.

Some of these machines live longer than others – for instance, a periodic or triggered location request state machine has a longer TTL (Time To Live), since the request is generally alive until the client issues a 'Stop' directive on it. A one-shot location request, on the other hand, has a relatively short lifetime.

We consider only state machine (a) here in greater detail, since it is directly related to the content of this section.

As can be seen from Figure 9.3, it is relatively easy to draw this state machine for a simple SCS implementation (with no nested SCFs, no dual SCS agglomeration). Now, let us see how these other cases may be addressed.

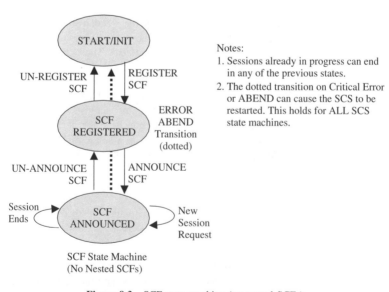

Figure 9.3   SCF state machine (no nested SCFs)

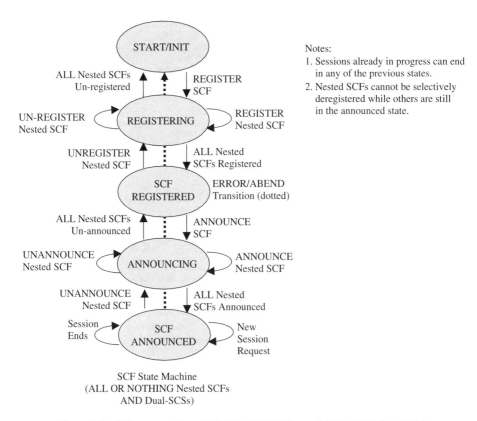

Notes:
1. Sessions already in progress can end in any of the previous states.
2. Nested SCFs cannot be selectively deregistered while others are still in the announced state.

SCF State Machine
(ALL OR NOTHING Nested SCFs
AND Dual-SCSs)

Figure 9.4   SCF state machine (ALL OR NOTHING nested SCFs AND dual-SCSs)

In the case of the scenario of the CHAM SCS (described above), there are two different SCFs implemented within it, and two different service factories, each of which is only capable of serving up a single type of service manager (one for Charging only, one for Account Management only), but where the two SCFs behave as one logical entity in the sense that either both SCFs are simultaneously available, or in case of failure, neither is available. This FSM is as shown in Figure 9.4.

Finally, let us consider the UL SCS – a case where it has a single service factory that registers thrice with the Framework, once per nested SCF, but is capable of serving out three different kinds of service managers based on the parameters passed into the SILM at createServiceManager() time. The FSM for this case is as depicted in Figure 9.5. The state map essentially remains the same from a functional perspective, but one that indicates mappings to specific state instances of each type cannot be generated unless certain other constraining assumptions are made with regard to the operation and behavior of the associated SCS.

9.3.3 Distributed SCSs

So far we have discussed the Multiple SCS Clone pattern as more advanced architecture for high performance deployments. The pattern consisted of distributing multiple identical SCS clones.

A second pattern to achieve load sharing or increased reliability through redundancy consists of physically distributing components of a single SCS across different nodes that make up a gateway cluster and have each node take its share of the load. With this pattern, the Parlay Client Application is no longer tasked with the responsibility to distribute its service requests among various physical

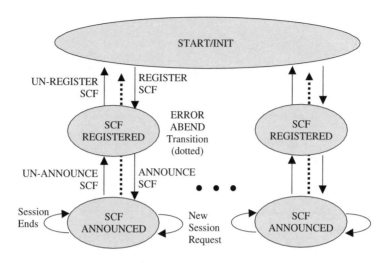

SCF State Machine with Nested, Independent SCFs

Notes:
1. Sessions already in progress can end in any of the
   previous states.
2. Nested SCFs can be selectively deregistered while
   others are still in the announced state.

Figure 9.5   SCF state machine with nested, independent SCFs

SCS components. Indeed, it may not have any knowledge of the SCS deployment architecture in the network, or may not even care. Let us take a closer look at a possible solution to support physically distributed SCSs. Physical distribution of SCSs should be transparent to the Parlay Client Application.

In the simple scenario, featuring a monolithic SCS, such an SCS would host a service factory, which creates service sessions based on service requests initiated by the various Parlay Client Applications. Nothing new here, we have already covered this ground in Chapter 5. As we are dealing with a monolithic SCS, the service instances are created on the same physical node as the one hosting the service factory.

In order to support physical distribution of SCS components across different nodes, we introduce the notion of a primary SCS. The other nodes available for distribution are subsequently referred to as the secondary nodes. The primary node is distinguished from the secondary nodes, as it is the node that hosts the service factory spawning the service sessions based on service requests initiated by the various Parlay Client Applications. However, this time the service instances are not created on the same physical node that hosts the service factory (i.e. primary SCS), but distributed across the secondary SCSs. Transparency towards the Parlay Client Application is ensured, as a service session can exist only between a single application and a single (secondary) SCS. However the load of the total number of service sessions for applications running on a Parlay Application Server can now be distributed across multiple secondary SCSs.

As it is the primary SCS node that hosts the service factory, it is the primary SCS node that performs Integrity Management, including load balancing. Integrity Management is still performed at the granularity of a service session, but now the service sessions, and hence service instances, are physically distributed. This is illustrated in Figure 9.6.

Distribution of service sessions is based on some distribution policy hosted on the primary node. The service factory executed on the primary SCS node may evenly allocate sessions among the

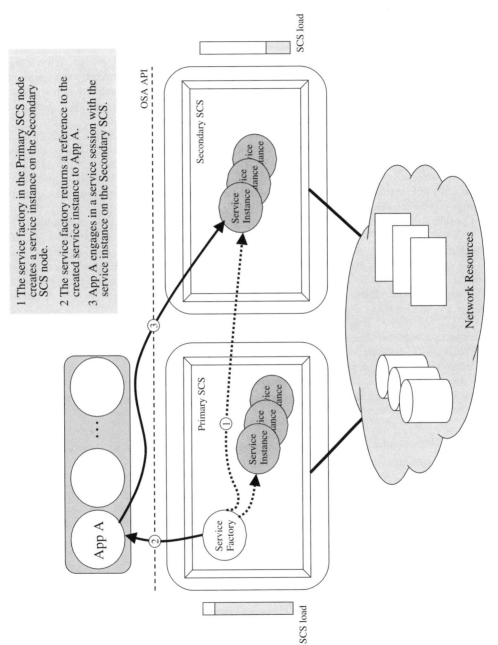

1 The service factory in the Primary SCS node creates a service instance on the Secondary SCS node.

2 The service factory returns a reference to the created service instance to App A.

3 App A engages in a service session with the service instance on the Secondary SCS.

Figure 9.6 Distributed SCSs

available secondary nodes, for example, using a random or round robin mechanism. With some added intelligence, however, more sophisticated policies can be envisaged. For instance when allocating new service sessions, the service factory could temporarily skip the secondary node which is operating above a certain capacity threshold in an attempt to prevent overload and hence failure. Graceful service degradation can be supported by re-allocating the service sessions of a failing secondary node equally among the remaining available secondary nodes still operating within their specified load constraints.

The solution of physically distributed SCSs presented above provides transparency towards the Parlay Client Application, as it conforms to the basic Parlay system architecture, and does not require any modifications to the specified Parlay APIs. The distribution of SCSs is not visible to the Parlay Client Application, but the Parlay Client Application does benefit from increased perceived service availability and fault tolerance.

As service sessions with the Parlay Client Application are distributed across the secondary SCS nodes, the granularity of the load balancing mechanisms is limited to the level of service sessions. That is, there are no mechanisms for the primary SCS node to allocate individual service requests to different subtended secondary SCS nodes. Load sharing on a per request basis is not supported. In some cases this level of control will be sufficient, however, there are examples where finer grained control may be required.

Consider for instance an application that is engaged in a service session with a User Status SCS in order to monitor the status for a set of end-users. The application has subscribed to triggered status reports for a large set of users. Both the subscription as well as the triggered reports are part of the same service session. During normal network operation and end-user behavior, the triggered reports, reporting a change in status, will be generated by the network and hence processed by the SCS spread over time fairly equally or in some predictable pattern. But as we all know we cannot design a system for sunny day scenarios only. When the network becomes operational again after an unplanned outage, the status for all end-users that are being monitored will simultaneously change from 'not reachable' to 'reachable'. This could cause an overload situation when all triggered reports are processed by a single SCS component. In such a scenario the Gateway would benefit from a distribution scheme where individual service requests can be shared among SCS nodes. Whether the poor Parlay Client Application will be able to deal with the large number of triggered status reports coming in from a number of SCS nodes, neatly distributed according to some sophisticated load balancing scheme, is of course an entirely different matter. But we will worry about that in another section.

9.3.4 Tiering of Multiple Cloned SCSs

The Multiple Clone SCS and Distributed SCS patterns can be applied irrespective of the architecture alternative chosen. That is to say, both patterns can be applied in order to build an SCS, be they deployed using the embedded, or gateway, or hybrid approach. The possibilities however for an SCS to distribute for instance load (or service requests) across various network elements, in the case where the SCS is embedded with one specific such element, obviously has its limitations. Also, supporting multiple instances of the embedded SCS on the same physical element for reasons of reliability or performance does not overcome the single point of failure of the physical element itself. The Gateway approach on the other hand provides more promising opportunities for further architecture decomposition in order to validate the support for real-time, high performance, and scalable network deployments. In Chapter 14 we will further explore the possibilities of the Gateway approach, and introduce a new architecture pattern that can be considered a combination of the Distributed SCS and the Multiple Clone SCS patterns – the Parlay Proxy Manager.

The most distinctive feature of the distributed SCSs is the existence of a primary node, hosting the service factory, and a number of secondary nodes. A single service factory is supported in this set-up. We have seen that this implies that Integrity Management is performed on a per service session basis. A logical next stage would be to take this model one step further. Imagine the case where all

SCS nodes support a service factory. A straightforward service session is set up between the Parlay Client Application and the primary SCS node, just like in the basic scenario (the reader might also want to refer back to Chapter 4 for more details). In fact, so far this architecture is the same as the Multiple Clone SCS pattern. Now here comes the novel part; the primary SCS node subsequently sets up service sessions with one or more of the secondary nodes (which are SCS clones), rather than creating service instances on these secondary nodes, as would have been the case with the Distributed SCS pattern. This second tier of service sessions is used to distribute individual service requests within the context of the single first tier service request between application and primary node, across the various secondary SCS nodes. The service factory hosted on the secondary SCS node will then create a service instance on that secondary node. As we can see it is now possible to perform load sharing at the granularity of individual service requests, rather than at service session level as was the case with the distributed SCS case. We shall refer to this architecture option with service factories on all SCS nodes as Tiered Multiple Clone SCSs. The secondary SCS nodes in the tiered multiple clone SCS configuration are called leaf SCSs, whereas the primary SCS node is called the Parlay Proxy Manager SCS. We chose the name Parlay Proxy Manager because in the multiple clone SCS configuration the PPM SCS acts as an ordinary SCS from the point of view of the Parlay Client Application, and conversely as a Parlay Client Application towards the leaf SCSs.

The Parlay Proxy Manager will be described in greater detail in Chapter 14.

9.3.5 Getting Practical with Architecture Patterns

We have looked at architectural alternatives naturally from an architectural perspective, discussing architecture deployments and design patterns to achieve certain improvements in performance. That is, we have approached this as an engineering problem using theoretical arguments. Let us look now at the Application Programming Interfaces and see whether our elegant architectural solutions hold by looking at some practical use case scenarios.

When discussing load distribution of individual service requests, we have made an implicit assumption that all service requests are both alike as well as independent of each other, and hence can be distributed without consequences. For some Service Capability Features such an assumption may be more valid than for others. For instance, one could envisage a requirement where all method invocations pertaining to a single charging session should be distributed to the same leaf SCS. Such method invocations have a functional relationship. Service requests or method invocations could also have a logical relationship, for instance a request to stop notifications should be issued on the same leaf SCS to which the notification start request was dispatched, as state information is maintained on that leaf SCS node. So we see that logical or functional relationships between service requests may impact whether certain requests can be freely distributed across SCS nodes.

Another implicit assumption we have made is that the distributing entities have knowledge of the load impact of individual service requests. Such knowledge may require complex intelligence on the part of the distributing entity. For instance, will a distributing entity recognize the difference in load demands between a request to arm triggers for address range ' + 1 61*' and address range [' + 1 613 1000' − ' + 1 613 9999']? And how about even more complex behavior. Do we allow a primary node SCS to break up a request to arm triggers for address range [' + 1 613 1000' − ' + 1 613 9999'] into range [' + 1 613 1000' − ' + 1 613 4999'] and range [' + 1 613 5000' − ' + 1 613 9999']? Preferably, such a procedure would not impact the application ID for the trigger arming request to the Client Application. The two separate trigger arming requests would exist only locally between primary SCS and leaf SCSs and then the primary SCS would have to aggregate resulting event notifications towards the Client Application.

These examples show that the theoretical performance benefits from alternative architectural options and design patterns, though providing significant performance enhancements, may not be fully achievable in real-life deployments.

9.4   Summary

In this chapter, we have validated the flexibility of the Parlay system by outlining several architectural alternatives as defined in the Parlay standards. We have then explored increasingly more advanced architecture patterns to address issues like performance management and reliability of implementations. The concepts introduced in this chapter have also laid the ground works for an even deeper dive of architecture considerations in Chapter 14 on the Parlay Proxy Manager, pushing the envelope of architectural possibilities of the Parlay standards even further.

10

Considerations for Building 'Carrier-Grade' Systems

10.1 Introduction

This chapter is organized into six separate sub-chapters[1], each focusing on a single aspect of 'surround' capabilities that are normally assumed to be part of any carrier-grade or telecommunications grade equipment or solution.

As we have seen in previous chapters, Parlay/OSA technologies do in fact provide an excellent means to leverage Internet technologies and toolkits as one builds newer and more exciting services and applications that use such services, leveraging network context information in transaction processing logic. However, it is not sufficient to have services that are new and exciting. After all, the investment that service providers make in new technologies is with a view to tapping into as yet unrealized revenue potential. This can only happen when the infrastructure hosting these services and applications is sound and operates to the expected guidelines of performance, reliability, scalability, overload control, etc. that all parties – the service providers, and subscribers, among others – have come to expect of telecommunications systems.

As technology gets more seamless and ubiquitous, one tends to expect that it works more reliably without failure. For example, you get rather annoyed when your landline phone for some reason, offers no dial tone. As cellular technology gets more prevalent, subscribers will be more unforgiving of lost or dropped calls during cell handoffs, roaming or other scenarios. As technology crosses the proverbial chasm [Moore 2002], and sees more widespread use, these 'surround' aspects have to be addressed as well, or subscribers will reject the solution (even a very good one in theory) and look for alternatives that better meet their needs.

This chapter focuses on some of what are considered '*-ilities' in common telecom parlance. The term originates with sometimes overlooked but very important elements such as scalability, reliability, etc. that all have the common '-ility' suffix. Other elements that fall into the same category and are of interest here (but lack the suffix that lends its name to this category) include performance, error handling, and upgrades.

We emphasize that this chapter merely addresses some of the important issues, and is not an exhaustive listing of all possible items worthy of consideration in the design of telecommunications

---

[1] The organization of this chapter presented its own unique challenges. Several topics need to be covered here, but they are not so closely related as to be sections within a larger chapter, and not so disparate and disjoint that they merited their own individual chapters. A compromise structure, using sub-chapters, therefore seemed the most appropriate to use.

---

systems. We do try to give the reader an appreciation for the kinds of things that merit closer inspection, and offer up some ideas in each case as candidate architectures are analyzed, but do not provide specific answers in most cases.

The reason for this is twofold:

1. Many of the issues discussed are critical to the development of telecommunications systems in general and service mediation gateways in particular. Specific choices or sequences of choices made can make or break a product and can either give you or take away capabilities for competitive differentiation.
2. Choices one makes as one builds more of the product depend on choices made previously, and many of the decisions need to factor in non-technical aspects such as the amount invested, the timelines for development, the staffing requirements etc., and to some extent, the culture of the company. One size certainly does not fit all.

This chapter is focused on the service mediation gateway component. However, generally speaking very similar, if not the same, conditions apply to client application architecture, design and development as well. We have tried to point out applicable elements wherever possible, in the narrative. We do expect that the astute reader will extract and also apply similar patterns to client application design.

With this introduction in mind, let us go through some of this material with a view to educating ourselves on considerations for building efficient Parlay/OSA gateway solutions.

10-1

Reflections on the Performance of Implementations

10-1.1   Introduction and Scope

In this sub-chapter, we study issues associated with the performance of implementations. A good comprehension of performance is important for many reasons: it enables people actively and consciously to factor concepts and ideas relating to maximizing performance into their designs; it gives them a view into optimizing implementations already in existence; and more generally it enables them to grasp more efficiently which sub-processes in transaction processing take what amounts of the total time, and what their contributions to the overall latencies tend to be.

Performance engineering, like all other important subjects, is both an art and a science. The scientific aspects are codified in reusable patterns (both architectural and mathematical) that are applied over and over again, in various projects, as engineers, designers and developers attempt to squeeze out every iota of slack from code to make it as taut and efficient as it can be. It is an art, in that sometimes the beauty of an improvement comes not from local optimizations, but rather global design changes that may, in some cases, even seem counter-intuitive at first glance.

High-performing systems are built that way from the ground up. Trying to engineer in performance as an after-thought typically does not work. The reader would do well to factor in aspects of performance engineering as he/she designs the Parlay-compliant SMG or application element.

This sub-chapter cannot claim to do justice to the topic of performance optimizations in the general sense, nor completeness in coverage of either the art or the science aspects of this important topic. We merely attempt to give the reader a good understanding of some of the sound principles surrounding performance engineering of Parlay/OSA-based systems, including SMG components and Parlay-compliant client applications.

10-1.2 Performance Aspects

There are at least two aspects of performance of which an engineer should be aware, as he or she takes on this exercise. The first deals with defining models to compute the expected performance of a given piece of software being written, and identifying optimizations, while the second is more focused on ensuring the implementation actually performs within the bounds prescribed by the model.

Before one undertakes the first step, however, one has to decide clearly what one wants to achieve. What does success look like? What are the items worth measuring? What are the prescribed boundaries for parameter variations? And are these reflective of reality?

As can be seen, some of these can be answered using knowledge and experience accumulated from previous projects, but modified to apply to peculiarities of the Parlay/OSA context. How performance engineering is actually carried out varies from person to person, and from one corporation to another, with success being determined by the processes followed, the experience one brings to the task at hand, and how rigidly or flexibly the models used have been designed.

Typically, two parameters that are factored into performance computations include the transactions per second (TPS) or busy hour call attempts (BHCA) rating of the software in question, and the end-to-end (e2e) latencies observed when typical transactions are run. The TPS or BHCA ratings are both representative of what may more generally be called Xpd (Transactions per duration) parameters, since they each specify, using different units, the number of transactions the software in question can process, per specified time interval.

Other performance engineers like to work with metrics such as 'milliseconds per transaction', which is an example of a Dpx (duration per transaction) parameter.

A computing platform's CPU performs important functions like provisioning support, operations and administrations support (sometimes collectively referred to as OAM&P – Operations, Administration, Maintenance and Provisioning) etc., among other things. In addition, a CPU needs some spare room to be able to handle overloads gracefully. It is therefore unreasonable to expect that the entire 100% of the CPU capacity would be dedicated to processing transactions. The engineer for that reason picks a suitable allocation of CPU cycles for transaction processing and proceeds to compute the available milliseconds from that, and can then calculate the TPS from the ms/transaction rating for the software in question.

Let us illustrate this by means of an example. Consider any quad processor symmetric multiprocessing system. Since this has four CPUs, the total processing capacity is the equivalent of 4000 ms per second of real time. Now, symmetric multiprocessing overheads account for some percentage of the CPU cycles (let's say 15%), which leaves 3400 ms/sec for use by applications. As we said before, some percentage of those are needed for other tasks such as OAM&P and overload control, etc., so we allocate say 20% of the remainder for that. 2720 ms are now available. Next, we allocate some 10% (say) for future growth and expansion. This leaves around 2448 ms/sec for actual transaction processing[2]. We call this the CPU Transaction Processing milliseconds per second of processing time (CPUTP).

---

[2] Note that the percentages used here are made up values that serve to drive home the points we are trying to make. In reality, there are usually guidelines individual corporations or engineers use as they design systems. Conservative estimates and allocations to transaction processing will doubtless yield lower TPS systems, but these ratings will be more stable, and almost certainly achievable in real deployments. Very aggressive models will yield systems that look good on paper, but which may fail to meet expectations in the field.

But all this raises a fundamental question: What exactly is a transaction? For if we do not know what it is, how can we even begin to measure it?

Transactions can mean different things to different people, so what appears to be a low TPS system in one context may actually be a high throughput system in another – it may merely be counting transactions differently. When comparing two systems, one must necessarily ensure that the two are using the same, or at least, a very similar, definition of transactions for it to be an apples-to-apples comparison (as opposed to an apples-to-oranges or an oranges-to-basketballs one).

Unfortunately, the standards documents provide no guidance on the definition of transactions in the Parlay/OSA space. So we come up with our own definition(s). We define the following terms:

Definition: Transaction
Given that Parlay/OSA supports both synchronous and asynchronous responses to requests, and also asynchronous notifications to applications on subscribed events, we define a transaction as comprised of:

1. a synchronous request response pair; or
2. an asynchronous request or notification and its associated asynchronous response, even if a synchronous answer is returned to the original request; or
3. an asynchronous notification, for which no response needs to be accounted.

Definition: User Transaction
This is a transaction that starts with the issuance of a user request to the application software, and ends when the user equipment receives a final response to that request. The Parlay/OSA client application may, or may not, perform additional processing after the final user response for the transaction has been returned.

Definition: Application Transaction
This is a transaction that starts with the issuance of a user request to the application software, and ends when the application software had completed processing that request. The application may, or may not, keep transaction state after the transaction ends.

Definition: SCS Transaction
This is a transaction that starts with the issuance of an application request to the SCS software, and ends when the SCS software returns a final response to the client application. The SCS may, or may not, keep transaction state after the transaction ends.

Let us illustrate these three definitions by means of examples (Figure 10.1). Alice connects to the Weather application and requests the weather forecast. This is the beginning of both the Application and User Transaction. The application formulates an appropriate request using some form of identifier (such as MIN or MSISDN) for Alice's terminal, and passes along this request to the User Location SCS on the SMG. This starts an SCS transaction.

The UL SCS processes the request by interacting with network elements such as GMLCs/MPCs and provides the response back to the client application. This ends the SCS transaction on UL. The application then takes the provided location fix, computes location specific content with that, and sends it to Alice's terminal. This ends the user transaction.

The application then (if it does not support a subscription-based billing model, but more a pay-per-use scheme) makes a request to the Charging SCS to debit Alice's prepaid account for services rendered. This is another SCS transaction that ends once the appropriate action has been taken. Once this is completed, the application-transaction for Alice's request ends. Note that the

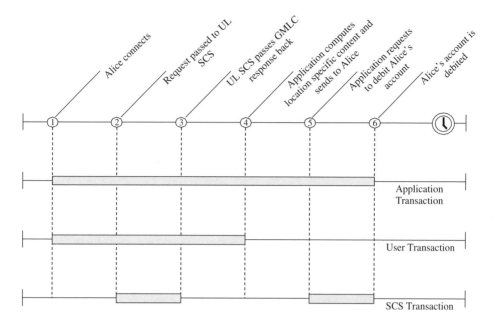

Figure 10.1   Performance aspects – transactions

application could have done this before sending Alice the requested content, thereby requiring two SCS transactions in processing a single user transaction.

As can be seen, one or more SCS transactions may be invoked by client applications either within or outside a single user transaction. But they are always invoked within the context of a single application transaction (asynchronous notifications are an exception).

If Alice were to request tracking of her handset on a map, the application transaction would begin with a periodic location request to the UL SCS, and end with an acceptance of that request by the SCS. The user transaction has the same lifetime, and ends once the application informs the user that it will perform the requested operation. At periodic intervals, the application would receive notifications from the UL SCS, each an SCS transaction in itself, and each tied to its own application transaction (and updates sent to Alice's handset would constitute their own user transactions). Once the user requests monitoring to cease, this opens a new user transaction, a new application transaction as the application turns off the periodic location reports on the SCS, and a new SCS transaction as this de-activation is completed. As responses are sent by the SCS to the application, and from the application on to the end-user, the SCS, application, and user transactions are closed.

Now that we have our terminology defined, let us look at how performance may be computed.

10-1.3   Performance Computation – Flow Composition

Each call flow, from the user to the application, the application to the SCS, and the SCS processing itself, can be composed into its constituent elements. These can then be timed or estimated based on prior experience. Which aspects of a call flow are considered for this exercise may vary based on what the engineer is trying to do. An SCS designer and performance engineer will likely focus more on SCS transactions, leaving the application designer to worry about the timing of application and user transactions.

Let us say the transaction in question requires three database dips, invokes two procedures for pre- and post-processing of the initial query and final response parameters, and performs one other special function. Once this is known, and suitable millisecond values assigned based on CPU time to each of these elements, and a transaction schedule (that accounts for parallelism across these constituent processes) has been put together, a total ms cost for this transaction can be computed.

Thus, flow composition helps one assess where the time is being spent in the transaction, and how long a transaction should take in terms of CPU cycles. Once this is known, TPS can be easily computed as CPUTP/(total milliseconds needed to process a single transaction).

10-1.4   Performance Computation – Transaction 'Mix'

As discussed in Chapter 6, each SCS supports several different kinds of transactions, each with its own processing steps. Some of these involve more processing than others and consume more CPU cycles. We need to account for this in our modeling.

This may be done in two ways:

1. We select the simplest transaction for the SCS, or across all SCSs, as a benchmark, and assign a 'complexity index' to all other transactions as a multiple of that, using the closest whole numbers wherever possible. Then, we compute the processing capacity of an SCS and state it as so many benchmark transactions per second.

   If we are then asked, how many transactions of types A, B, C etc., can be supported, this is easy to compute, by setting up an equation of the form

   $$a * 3x + b * x + c * 4x = \text{CPUTP}$$

   where a, b, c = number of transactions of type A, B, C respectively that can be accommodated; x is the milliseconds per transaction for the benchmark transaction (in this case, B); and 3 and 4 represent the multiplicative index indicating how much more complex A and C really are than the benchmark.

   As can be seen, this equation has no single solution, but a range of suitable solutions can be obtained. Note that the above analysis assumes the entire transaction processing capacity is available to the SCS in question. If not, the appropriate fraction of CPUTP should be used on the right hand side of the equation above.

2. We choose suitable 'mixes' of transactions, say 25% of type A, 55% of type B, 20% of type C, that reflect most closely the estimated proportionate realistic use of this SCS, and then compute the number of each type that can be supported given the CPUTP rating. One may use flow compositions for each transaction type or a complexity index as indicated above to compute the required value.

   Using flow compositions, and assuming x ms per transaction of type A, y for each of type B, and z for C, the equation becomes $a * (25x + 55y + 20z) = \text{CPUTP}$, where a indicates the multiplier. Solving this equation, we conclude that 25a transactions of type A can be supported at the same time as 55a transactions of type B, and 20a transactions of type C. Use of complexity ratings in performing similar computations is left as an exercise to the reader[3].

   The above procedure may need to be repeated for different mixes, as different types of applications with different behaviors are supported by the same underlying SCSs over time.

---

[3] This chapter is intentionally vague on some of the details relating to performance optimizations. Specific processes used contribute to competitive differentiation, and are unlikely to be revealed by any developer of either Parlay applications or Parlay gateways.

10-1.5   Performance Computation – Abstract Models

The above discussion sets the stage for a slightly tangential discussion on the mathematical models themselves. This section is meant to give the reader a better appreciation for some of the complexity involved in the computations, and to help him or her visualize the procedures being carried out.

Consider an N-dimensional hyperspace, with each axis indicating a given transaction type for a given SCS type (for example, an implementation of the User Status API in an SCS could support the 'single-shot' request for the status of a terminal as well as the 'triggered reporting' mechanism for terminal status – each is an example of a different transaction type supported by that same SCF API). The methods outlined in the earlier sub-section serve to establish a kind of 'convex hull' if you will, a hyper-surface that bounds the capacity of the SCS in various dimensions (i.e. in terms of transaction types). Figure 10.2 demonstrates these concepts pictorially.

So long as the true load experienced by the SCS is within those bounds, it will operate as predicted by the model. Behavior outside those bounds is indeterminate. Since software needs to have fail-safes built into it, especially in carrier-grade implementations, engineers typically tend to build-in safety factors in various ways. These, when accounted for in our model, may appear to support another convex hull or hyper-surface, parallel to the first (though not necessarily so, since the tolerance values can be different along different dimensions), but contained completely within it. Typically, if load exceeds the inner hull, but lies within the outer boundary, overload control and throttling mechanisms would ensure that the outer hull is not 'breached', and the system performance remains within desirable ranges and behavior continues as specified. Similar considerations apply to client applications as well.

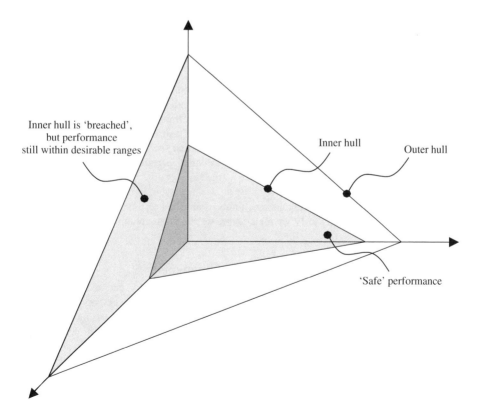

Figure 10.2   Performance computation – abstract models

10-1.6   Performance Computation – Round Trip Times

Round Trip Times (RTT), also referred to as Turn-Around-Times (TAT), specify the duration it takes in real time, not CPU time, in getting a transaction through the system. In traditional telephony, typically components are designed to satisfy a 300 ms latency requirement in terms of a 'last bit in, first bit out' metric. This metric is derived from the maximum delay tolerance in voice connections. A delay in excess of 300 ms is audible to the human ear and is experienced as a hindrance in voice conversations. Actual end-to-end latencies may vary based on network transmission delays, the number of components strung together in the call flow, etc.

In Parlay scenarios, one does not have specified latencies per network element, but it stands to reason that carrier-grade systems would mirror these requirements closely, though perhaps, due to the fact that these are not, strictly speaking, telecommunications equipment, the requirements could be relaxed a little.

Call flow composition in an end-to-end manner along with a sharp focus on individual elements and load assessments under stable state could help compute these figures, and then appropriate plans for tuning and optimizations can be made, as in the case with processing times. Some degree of variability typically exists in RTT measurements, due primarily to the fact that these times vary based not just on the conditions of the specific elements involved in the processing path, but also more generally on network related factors such as load on particular LAN segments or across the WAN as packets are transmitted back and forth.

10-1.7   Performance Verification and Validation – Tuning the Code, Measurements

Application behavior may be modeled, and the 80/20 rule applied, where a certain type of transaction is processed 80% of the time. Similar rules may be applied to each SCS, where a certain transaction is executed in an overwhelming majority of cases as opposed to others, or in SCS flow compositions where a certain constituent operation is invoked in the vast majority of all SCS transactions.

These 'most often used' sections of code, or transactions, are those that need to be optimized first. Application developers and performance engineers would be well advised to focus their (often times limited) resources in performance tuning and optimization on those aspects. Later, the rest of the code could be made more efficient as well, as time permits.

In general, it is considered good practice for the makers of carrier-grade products (both software and hardware) to perform performance tests including load and stress tests of their software, along with stability tests for memory growth, etc., over extended periods of time, and compare the measured performance per transaction type in the stable state against that predicted by the models in question.

These numbers should be within the tolerance limits established by the models. If they aren't, one needs to revisit the models, and verify what elements, so critical to computations, were mistakenly ignored.

10-1.8   Performance Engineering for Deployments

While deploying gateways or applications in service provider networks, one needs to go the extra mile beyond engineering considerations for individual SCSs. This process involves computing expected utilizations per service type across the deployed set of applications, using expected transaction mixes, and then scaling the SMG component deployments to meet these service provider needs for the required throughput. The models discussed previously in the section on transaction mixes can be easily applied with minor modifications to support this process as well.

Communications Networks are always designed with some judicious performance assumptions. For instance, the PSTN is designed with the assumption that no more than a certain percentage of all phone lines will be in use at any given point in time. If half the phones in the world were to call the other half at the exact same time, the system would not be able to cope with the load, but this is not typically a problem, and overload controls are built in to handle some of these pathological situations. Building to the worst case often results in unaffordable solutions.

Similarly judicious assumptions need to be made in Parlay capacity engineering. Different applications have different behaviors, and offer different services to their end-users. They also have different 'busy hours' – the times during the day that they are most utilized and offer the most load to the underlying SMG element. Where this information is available, it can be used to engineer the networks to support the required capacity in most non-pathological situations. For the rest, suitable overload controls can be built into the products in question to provide graceful service degradation, or rejection of new requests. These, and other factors, are discussed in the next sub-chapter of this set.

We leave the reader with a simple example on SMG engineering. Let's say Freedom Wireless wants to deploy a Parlay gateway and five applications. Their details are as indicated in Table 10.1.

The 2nd column in Table 10.1 indicates not just the peak hours where 100% of the capacity needed is expected to be used, but also the average load estimated at all other times during the day. Note that these patterns may vary from one day to the next, or on weekdays versus weekends. The performance engineer must account for that.

Using this information, the engineer then constructs a graph of the expected load per application during the day, and computes the peaks per service type, using the superposition principle (adding the waveforms for individual services across all the applications). The cumulative waveforms for each Service Type (recall that in Chapter 5, we have discussed how each SCS registers and announces itself using one or more of the standards-defined ('P_') or proprietary ('SP_') service types) indicate the total maximum Xpd capacity needed across all the applications for that service.

Next, he factors in a multiplier, to account for growth, say 20%, and computes the Xpd rating needed per service, for the applications to be deployed in that network. Once he knows this value, and the Xpd ratings in the same units for individual SCSs, it is a simple matter to factor in the number of clones (recall the discussion of this in Chapter 9) of each service, or the scaling factor needed for the deployed SMG to support the service provider required configuration satisfactorily.

Let us consider a simpler, more intuitive example for this as well. Here, there are two applications, A and B, that each use SCS X. These are the only applications in the network that use that SCS. The usage characteristics of this SCS by these applications are depicted in Figure 10.3.

This Figure indicates the required number of benchmark transactions that need to be processed (on average) during each hour of the day. The engineer applies the superposition principle to come up with the composite total number of transactions needed for that SCS (service of type X). This is repeated for each SCS hosted by the Service Mediation Gateway in the network (recall from earlier examples in Chapters 6 and 7 how each application may use one or more services). From Figure 10.3, it is clear that the maximum Xpd rating needed is 55, and that this occurs around 10 AM. The engineer may now add a safety factor to this, and indicate that a capacity of around 80

Table 10.1 Freedom Wireless: application performance characteristics

Application	Busy Hour (Peak Period)	Service Types used	Capacity needed per Service Type
Application 1	7–8 AM, 10% average	A, B, C	A – 80 Xpd; B – 100 Xpd; C – 50 Xpd
Application 2	8–10AM, 10% average	B, D, E	B – 150 Xpd; D – 20 Xpd, E – 10 Xpd
Application 3	9–11AM, 10% average	A, C, E	A – 10 Xpd, C – 200 Xpd, E – 5 Xpd
Application 4	4–7 PM, 5% average	D, F, G	D – 10 Xpd, F – 10 Xpd, G – 10 Xpd
Application 5	6–9 PM, 25% average	F, G, H	F – 30 Xpd, G – 200 Xpd, H – 35 Xpd

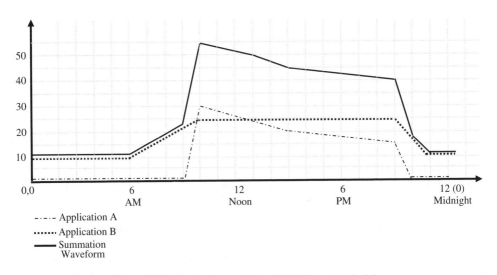

Figure 10.3   Usage characteristics of SCS X on a typical day

Xpd is needed. If each clone of SCS X can handle 30 Xpd, the engineer concludes that to provide the required capacity, 80/30 or 3 (rounding up) clones need to be deployed[4].

Of course, as new applications are added, or as existing applications are enhanced to support new capabilities and therefore new call flows, some or more of this analysis may have to be repeated based on usage and take rates in each case.

10-1.9   Summary

In this sub-chapter, we have illustrated some of the performance engineering guidelines for Parlay-components and deployments with a view to enabling the practitioner to develop reasonable procedures to maximize the performance of Parlay-compliant implementations.

10-2

Overload Handling Considerations

10-2.1   Introduction

Growing up, children are told a folk tale where a donkey, used to carry loads, was once burdened with sacks of salt. As it was crossing a stream, it ducked into the water, and its load got lighter. It

---

[4] Please note that this computation only considers capacity and not the reliability or high availability requirements that may also apply to the service. That aspect is addressed in sub-chapter 10-3 on scalability and reliability aspects of deployments.

repeated this over and over much to the owner's chagrin. Then the owner once filled the donkey's bags with sponges and thereby taught it a lesson – sponges soak up water thus gaining in weight, unlike salt that dissolves to lighten the load. Henceforth, the donkey was careful while crossing the river.

Overload control in deployed network systems has some similarities. Systems that are not well designed exhibit performance that drops precipitously when they are heavily loaded; everything seems fine under light or moderate load, but things fall apart when the presented load crosses a certain threshold.

Overload handling routines may also be viewed to be designed as a means of 'self-protection' for the system in question. This is somewhat similar to Asimov's Third Law of Robotics, which indicates that 'A robot must protect its own existence as long as such protection does not conflict with the First or Second Law'. Functionality is paramount, but while it services requests, the system should also try to take care to ensure that its continued well-being (and ability to service more requests into the future) is not adversely impacted.

In this sub-chapter, we shall examine some considerations involved in designing systems that behave more predictably and deterministically during times of heavy load. As with the other companion sub-chapters in Chapter 10, we will discuss issues of relevance here with a view to giving the reader things to think about. Specific answers to individual questions or design choices that work for particular situations are not presented since there are competitive advantages to be gained through use of these techniques, and as such, are closely guarded secrets by various telecom equipment vendor manufacturers[5].

The focus of this sub-chapter, as with the other adjoining sub-chapters, shall be on service mediation gateways, though the astute reader will note that similar considerations will also apply to the design of Parlay-capable or compliant applications.

10-2.2   What is Ideal, What is Practical?

The study of overload conditions is very involved, and merits a book in its own right. In fact several books abound [Hanmer 2000] on this very topic, and the treatment tends to be very mathematical factoring in elements of queuing theory and other related fields. We take a simpler, more intuitive, though not quantitative approach to the matter at hand. The intent here is to give the reader an appreciation for what overload is, some associated design patterns to think about for alleviating these conditions in systems to be built, and raise awareness of the importance of overload considerations from a performance and capacity engineering standpoint.

Addressing overload involves three steps – those of overload detection, overload reporting, and overload handling. The first of these deals with detecting that overload has occurred – to achieve this, some process or element in the system must keep track of the current load of the system at every point in time (or some suitable intervals), and the threshold beyond which the system may be considered overloaded (for how else would one know that overload conditions were prevalent?) or by some similar means. The second deals with reporting these overload conditions to other processes, administrators or users who may be able to modify their behaviors or expectations appropriately, or take suitable steps to alleviate the condition. The third of these deals with empowering the system itself to take steps to defuse the situation and bring the system back to a more stable state.

It is important to note that overload should not be a commonly occurring condition in the system. If overload conditions occur very frequently, it is typically an indication that the system has not been engineered appropriately for capacity or performance or both. It may also happen that the system was initially well-engineered with those metrics in mind, but as more and more services

---

[5] Just like how the secret formula behind a well-known caffeine based soda-pop is a trade-secret worth a lot of money to that company, or how some cooks try to keep back 'magic' ingredients from their soup so it makes it harder for rivals to equal them.

and applications were deployed, the infrastructure was not upgraded to keep up with the growing demand. In such situations, overload control, although still helpful, cannot resolve the forces that cause these repeated conditions. A more permanent solution would have to be identified.

Catastrophic failures during overload do happen every so often in systems where reliability is expected. In the case of natural disasters or acts of terrorism for example, many people may want to call their friends or loved ones in affected areas, and the telephone system may not be able to cope with the presented load. The PSTN is designed with the expectation that no more than a certain percentage of supported users will want to make a call at any given point in time. If all users in an area were to go off-hook at the exact same instant, it is unlikely that the poor switch at the central office would be able even to provide them with simple dial tone. But the design principle is a valid one – cost-effective network design requires that one make judicious choices in terms of capacity engineering factoring in aspects such as cost, capacity, and ability to meet the busy hour loads presented to the system.

So, given these constraints, what does one want ideally in a well-designed system? A good requirement would be that the system should not demonstrate a precipitous fall in performance as load crosses a certain threshold, and that it should throttle sources equitably when under heavy load. Several design patterns have been defined in literature today for use in systems design in dealing with overload considerations. Some of these are explored in sections that follow. Another requirement would be that as load increases, the transaction-processing rate remains as nearly constant as possible. All real world systems will see some degradation, but the ability to degrade gracefully (and not precipitously) goes a long way towards alleviating overload conditions.

10-2.3   General Patterns for Overload Control

As we said before, overload control and issues relating to it are well understood and mechanisms for the same are widely documented in literature as design patterns etc. In this section, we discuss a couple of these, though merely to give the reader an appreciation for what load management entails. The interested reader is referred to [Rising 2001] for more on pattern languages for communications software. We also note that this treatment is not exhaustive and that several other equally interesting patterns are also extant. What follows is merely a high-level description of some of the patterns that are applicable – for a more in-depth treatment including the details of the pattern, the pattern language, relationship to related patterns, the forces resolved, application context, etc., the interested reader is directed to [Hanmer 2000].

10-2.3.1 New Work Before Old

In telephony systems, delays in the processing of a request would result in a user abandoning the request and then retrying. Therefore, the more times a request has already spent in a queue, the more likely it is that it would be abandoned. Similar patterns also apply to systems, which have well-defined timers and timeouts. If a request has waited too long, perhaps the server is better ignoring it in favor of a more recent request, for processing it and having a ready answer for the client after the client timer has expired is indeed worse than not having processed the request at all. The client cannot use the response, the system resources utilized in processing the request are completely wasted, and the client may re-issue the exact same request again when the pre-computed response may no longer be usable.

10-2.3.2 Shed Load at the Periphery

If multiple systems are involved (say) sequentially in the processing of each transaction, then overload at one or more points along the chain would lead to traffic backing up at other places as well. As more traffic backs up, more nodes get overloaded, and this is a recipe for overall system failure. This design pattern advises systems to shed load at the periphery. If the external interface systems were to start rejecting requests or throttling them when they saw that internal systems were overloaded, and only passed along messages associated with transactions already in progress, thereby enabling those to complete (completion of existing transactions contributes towards lessening load), this helps handle the overload condition.

10-2.3.3 Evaluate Overload Globally

Often, it makes no sense for a system of cooperating processes or entities to take on new work in an overloaded state when one or more of the component processes is overloaded. Stated another way, since a chain is only as strong as its weakest link, and the overloaded nature of one component may very well propagate across the system to translate into overload conditions in neighboring components and so on, there is some logic to evaluating overload globally across the system as a whole and then taking steps to manage these conditions also on a global basis if possible.

10-2.4   Overload and Parlay Gateways

With a general introduction of overload considerations and patterns for overload control under our belt, let us now explore how these concepts apply to Parlay Gateways.

10-2.4.1 Overload Detection, Reporting, and Handling

As we have said before, overload detection deals with determining whether an overload condition exists, and is typically implemented through the use of a monitoring process that tracks system resource consumption across various contributing elements like memory, CPU cycles, I/O channel consumption, etc. When a pre-defined threshold is crossed, the system may be configured to raise alarms or otherwise report overload conditions to administrators or communicating processes.

Many protocols include in their design some overload handling routines. For instance, some IN protocols support the notion of an Automatic Code Gapping (ACG) component that the SCP can issue to the switch telling it not to send any more requests for a certain period. Other systems support yet more mechanisms for overload handling by responding to only a certain percentage of received requests – say K out of every N for a given duration from each source. These mechanisms, where supported by the underlying protocols, must be leveraged by the systems to throttle sources that present excessive load.

Some overload handling routines are more equitable than others. In SIP, for example, one may support source throttling through the use of a final response (something suitable like 480 Temporarily Unavailable, 486 Busy, 503 Service Unavailable, etc.) along with a 'retry-after' header that specifies when the source of the transaction may retry the same to get a processed response. Now, this throttling could be applied either in a context sensitive manner (i.e. depending on the source) or in a completely context agnostic manner (throttle all sources equally during overload). Some may be more equitable than others, and as Andrew Tanenbaum says in his excellent text on Computer Networks [Tanenbaum 2003] 'fairness and optimality are like motherhood and apple-pie', and the details of how this is supported may vary from one SIP implementation to another. Thus, the protocol itself gives the developer the flexibility to build in overload controls, but also leaves the door open for the developer to differentiate her implementation from someone else's. This is as it should be, for standards are meant to promote interoperability while encouraging creativity, yet not stifle innovation.

10-2.4.2 Parlay Gateway Related Considerations

Parlay gateways present some unique challenges to the overload control. From a network perspective, SCSs need to be able to react to throttling from the underlying network elements whenever the protocol mappings support such capabilities. Since this determination needs to be made on a case-by-case basis, we do not explore this issue any further in this chapter, though we acknowledge that these kinds of considerations are of significant importance.

From an API perspective, load control and throttling mechanisms are supported, though the controls and notification mechanisms are distributed within the Integrity Management APIs defined for load management along the SCS to Framework and Framework to Client Application interfaces. In addition, the Service Agreement digitally signed by the Framework and Client Application that could contain within it the governing parameters for the load-related contractual agreements between the two communicating parties (namely the Client Application and the SCS) may be used to enforce some degree of load control.

For example, the client X may be permitted to run 30 transactions per second (TPS) to the CH SCS, 20 TPS to the UL SCS, and 5 TPS to the PAM SCS. Client Y may have a different profile and be permitted to run 20 TPS to PAM and 2 TPS to the AM SCS. It would be wise to design the SCSs such that when new service managers were created to handle specific service sessions, they each try to honor the specific service contracts associated with the client application they are dedicated to serving. All excessive load incident on the service manager, regardless of spare capacity on the rest of the SCS, could be throttled[6], and a suitable exception such as one indicating 'resources_unavailable' may be returned. Since this throttling takes place on a per-client session basis, it remains equitable.

If a Parlay gateway were not engineered with adequate capacity, situations may arise where none of the individual service agreements were violated with an overload condition, but that the overall cumulative load on the SCS in question is still too large for the poor clone to handle alone. In such situations, an equitable scheme might be to throttle individual applications in proportion with their allocated service agreements. For instance, in the example from the previous paragraph, if PAM were to be in overload, throttling in the ratio of 1:4 would be applied across the inbound traffic from applications X and Y respectively to be in keeping with their service agreements. Note however that such conditions should be remedied by growing the capacity of the gateway – adding more nodes, more clones of particular services – for it is not a good idea to violate contractual agreements through proportionate load throttling just to satisfy immediate overload concerns. All attempts must be made to meet contractual obligations in all cases.

Similar situations may occur in deployments where multiple SCSs were deployed together on a common hardware platform because their load characteristics were thought to be well understood (until the next new application that causes subtle but interesting changes in these is innocently introduced). The reader is referred back to the graph from the previous sub-chapter that shows such a load characteristic across a 24-hour period. Or consider the case where a new SCS is deployed on an existing node that was thought to be under-utilized: one may suddenly be faced with a situation where each SCS appears to be well within its configured load parameters, but the node as a whole is in overload (running out of CPU cycles).

Throttling may need to be done intelligently for it to have the desired effect. For instance, there are some transactions defined in Parlay SCFs whereby the transaction is long-lived, and is only closed when the client issues such a 'closure directive'. An example of this is the periodic user location request in the User Location part of the Mobility SCF. The transaction starts when the client issues a periodicLocationReportingStartReq () and remains open (with the client receiving notifications periodically) until the corresponding periodicLocationReportingStop () method is invoked. Thus, although SCSs can implement patterns such as 'Shed Load At The Periphery', exactly which methods to reject needs to be carefully factored in, as the processing of certain methods (such as the periodicLocationReportingStop () mentioned above) can actually reduce overall traffic and help with overload mitigation.

As was previously indicated, Parlay also provides load management APIs that enable clients to query the load of particular SCS clones or for the SCSs themselves to report their load to the

---

[6] Alternatively, service providers may choose to support this overload if spare capacity beyond the negotiated parameters was available, but then bill for these transactions at premium rates.

Framework. These load management API methods (as described in Chapter 5), contain a parameter that permits the SCSs to specify their load in terms of a percentage or a number indicating load level (0 for normally loaded, 1 for overloaded, 2 for severely overloaded). These load reports are generated per service session, not per the SCS as an aggregate, though such an aggregate can be computed by the Framework (which knows about all the service sessions anyway) if need be.

In implementations, each service manager on the SCS could be created with knowledge of the maximum load a client can present per the pre-negotiated service agreement, then compute the load level threshold values as pre-defined percentages of the total value. Anti-hysteresis mechanisms may need to be built into the load control scheme dealing with load level transitions so as to prevent repeated and frequent load reports and level changes that result when a particular client presents heavy load, gets throttled, falls below the overload threshold, then presents the same load again, rises above the threshold and repeats this cycle in a loop.

The aggregate load reports if computed by the Framework, could be used by it with other heuristics or algorithms to distribute incoming service session requests across the different SCS clones for that service, thereby ensuring a more equitable loading of the various servers that constitute the gateway.

Thus, the Framework can apply the 'Evaluate Overload Globally' pattern, albeit to a limited extent, since it controls coarser granularity control (at the session level, in being able to assign particular service sessions to particular SCSs). There is no finer granularity mechanism defined in the standard in support of this pattern. However, implementations are free to explore creative means to achieve this end.

For a more quantitative and simulation-oriented treatment of Overload related considerations in the Parlay/OSA space, the reader is referred to [Andersson 2004].

10-2.5   Summary

In this sub-chapter, we have examined some of the issues relating to overload monitoring and detection, reporting, and control or handling. Specific ideas relating to the Parlay context, and how these apply to Parlay gateway implementations, were also discussed.

10-3

On the Scalability and Reliability of Implementations

10-3.1   What are High Availability and Reliability? Why Consider Scalability?

Telecommunications systems, as previously mentioned in Chapter 1, have become central to the operation of a free society. One relies heavily on this basic infrastructure, and almost takes it for granted. There is no doing without it.

Given this kind of critical role that telecom plays in people's day-to-day lives, service providers absolutely require the highest levels of availability and reliability of their systems deployed to support these basic communications services. These characteristics are typically expressed in a rating of '9s'. A five 9s system is 99.999% available; that is, planned and unplanned service downtime (i.e. the time when service is not available to end-users) should never exceed 0.0001% of the total system time. To make this characteristic more tangible, 0.0001% of the total system time equates to 5 minutes a year. Similarly, 99.99% (or a four 9s reliable system) permits you to have a downtime (both planned and unplanned) of 50 minutes per year. As can be seen, the requirement for higher nine systems (five 9s and above) are very stringent indeed, and rather difficult to meet in real-world systems (but almost all telecommunications equipment and services are in fact designed to these exacting standards).

The computation above is performed as follows:

There are 365.25 days in a year (accounts for leap years) and 24 hours in a day, each of which consists of 60 minutes. There are therefore 525960 minutes in each year. Now, 0.001% of that is ~5 minutes of downtime per year (for a five 9s system) and so on.

Purists will object to our description in the previous few paragraphs, and rightly so. High Availability, or HA as it is sometimes called, is distinct from reliability from a pure technical perspective. It can be argued that the '9s' referred to above are a measure of high availability, not of reliability. This would be an accurate criticism.

One can have very high availability services that are not very reliable, and vice versa. Imagine, for instance, a service that fails every thirty minutes, but recovers within two seconds each time. The service is available a large percentage of the time (unavailable only two out of 1800 seconds, or 0.11% of the time), but not very reliable. Consequently, it cannot be called 'carrier-grade'. In this chapter, we are concerned with both aspects of Parlay services – namely both the reliability and high availability characteristics. We also briefly cover aspects relating to the scaling of services supported by Parlay gateways, and related concepts. Although scalability aspects are not directly related to availability, one must take care to ensure that systems that are designed should not suffer from a degradation in availability and reliability characteristics as the deployment is scaled up or down to meet the required capacity needs.

In what follows, we first discuss, at a high level of abstraction, some important issues relating to the availability, reliability and scalability of Parlay components, including engineering aspects for deployments. Then we delve down deeper into standards patterns that can be leveraged to build some of these capabilities into real-world implementations. This sub-chapter is slightly different from the others in Chapter 10 in that while the others focus more pointedly on engineering considerations, here the discussion also factors in some aspects from the standards to present a more complete picture of what is required.

10-3.2   Reliability and High Availability of Parlay – Applications and Gateways

The stringency of these aforementioned requirements directly indicates that the hardware and software elements required to support services for communications networks need to be very carefully designed indeed. Similar considerations apply to Parlay/OSA systems as well – more directly to the Parlay-compliant SMGs that are deployed directly into the service provider networks, and perhaps less so to the application servers hosting Parlay-based applications in the enterprise domains (after all, this latter set of elements are Internet domain components that are not directly influenced by the same set of restrictions). We do note however that some Parlay/OSA applications may be directly hosted within the service provider domain itself, or the service provider may have sub-contracted a service bureau type arrangement with the enterprise domain hosting the application, and in either of these cases, since the application contributes directly to the expected service provider revenue streams, stricter requirements on the operation of these application elements may be imposed.

This typically means that both the physical box[7] on which the SMG is built, and the middleware platform on which it runs, need to be carefully selected. A resilient system is only as strong as its weakest component. In addition, systems built to satisfy such stringent requirements are built with off-board or on-board monitor processes that continually track their health and sanity, and restart processes when they appear to be malfunctioning. And there is an escalation scheme built in where critical failures of a node would result in it being restarted to bring back the service in a form usable to subscribers just as soon as possible.

It perhaps needs to be emphasized that high-availability (HA) and reliability are quite different concepts, though frequently used interchangeably in informal discussions. Software can be highly available, without being too reliable (refer back to our example from the previous section). Consider the case of software that fails often, but where the Mean Time To Repair (also called Mean Time To Recovery or MTTR – the average time between a failure and its repair) is very short. It is therefore available most of the time, but not very reliable.

Some terms used to gauge the reliability and availability of software include: MTTR (described above), and MTBF (Mean Time Between Failures – the average time between successive observed failures of an element). Generally speaking, the larger the MTBF, and the smaller the MTTR, the more reliable and highly available the system is. Of course, sufficient field data need to be available before these values can be computed. [Shooman 1983] gives a good mathematical background in availability and reliability computations for different kinds of systems.

10-3.3   Scalability and Reliability

Scalability refers to the ability of a component to support very small and then very large deployments either in terms of subscriber capacity, or in terms of throughput requirements. From the definition, it is apparent that scaling small (also sometimes called 'scaling down') is just as important as scaling up. The former permits an equipment- or application-vendor to provide a cost effective low-end configuration, while the latter enables them to extend already deployed systems to take on greater volumes of load in as transparent a manner as possible (sometimes this may involve adding additional boxes, new software, or the removal of an old box and its replacement with a new box with greater capabilities – what is sometimes referred to as a 'forklift upgrade').

Scalability and reliability go hand in hand. In the sub-chapter on SMG performance, we described how the SMG can be designed to handle the load requirements of the applications supported by the service provider network. This is typically done, as described in that sub-chapter, by considering the aggregate load across all applications to be supported, on a per service basis, and then using the TPS metrics of these service types to figure out the number of clones of each service (say N) needed to meet the throughput requirements. Scalability calculations seemed simple enough.

But those calculations assumed that the services themselves, and the nodes that host them, are always operational. Sadly, however, the real world does, at times, operate in keeping with Murphy's law: 'Anything that can go wrong, will go wrong'.

10-3.3.1 Engineering for Scalability and Reliability

When services or the nodes that they are hosted on fail, or the network becomes inaccessible, or a data server is no longer reachable, or some other catastrophe occurs, one still has to have a means to provide as reasonable access to service as the service provider requires. Thus, scaling just to meet the needs of peak hour transaction processing is no longer sufficient. One has to anticipate, and plan for some level of failure, and factor that into scalability assessments.

---

[7] Hardware fault-tolerance is also of great importance. NEBS-compliant (NEBS stands for Network Equipment Building Standards) elements are typically used in telecom equipment design. The details of hardware aspects are outside the scope of this book. See [NEBS 2002] for more.

This is typically achieved through Markov Chain modeling. What one does here is assume a certain probability for failure of each individual element and then derive a mathematical function that enables one to compute the level of redundancy needed to provide a certain guaranteed availability assuming K concurrent failures. [Norris 1998] provides a good overview of Markov Chain theory and the associated models.

Once this analysis is complete, in conjunction with the scaling exercise based on performance metrics previously described, one gets a good sense for both the N and the K values for the number of replicas of a given type of element needed in a deployment (N) to support the throughput requirements for a service at a given level of availability. The output of this exercise looks something like: 'To provide 99.97% availability of the Call Control SCS at the required 2800 TPS system rating, we need to deploy five replicas or clones of this service, each capable of processing 700 TPS, each with an availability rating of xx.xx%.'[8] Here, N is 4 (derived as 2800/700 and rounded up where necessary, since one cannot deploy fractional systems in the real world) and K is 1 (intended to provide resiliency with no capacity degradation after one failure).

N is 4 here because a minimum of four systems each capable of processing 700 TPS are needed to support an overall capability of 2800 TPS for the deployment as a whole[9]. The K is 1 because, even though four boxes are together able to meet the capacity requirement, we need an extra box to support the availability requirement for the service. This resultant configuration is thereby able to support a total of one failure and a guaranteed capacity of 2800 TPS even under those conditions. It can more easily support 2800 TPS in situations where there are no failures.

In N + K architectures, all the N + K nodes concurrently process traffic. Thus, the cluster as a whole has some spare capacity when all nodes are active and available. But even after there is a failure, the total capacity of the system as a whole does not go below the required 2800 TPS for which it is engineered. If more than one concurrent failure occurs, degraded operation may result with a loss of capacity, and if greater resilience is required, one needs to simply increase the number K (with a proportionate increase in the cost of the deployment).

The astute reader will note that there are two levels of availability that need to be factored together to provide the metric used in the above calculation – one is the availability of the node/box/physical hardware itself, along with its middleware platform, the second is the availability of the application or service that runs on that hardware element. The cumulative availability of the two elements together is what needs to be considered as Markov Chain related computations are made.

10-3.4 Parlay Considerations

In Chapter 5, we have studied how applications react to failures of SCSs with which they are connected, and how they can detect and report these failures to the Framework using the appropriate Fault Management methods. We also saw how applications can be monitored by the Framework using the Heartbeat Management interfaces provided as part of the standards.

Client applications can also be designed to exacting availability standards if the enterprises so choose, or if the service providers enabling network connectivity so require. Application providers may persist service references in a data store, make them accessible across replicas of the application, and resort to other such tricks. So long as the various replicas appear as one application

---

[8] The computation referred to here (particularly the one pertaining to how K is calculated) is interesting, but is beyond the scope of this book. The value of K is computed while factoring in the kind of model in use (e.g. the Parallel Redundancy model), the number of repair facilities, etc. (since these contribute towards how quickly a failure can be repaired on average). [NIST 2005] provides a nice brief description of the ideas involved here.

[9] Depending on implementation, sometimes, SCS clones that provide the same service, or SCS clones that provide different services, may need to share information amongst themselves. This inter-clone messaging also has a potential impact on the performance and capacity of the system as a whole, and needs to be carefully factored in as these computations are done. Given this depends on particular vendor architectures and implementations of the gateway components in question, a generalized model cannot be presented here.

instance to the Parlay components on the SMG, this is all well in keeping with the defined standards, and offers yet another avenue for application providers to differentiate themselves from the competition as Parlay technology takes off, and the marketplace gets inundated with a large number of Parlay applications.

10-3.4.1 Building HA Parlay Applications

As we have seen first in Chapter 5 and then in more detail in Chapter 6, the Parlay standards define interfaces whereby the client application can register a callback interface and issue asynchronous requests to the server. Once the server completes the processing of the request, it can issue a callback with the response, on the previously registered callback interface.

The standards permit the client to register more than one such callback address to which the response may be sent. Where multiple alternative callbacks are registered, the last registered or most recent one is used first by the Parlay server or SCS, and if some kind of failure indication results, then the server may attempt to use the other back-up callback addresses to transmit the response before giving up with some kind of failure indication (if necessary, even to the Framework).

This mechanism can be very cleverly utilized by client applications to support HA constructs, see Figure 10.4. If we explore the concept of 'application replicas' alluded to in passing from the previous section, we can conceive of situations where multiple application replicas each register: a) their own local callback reference as the primary and then b) a replica's callback reference as the secondary callback address, with the SCS to which the primary has a service session.

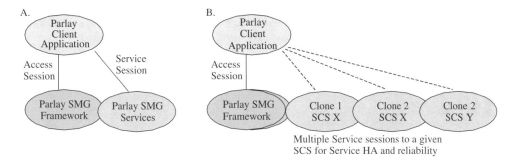

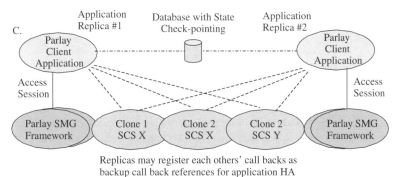

Note: Other configurations are possible, the above is merely illustrative of possibilities.
A. Simple figure of a Parlay client interacting with a Service Mediation Gateway (SMG)
B. Figure showing details of multiple clones, configuration for Service-level HA
C. Figure showing Application Replicas for Application-level HA

Figure 10.4   Examples of Parlay configurations in multi-cloned SCS scenarios

In such a situation, one very elegantly achieves higher availability of the application entity for, if the primary application replica were to fail, or if its callback address were to become suddenly unavailable, the SCS would notify the secondary replica of the response or of an asynchronous, subscribed network event. In either case, the application as a whole could update its transactional context and function in spite of the failure of some subset of its registered replicas.

Admittedly, the viability of such a strategy depends on many factors, not the least of which is the level of context sharing between the various application replicas, how frequently they communicate with each other, how the transactional state is stored, etc. But the standards do provide a mechanism for high-availability support as a beneficial side effect of this mechanism of enabling multiple concurrent optional callback addresses.

Although this standards-specified mechanism for achieving HA in the application domain is elegant and simple, it is not a firm requirement. Implementations (both of the client application and the server or SCS) are free to use whatever options or alternatives they deem most appropriate. In CORBA-based implementations, for example, Parlay components may be built to use fault-tolerant CORBA and related constructs. We do note, however, that although various technology-specific options for HA exist, the standards do in fact provide some capabilities that are generic, useful and technology neutral, for achieving the desired goals.

10-3.5 Summary

In this chapter, we briefly considered some aspects relating to scalability, reliability and high-availability aspects of Parlay deployments. The focus here, as in the other adjoining sub-chapters of Chapter 10, has been more on providing the reader with things to think about, rather than prescribing particular solutions to the issues raised. There is a wide variability in applicable solutions to problems in this space, with ample scope for competitive differentiation as desired.

10-4

Failure Handling in Parlay/OSA Environments

10-4.1 Introduction

As more applications are built to comply with Parlay/OSA technologies and as gateway deployments become more pervasive and prevalent, smooth interoperation of products from different vendors in Multi-Vendor Environments (MVEs) becomes increasingly important. Transparent and efficient identification and resolution of failure scenarios takes on more importance and gets more attention, deservedly so. Careful planning in this regard, on the part of the standards bodies, and those that build the gateways and applications in question would enable the Parlay/OSA technologies to cross more easily the 'technological chasm'[10].

---

[10] Reference is made here to the book 'Crossing the Chasm' by Geoffrey Moore [Moore 2002]. In that classic, in a chapter focused on the maturation cycle for various technologies, Moore argues that there is a certain point in the technology lifecycle where the technology gathers a mass appeal and wide following, and this causes a sudden upsurge in interest and demand in it, which causes it to become very widely used and

The ETSI standards body, with a view to 'greasing the skids', so to speak, and enabling easier interoperability of implementations (which theoretically should be seamlessly interoperable anyway, since all compliant products are built to the same set of standards-defined interfaces), has conducted over the past few years, events called 'plug tests' (which previously used to be called 'bake-offs'[11]) for OSA/Parlay. At these gatherings, people from different companies could go in with their implementations of gateways, applications or other components that implemented the Parlay/OSA interfaces, and then test their components with those built by other vendors.

During these events, it is not uncommon for developers to discuss and argue interpretations of the standards themselves, and then make appropriate fixes collaboratively, or to come up jointly with standards change requests (CRs) for submission to the bodies that publish these documents, as they ensure interoperability between their respective implementations. In addition, these events are valuable because it is fairly typical for service providers to want to deploy products from different vendors in their network to hedge their bets, and in such cases, the degree of effort involved in testing interoperability and compliance of implementations to the standards in a service provider setting is simplified if the products involved have been interop tested at events such as plug tests.

Both in live networks, and in situations where interoperability is merely being tested in a lab-setting, it is often valuable for the product to return appropriate failure indications so that the application, operator, administrator or user (as the case may be) is aware of what is happening, and why, and which, if any, corrective action may need to be taken. The error indications themselves may be of different types, and at different levels of detail, depending on who is expected to react to them.

For instance, a user presented with a 'blue screen of death', and an error which says 'Error 231X in memory location 231F:123E:21BC:0AFD, press ¡Enter¿ to reboot', is rather helpless when it comes to de-bugging the application or divining the cause of failure, unless he or she is a programmer or has access to a manual. (Users do not like being told to 'Read The Forgotten Manual' – the infamous RTFM fix for all non-sunny-day scenarios.) This error is in turn better than one where the computer reboots itself without giving an error indication at all.

On the other hand, an application dealing with a gateway would prefer to receive an error indication or error code it can understand (e.g. 231X) without too much supporting text or explanation, for the program logic is, more often than not, keyed off the error indication or error number received. The explanation, if any, is normally either used offline by the concerned administrator (who examines the application logs), or is passed on to the user, as the case may be, though the application itself probably makes best use of just the error code.

OSA/Parlay products, like all other software, will, from time to time, face error situations[12], and are expected to gracefully recognize, isolate, react to, log, and then recover from, these kinds of situations. Errors can be of many types, and may occur at different layers of software and different points in the operation of systems. Here, we study some of them with a view to giving the reader a somewhat finer appreciation of what exactly is involved in designing a reasonably stable, usable

---

also very successful. Technology itself may be very attractive but marketing can only push it so far, beyond that, it is up to the demand-pull that is exerted by consumers that really makes or breaks it.

[11] It's rather interesting to note that the term 'bake-off', which was widely used for a number of years to denote events of this type (especially with regard to SIP implementation interoperability), had to be changed to 'PLUGTESTS™' because the term 'bake-off' was apparently copyright-protected by a popular pastry/baking products company, and its use in the technical arena was discouraged by them in an effort to protect their copyright.

[12] There is no such thing as 'error free' software. Testing can be extremely methodical, follow all the right techniques, and try to ensure that all the normally used program legs are error free. But programs may behave differently in different situations, under different kinds of load, when security attacks are in progress, when they interact with different kinds of servers or clients, when they see different unrecognized codes or responses, etc., and any scenarios not considered during testing can cause problems later on. Software engineers are frequently surprised when they hear laymen talk about 'perfect' software.

system from an error handling perspective. Although this chapter appears in the section on gateway design, the contents here apply equally well to the design of Parlay/OSA client applications.

10-4.2   A Layered Software Architecture – Again. . .

Figure 10.5 indicates the layers of software that are typically involved when one builds a carrier grade system. It must be emphasized that the figure merely indicates what is typical, and the actual product architecture can vary widely from one vendor to another, based upon factors such as history, design choices previously made, company technical culture, reuse of existing assets factored in, reuse of legacy company platforms or interfaces to such components sold together in an offer, etc. We have examined this layered software architecture, rather informally in previous chapters, but take a closer look here, to set the context for our discussion of failure handling.

In Figure 10.5, the lowest layer represents the hardware platform and associated software drivers etc. This layer also includes operating system software, protocol stacks 'plugged into' the system and so on. Capabilities and services provided by this layer are used by the layers[13] above it (the astute reader will note that these concepts are similar to those in the OSI layered protocol model, where services from one layer are made accessible to the layer above via Service Access Points or SAPs).

The next layer up contains the platform and services middleware. Vendors typically tend to use the platform middleware layer to give their products the robustness and other carrier class capabilities for reliability, availability, redundancy, failure detection and recovery, etc. This middleware provides the supporting standard platform capabilities, interconnection mechanisms, binding to operating system capabilities, and related features to upper layers. Platform middleware with more years of field data from actual deployments (giving MTBF, MTTR and other such information) is typically considered more hardened and this serves as more of a metric of the reliability of the implementation than any claims the vendor might make.

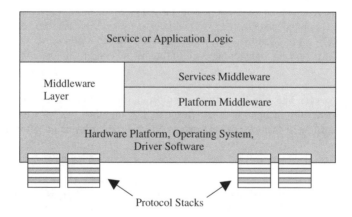

Figure 10.5   Generic layered architecture for carrier-grade systems (nodal view)

---

[13] The novice reader can be forgiven for wondering: 'what is it with engineers and these layers in every model we see?' Engineering is often an exercise in management of complexity. Layering is a tool that enables the engineer to partition the problem, simplify it into a view of smaller, interconnected problems at different levels of abstraction, solve these sub-problems (even share sub-problems amongst team members, one or two each), then put the whole thing back together. If the problem is amenable to this kind of treatment (and it often is), the solution works. In other words, layering is a kind of 'divide and conquer' technique, applied simultaneously across different levels of complexity.

Services middleware, on the other hand, is used to provide the base infrastructure for the actual service logic that runs on the node. For instance, Parlay/OSA gateways or applications could be built to use any one of several service middleware technologies such as CORBA, SOAP/XML, Microsoft MIDL or DCOM, COM etc. These would constitute the services middleware layer. One or more middleware technologies could be utilized by any given implementation.

Finally, at the top of the stack, comes the services or application logic itself. This is the software intelligence that actually runs the node. One or more services or applications can co-exist (of course, the engineer must, needless to say, factor in other considerations such as capacity, performance etc. while doing this) on a given node, and reuse the same underlying infrastructure. If we consider Parlay/OSA applications, this layer would implement the standards-defined interfaces, the programmatic APIs, and the logic that lies behind them, including that needed for handling error cases.

10-4.3   A Layered View of Errors?

Now that we have looked at the different layers of an implementation from a software perspective, let us also look at the different kinds of errors that might be reported at each of these.

From a bottom up perspective, the first kind of error we encounter are simple protocol errors. For instance, failures of TCP connections to remote nodes, buffer overruns, timing errors, etc. Some of these may be reported to application logic so appropriate corrective action may be taken, while others may be localized problems in the protocol stacks or driver software, and need fixes or patches from the vendors that provided those components.

The platform middleware might report errors in inter-process communication infrastructure, failures of critical processes, and so on. In such cases, the middleware may itself take corrective action such as trying different strategies for recovery perhaps escalating eventually to a nodal reboot after termination of resident applications: all this, in the interests of keeping the node healthy and operational to the best of its abilities. Or it may simply generate alarms and ensure that some human administrator would eventually notice and initiate such action manually. Generally platform errors are not propagated to the application layer, especially for hardened, thoroughly tested applications, and are few and far between. But this depends, as we said before, on how hardened the platform software itself is, and how capable it is, in terms of dealing with the kinds of real-world network conditions it may be subject to in the field on its own.

Services middleware, being somewhat less resilient and less tightly coupled with the platform than the platform middleware, may see errors from time to time, and will generally pass these errors on to the service or application logic that executes on the node. For example, it is not uncommon for an OSA application to see occasional CORBA errors if, for instance, it were to invoke a method from the defined standards interface on a server, where the server did not implement that method and does not even support it in its view of the same interface. Section 13.6 in Chapter 13, which discusses the least cost testing of Parlay interfaces, considers these issues in greater detail. Normally, it is recommended that clients and servers implement all the methods in the standards-defined interfaces if they implement any method in said interface at all – if nothing else, a dummy implementation with no real service logic is still better than no implementation.

Perhaps the most important kind of error is that which occurs directly at the application or services layer. The programmer must ensure that he or she deals with these errors in the code, and that every expected situation is provided for programmatically, and that there is at least some logic to handle unexpected errors in strange situations. The standards help to a large extent here, in defining the error conditions that may arise when these interfaces are used, along with error codes, associated cause codes, and diagnostics. The standards also provide flexibility to define user- or administrator-readable error text that can be put into logs by the recipient for later use.

In the Parlay APIs, such application layer errors are dealt with either by exceptions or by explicit error return methods. Exceptions can be thrown by any method. In order to structure exceptions and minimize the need to define exceptions exhaustively for each method, an exception hierarchy is introduced. The hierarchy consists of common exceptions and service specific exceptions. The common exceptions, as their name suggests, are applicable to every method (e.g. P_METHOD_NOT SUPPORTED in case an invoked method is not implemented on the server, or may not be invoked according to the Service Level Agreement). Service specific exceptions are limited in scope to a specific service only (e.g. P_REQUESTED_ACCURACY_CANNOT_BE_DELIVERED in case a location report request is issued with an accuracy of location details that is not supported in the network). In addition to the standards-defined exception types, each exception may provide additional information to the application programmer, in the form of free-format text strings. For example, when the exception P_INVALID_STATE is thrown by the service to indicate that a method is invoked in an unexpected or invalid state of the state machine executing on the gateway, the gateway may provide additional explanation on that current state.

Errors can also be reported in the asynchronous response to a specific request. For example the method locationReportErr in the IpAppUserLocation interface returns an error cause and diagnostic towards the application. In case certain privacy consideration are enforced by the regional or national regulatory body, a location report request might fail with error cause P_M_UNAUTHORIZED_ APPLICATION and diagnostic P_M_DISALL_BY_LOCAL_REGULAT_REQ.

Admittedly, although the programmer may do his or her utmost to deal with all kinds of error situations that may arise, some will be overlooked; which is why the default error-handling clause (i.e. some default logic so the code does not choke when it sees a new kind of error arising in a particularly infrequent condition) is necessary. Any client server system deals with communicating state machines – one at the client side, and the other at the server end, and each end essentially performs its functions based on its view of its own state and its perceived view of the state of the other interacting component. The standards-defined interface and (real or perceived) behavior underlying it govern the interactions between these communicating components, and the software at either end must be in a position to handle situations when perceptions of their state as viewed by the other parties may not agree with their own[14].

Contrary to popular wisdom, doing the same thing over and over, when dealing with software systems, may actually result in different outputs. The output is, in some cases, not just a function of the input, but also of the state of the other element with which one is interacting. Of course, if the output is not what is desired, and the user (person or application) has no view of the internal logic or states of the other entity, this knowledge is of little consolation.

The astute reader will note that errors in lower layers, if they are not caused by the controlling logic in the highest layer, will be caught and ironed out more quickly, since these lower layers are utilized more. Stated another way, the more particular software is used, the more likely that obscure errors will be caught and reported, since this software is tested under varying conditions more frequently, in different deployments. This also explains why the number of hours of field data collected on lower layer software components, particularly those that deal with platform reliability and availability, etc., contribute directly to how hardened those components are.

10-4.4 Summary

In this chapter, we have studied the different kinds of errors, a possible classification of these from a layered software perspective, and then examined how software can be better constructed to handle error situations that occur, with a focus on Parlay/OSA applications and services.

---

[14] People, in similar situations, either argue, or agree with each other, though each person's view of the agreement is different (setting the grounds for later arguments). Software at a node either throws immediate errors or starts diverging from its peer in its view of the other's FSM state, which could lead to errors later on in their communications. Just like there are no perfect people, there is no perfect software. Sometimes, the errors cannot be localized, but some blame attaches on both ends.

10-5

Security Aspects

10-5.1   Introduction

Security is of paramount importance in various communications networks in general, and in telecommunications networks in particular. Historically, telecommunications networks have had a higher barrier to entry than packet-based networks such as the ubiquitous Internet and have used more esoteric protocols and variants carried over SS7 that restricted connectivity to a smaller set of nodes capable of speaking these protocols. Telecommunications service development also used to be an arcane skill, almost a kind of 'black art' requiring deep expertise of very specialized domain knowledge.

As we have seen in previous chapters, with convergence, and service-centric network support, and the need felt by service providers to bring in ever newer and more exciting services and applications and make them available to end-users to retain subscribers and grow subscriber bases, this picture is rapidly changing.

Open standards like Parlay and OSA are leading to service provider hosted service capabilities being exposed to third party client applications, developed with Internet toolkits and technologies, that can effectively leverage these to enhance the end-user experience in new and interesting ways. Even service provider supported SS7 capabilities are now accessible (albeit indirectly) to applications via IP-based API interfaces via network elements like Service Mediation Gateways.

Openness is good, but service providers want assurances that this new model does not compromise existing security and safety requirements. After all, if new services are made available to their subscriber base, but they result in serious heavy spamming or the download of viruses or other malware to their handsets, or in the exposure (accidental or otherwise) of end-user privacy information, or in billing/charging fraud, these so-called 'improvements' would be counter-productive. A good security model goes a long way towards addressing these concerns. In this and the following sections, we discuss some ideas in this regard.

10-5.2   Security and Service Mediation Gateways

In earlier chapters we have discussed how policy management may be supported to provide greater dynamism and flexibility in the processing of requests received by the SMG element deployed in a service provider network. The Finite State Machine or FSM that indicates the processing to be carried out for each received request on the server element, may be enhanced to support a policy query, so that a rules engine (supported by the architecture for example) may be requested to render a decision, and this decision is then enforced by the SCS that now acts as a Policy Enforcement Point or PEP.

Policies themselves can encompass a wide variety of applications, and one such application may be security associated with the authentication and authorization of applications, and privacy enforcement so end-users' rights are safe-guarded and only authorized application elements are given access to their information.

Note however that policies are useful once an underlying basic infrastructure exists, and then this infrastructure is enhanced to become policy-conscious. Thus, although policies themselves are very useful, core elements supportive of security must already be present. SMG implementations may address security from two distinct standpoints:

1. standards-based support for SMG security via the defined interfaces
2. network level security for SMG deployments utilizing 'surround' elements

Let us look at each of these in turn, in some detail.

10-5.2.1 Standards-based Security Support

In Chapter 5, we have briefly outlined in Figure 5.10 how security support is provided by the OSA/Parlay APIs. The access session, established between the client application and the Framework SCS, typically provides session-level security, and carefully regulates application access to service provider hosted services and SCSs. The defined interfaces support two kinds of authentication models:

a) P_AUTHENTICATION: whereby the actual authentication process may take place at the network level or utilizing mechanisms other than those defined explicitly in the Parlay API, but where the results of said authentication may be utilized within the Parlay application context; and
b) P_OSA_AUTHENTICATION: whereby the authentication procedure utilizes the capabilities specified within the API directly, for performing authentication and authorization functions.

The API indicates that the Challenge Handshake Authentication Protocol (CHAP), originally defined mainly for the authentication of computers over dialup lines [RFC 1994], should be used in client application authentication. As mentioned in Chapter 6, part 3 of the specification defines the Framework APIs or the Trust and Security SCF. This specification indicates how the CHAP packet is to be constructed, disassembled and parsed at the destination, and the identity validated. A hashing scheme (also called a trapdoor function, like MD5 or SHA1) is used along with a shared secret, to establish the identity of the parties being authenticated. The shared secret is established during an out-of-band pre-negotiation phase between the enterprise domain fielding the client application and the service provider domain hosting the SMG element.

Mutual authentication is supported, whereby the client and the server authenticate each other. This is useful because it ensures the client is not tricked into revealing security information to unknown malicious servers that attempt to masquerade as the SMG. Transparent re-authentication is also supported, whereby the Framework can periodically (based on a service provider supported policy, for example) require the re-authentication of client applications if these clients want to continue to avail themselves of the services offered via the SMG.

During the service discovery and service selection phases, the Framework may enforce security policies that are defined, to ensure that clients are only able to 'see' and select that particular subset of offered services that are in keeping with the service level agreement previously agreed between the enterprise operator offering the client application in question and the service provider. Also as previously discussed, Service Agreements need to be digitally signed between the client application and the Framework, re-affirming the terms of service usage between the two interacting parties. The specification supports the use of asymmetric cryptography and digital signature algorithms such as RSA and DSA for these purposes. Some implementations also support a NULL cipher-suite for authentication – it is not recommended that this be used in actual deployments unless there are other security mechanisms already in place, but this support for the NULL algorithm does provide a convenient means for interoperability testing at various plug tests and other such gatherings.

Other aspects of security, including 'surround' or related capabilities pertaining to key genera-tion, key storage, key exchange, key management, etc., are outside the scope of the Parlay/OSA specification and are not addressed by the standard. It is assumed that well-understood, widely deployed implementations of mechanisms from those domains are supported by individual SMG deployments. (As can be seen, although the standards do offer a lot of capabilities packaged neatly into APIs, there is adequate scope for differentiation and innovation among different vendors' products.)

Where the P_AUTHENTICATION model is supported, the discussion ties in rather closely with what kinds of network-level security mechanisms are utilized in particular service provider network contexts to provide assurances of security. This is the case because, as previously indicated, the P_AUTHENTICATION model utilizes the results of authentication carried out outside the scope of the Parlay/OSA APIs themselves. We shall study this in the next section.

10-5.3   Network-level Security Support

Whether or not security is built into the SMG element deployed in a service provider network, it is wise to support other security mechanisms, elements, or constructs so the design principle of 'defense in depth' is closely followed. Doing so ensures that the compromise of one layer of security does not lead to the total compromise of the system as a whole, since there are other safe-guards in place that can still protect the system as compromised elements are detected and fixed.

Thus, typically, SMG deployments would be shielded from the general Internet behind one or more firewalls (since the SMG will typically be hosted in the inner network, not in the de-militarized zone (DMZ) that exists between the inner and outer firewalls along the network periphery that typically hosts web servers and other such elements). Service providers could support VPN Security Gateways between their own networks and external enterprise domains that host applications, and utilize OSI layer 3 IPsec ESP tunnel-mode VPN tunnels between these different networks to guarantee the integrity, authenticity and confidentiality of traffic transiting these connections [RFC 2401]. Figure 10.6 shows possible security configurations.

It is to be expected that in such cases, the service providers will require all 'connected' enterprise domains to take some minimal set of precautions to ensure that their individual networks are not subject to compromise, thereby heightening security for the entire 'network of networks' as a whole. We have said it before: a chain is only as strong as its weakest link. Similarly, service providers may also require these enterprise domains to respect a common privacy policy with regard to use of subscriber information – any violations of the specified policy could be punished with forfeiture of service connectivity and the resultant loss of associated revenue through operations.

If so configured, or if alternative mechanisms such as TLS/SSL[15] are used, and credentials shared via digital certificates are utilized in the context of widely deployed cryptographic handshakes for mutual authentication, the Framework SCS may be instructed to permit client applications access to services without explicit API-level authentication.

10-5.3.1 Securing Service Sessions

Providing security to access sessions is nice, but is not necessarily sufficient. One may want to prevent some client applications from accessing particular methods from within the service interfaces. As indicated in Chapter 5, the parameter list passed by the Framework to the SCS SILM during the createServiceManager () invocation may be used to constrain the method set that is accessible to clients, in keeping with pre-negotiated service level agreement parameters.

Of course, the 'surround' elements supported for network traffic security between the enterprise and service provider domains apply to all traffic exchanged between these two networks, including

---

[15] It is to be noted that TLS v1.0 and SSL v3.0, the OSI layer 4 mechanisms, are typically more widely used in the context of Web Services and other technologies more amenable to their support. TLS and SSL are not as readily usable in CORBA contexts, since support for these typically requires the use of 'ORB-gateways' and 'SSL-packs' that are developed with vendor-specific features and capabilities that are not defined by the OMG standards for CORBA.

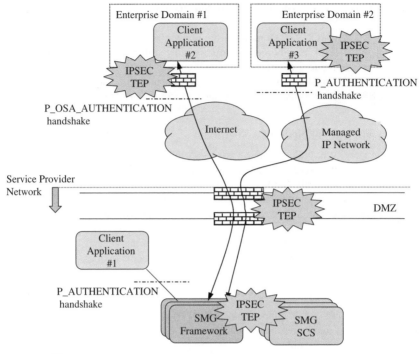

Notes:
1. TEP is a Tunnel End Point. IPSEC is used as an example of security protocol
2. Figure shows possible configurations only, other alternatives exist
3. Figure depicts access sessions only, security for service sessions may also be
   provided by implementations

Figure 10.6 Possible security configurations

the service sessions. In addition, service sessions connected to service managers hosted by policy-aware SCSs could dynamically enforce policy decisions rendered by a rules engine. The rules decision mechanism or Policy Decision Point may be accessed by the application or indeed by the SCS itself, via a separate service session to the Policy Management (PM) SCS. In such cases, the PM SCS may be a part of the same SMG hosted by the service provider network.

Rules may be service provider specific (e.g. enterprise applications require P_OSA_AUTHENTI-CATION and need to sign service agreements with 512-bit RSA), end-user specific (e.g. Alice does not wish to permit application X access to her location information), application specific (e.g. application X in an enterprise domain can only get some user data, application Y in the service provider network is trusted and has access to the complete user profile), or a combination of the above.

10-5.4 Summary

To sum up, security is important in SMG deployments. The newfound openness supported by the Parlay/OSA model needs to be tempered with regulated access and secure usage of hosted services. Some API support for security is embedded into the Parlay and OSA standards, but generally speaking, it is wise to utilize these capabilities within the context of a larger set of complementary technologies and mechanisms to ensure greater end-to-end security. In this chapter, we have briefly studied some of the issues involved in such deployments.

10-6

Upgrading Field-deployed Systems

10-6.1   Introduction

As described in sub-chapters 10-1 and 10-3, a Service Mediation Gateway could be developed as a distributed, extensible, modular system of N + K autonomous, spared and redundant Service Capability Server clones over a cluster of hardware nodes. Alternatively, it may be build to consist of extensible distributed SCS processes that subsume service managers that may be spawned on different nodes within the cluster, with these various nodes communicating with each other over some proprietary protocol, and the primary node supporting a service factory or service instance lifecycle manager (SILM) and interactions with the Framework.

In addition, a Service Provider may decide to deploy SCS and Framework components from different vendors into a true multi-vendor-environment (MVE). Also, since the SMG is a modular, extensible system, the service provider may choose to upgrade selective services or particular SCSs. Given that the SMG cluster may host not just Parlay/OSA SCSs but also proprietary, non-Parlay SCSs, it is important that a generic upgrade plan be made available that covers all the details of how such a process may be carried out.

An SMG, though a complete system in itself, does not operate in isolation. It is interconnected to network elements along the southbound direction and with Parlay/OSA client applications along its northbound interfaces[16]. It therefore stands to reason that when talking about upgrades, one must also consider client applications.

10-6.2   Upgrading an SMG

Let us first understand what is meant by the term 'upgrade'. As standards evolve, it is likely that more functionality would be added to existing interfaces, or that interface definitions may change from one version to another.

Since Parlay/OSA have base-lined the intended structure of the various SCS APIs in Parlay 3.1 and OSA R4 (what is sometimes referred to as the 'anchor'[17] release), it is very unlikely that any future changes to the API would result in incompatibilities with previous versions subsequent to this base or anchor release. In other words, the standards committees are working hard to ensure backwards compatibility with versions starting with Parlay 3.1/OSA R4. In fact, this is even a stated requirement.

Parlay 3.1, 3.2, etc. are essentially maintenance releases that fix minor bugs with the base release from Parlay 3.0. There have been significant IDL structure and mapping changes from Parlay 2.1 to Parlay 3.0, but again, given that Parlay 3.1 (with minor revisions introduced to Parlay 3.0) is the anchor release, only minor changes to existing APIs are permitted, though some existing method signatures may be deprecated/modified or replaced where errors are found. New methods

---

[16] Quite possibly, it also connects via some proprietary or standard interfaces to service provider OA&M systems; and all SCSs, we know, have to interface to the Parlay Framework.

[17] The term 'base release' applies to all the X.0 versions of the specifications. The term 'maintenance release' applies to bug-fix releases – the X.1, X.2, etc. The term 'anchor release' refers to a release version that is guaranteed to be the basis for all future work, while simultaneously promising to be a version with which all future versions will be backwards compatible.

and interfaces can, of course, still be introduced. And the Parlay body has finally put in place some well-defined rules relating to what kinds of changes can be made in what kinds of releases, and how method deprecations should be handled.

If the implementation of particular functions within a given SCS changes, so as to provide added functionality (or provide bug-fixes) while building to the same interface, this would qualify as an upgrade of the SCS being replaced with the next version. (Note that, in general, the SCS version and the version of the standard that it implements are two different things – for instance, v3 of a vendor's UL SCS is an implementation of the Parlay/OSA UL API, providing capabilities or performance above and beyond what was possible with v2; but both the v2 UL SCS and the v3 UL SCS may in fact build to the same Parlay 3.1 API, but with v3 building more of the functions in the API than v2.)

The above paragraph presents what is meant by the term 'SCS upgrade'. A 'Service Upgrade' refers to the case where ALL SCSs of a given kind are upgraded within a particular time frame. Note that as particular elements (of a given service type) are upgraded, the newer version of the service is available to client applications from those SCS clones. Once all SCSs of a given type are replaced with newer instances, the only version of the service available to clients is the one offered by the newer SCS version. This is what we refer to as a service upgrade.

In an analogous fashion, upgrading the Framework might imply replacing an existing Framework implementation with another version that provides a different, possibly better, or more efficient implementation. An 'upgrade' necessarily implies that some positive change in terms of added functionality or better client application experience results after the process is complete.

All the above processes (for Framework upgrades) work beautifully, in a manner totally independent of each other unless there are changes to the API or associated IDL for the interface between the SCS and the Framework. We define the term 'system upgrade' to cover this case. This is the most complicated of all the upgrade scenarios and essentially is expected to occur only when the SMG transitions from one base release of the specification to another, where there are significant changes between base releases of specifications from Parlay/OSA.

To summarize, upgrades always result in the deployment of a new SCS to replace an existing SCS, but may be classified into different types based on whether this results in:

a) changes to the SCS functionality but no changes to the supported interfaces (other than possibly the southbound protocol/API interface);
b) changes to the SCS northbound interface;
c) changes to the SCS to Framework interface; or
d) combinations thereof.

These cases will be referred to in the sections to follow as 'Type-X' upgrades, where 'X' refers to the letter from the above classification. Figure 10.7 illustrates these cases, and describes how upgrades may be carried out. The reader may refer to this figure while reading the rest of this chapter.

10-6.3 The Upgrade Process, and Addressing Inter-Component Dependencies

Upgrading traditional telecom system services is a non-trivial undertaking. To cite a completely different example, when upgrading pre-paid systems, one has to ensure that all the network elements involved in an end-to-end flow are identified, that their inter-dependencies in terms of both protocol operation, data storage, and changes are well-understood, and that very minimal (if any at all) service disruption be visible to subscribers who try to use the service during the upgrade period. Upgrades are also very carefully and meticulously scheduled to be completed in well-defined maintenance

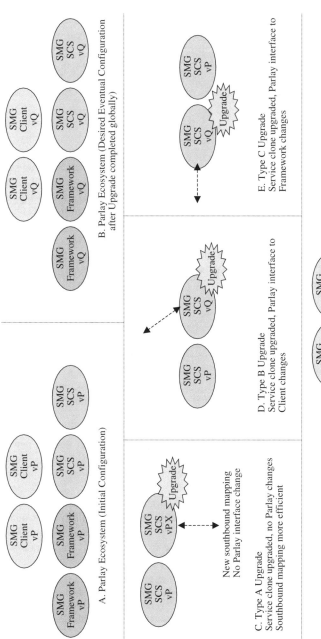

Figure 10.7 Upgrading field-deployed Parlay systems

windows[18] that are of fixed duration at well-established time-slots. These time windows are typically pre-scheduled by the operator during times of lowest network activity. This helps minimize any revenue leakage or loss due to calls completed as the upgrade is in progress.

Typically, once the upgrade process is completed on a given system, verified to be properly done, and the updates 'committed', the upgraded system is 'under observation' for a period of time, even as it handles traffic. After the service provider is convinced it is behaving as it should, the rest of the same type of systems in the network could then also be upgraded. Given upgrades have the potential to be revenue impacting if something were to go wrong, the whole procedure is very carefully designed, analyzed, tested, and only then applied – first to impact markets or systems with smaller loads, and then, as greater confidence is gained in the process, to other elements in the network.

10-6.4   SCS and Service Upgrades

For SMGs, we simply require that upgrades may be done at any time when the SCS has no service sessions connected to it. If an operator absolutely has to reset an SCS and upgrade it within a specified time window (due to other network element upgrades that result in an SCS change being necessitated), the operator could explicitly bring down the SCS by invoking appropriate commands on an administrative console. The operator must however note that this explicit shutdown of the SCS would not be transparent to the associated client applications, and would appear to them as a failure, and the associated service level agreements may be violated as a result of this action. Once the active SCS has been taken out of service, it may be deleted from the node being upgraded, and once this has been done, the new SCS package may be installed, configured, and made active[19].

In general, if an SCS is shut down explicitly as a result of operator action, existing service session data may be stored in a persistent data store for later recovery. Also, the SCS itself, if properly programmed, may send out a svcUnavailableInd() to the Framework for each service session instance before tearing down the associated service managers, and then un-announce and deregister itself with the Framework. Clients are thus informed of the service manager exiting, and can make alternative arrangements (e.g. by selecting another SCS and signing a service agreement with it) to avail themselves of the same service, with other SCS clones of the same type.

Once the new SCS software is installed and made active, it starts providing service by first registering, and then announcing its service reference (IOR, if CORBA, or some other appropriate reference if programmed to use a different kind of middleware technology) with the Framework, once per nested SCF it supports. Re-registration is required because we expect that the new SCS instance could register a new set of properties that are different from, and possibly a super-set of, the properties registered by the previous version of the same service (though the mandatory properties may remain the same).

Client applications that have been made aware of this change may then utilize these properties to identify and select new versions of the service for use. Other client applications would not factor these new properties into their selection. This raises an interesting and important issue. If the operator were systematically upgrading SCSs one by one in a given SMG configuration, it is

---

[18] A maintenance window is a period of time the operator or service provider agrees to take down a system briefly for routinely scheduled maintenance. Upgrades may also be scheduled to happen during this time. Operators would typically require that there be minimal service disruption during the time a given system is down, and this may be addressed by alternative routing to other clones of the system undergoing maintenance, or by other means.

[19] This section takes a very simplistic view of the procedures involved. Vendors may support more sophisticated procedures that minimize service disruption, and provide near transparent cutover from one version of the software to another (e.g. issues like database schema changes etc. need to be very carefully addressed). However, these steps tend to be carefully guarded secrets, since they provide for competitive differentiation between vendors' products.

possible that a particularly unfortunate client application would be bounced off (disconnected from) each new service session it tries to establish with SCSs of that type as these SCSs are upgraded one after the other, before finally settling on an already upgraded version that operates in a stable manner from that point on (since the upgrade was already completed on that node).

This is perhaps best explained by thinking through a simple physical model. Consider a set of N large red marbles. Each marble represents an old version of an SCS. Consider another set of small red marbles one or more of which are connected to each large red marble. These represent client applications of the same version as the SCSs. Now, as an SCS is upgraded, the large red marble is disconnected from the smaller marbles, and replaced with a large blue one. The smaller marbles previously connected to this then attach themselves with another large red marble. As more of the large red marbles get replaced with blue ones, the application marbles get shuttled to other old SCSs to connect with. An unlucky application marble may have to re-connect N-1 times as the upgrade is carried out across the set of SCSs. Gradually, for applications to retain their ability to utilize the service, the smaller marbles also need to change to blue ones to be compatible with newer SCS versions (but this depends on whether or not the blue version was backwards compatible with the red interfaces for client to SCS interactions). Do you still have all your marbles?

Of course, if the northbound SCS API interfaces change between versions, unless client applications are aware of these changes, the entire service (or selected portions[20] of it, depending on how the changes are integrated into the API) may become inaccessible to client applications.

10-6.5   'Type C' Upgrades

As noted in previous sections, these are the upgrades that result in the deployment of new SCS versions where the SCS-to-Framework interfaces also change, in addition to other modifications in behavior and/or implementation.

If the SCS-to-Framework interfaces change, then, in order for this kind of SCS to be able to register and otherwise communicate with the Framework, either a Framework that supports the new variant of the interface must first be incorporated into the cluster (through a Framework upgrade), or a proxy architecture (see next section) needs to be supported within the cluster before the SCS itself may be upgraded. Alternatively, the SCS clones may be upgraded first, but may be unable to register or announce themselves till a suitable 'new version' Framework becomes available (this latter case impacts capacity, since the upgraded SCS clones are now unavailable to client applications until the Framework is upgraded, and thus, this scheme may be undesirable).

This raises an additional issue. If there are also changes between the old and new versions of the client application to Framework interfaces, there is an inherent partitioning of the set of Frameworks into two subsets – one that conforms to the older version of the interface, and the second that conforms to the newer version. This partitioning also occurs even if there are no client application visible changes, if the SCS to Framework interface requires that the SCSs register with only particular Frameworks that implement the new version of the interface – so new service managers on these upgraded SCSs can only be created (or destroyed) by conformant Frameworks. This also means that these upgraded SCS instances may only be discoverable and selectable by client applications that have access sessions with the upgraded Frameworks (or that all Frameworks are no longer equivalent).

Changes of this magnitude could be done by taking the entire SMG down for the duration of the upgrade. Alternatively, if Framework changes were backwards compatible, one could first upgrade all the Frameworks, and then proceed to upgrade individual SCSs. If changes were not backward compatible, but a Framework implementation supported overloaded methods (using object oriented

---

[20] An example of this could be through adding in additional proprietary capabilities into an SCS implementation. The entire Parlay/OSA-defined service capability API could be supported, but additional proprietary methods can be seamlessly integrated into the API so as to be transparently accessible to client applications that are aware of these capabilities.

techniques such as polymorphism for instance) so that the appropriate method version would be invoked in communicating with selected services, again seamless upgrades may be achieved. Once the service upgrades were completed for all SCSs and across all service capabilities supported by the SMG, the Frameworks could then again be upgraded with cleaner implementations that supported only the new version of the Framework to SCS interface[21].

If there are changes in the Framework to client application interface across versions (an especially difficult class of 'Type B' upgrades), this is a much harder problem, and the 'mixed version' (MV) mode of operation would have to be supported over extended periods of time before the Frameworks could be upgraded again to cleaner instances that implement just the new version of the interface. This is because it is non-trivial or even undesirable or impossible to transition over the entire set of third party applications from one version of the interface to another.

📖 **Rule of Thumb:** The set of Frameworks MUST be stable across SCS and service upgrades. If necessary, in cases where there are significant changes to interfaces including possibly the interface between the SCSs and the Framework, the entire set of Frameworks may be upgraded before the SCSs themselves may be upgraded. There may be some service disruption in such cases.

10-6.6 Supporting Different Service Versions Simultaneously; The Proxy Architecture

Some service providers may want to support multiple versions of the API simultaneously: one method has already been discussed in the previous section, involving support for interim implementations of the Framework that implement overloaded functions.

Another, more transparent solution involves supporting a proxy in the communications path between the Framework and the SCSs and possibly also between the client application and the Framework. This way, Framework upgrades may be performed in a manner that is de-coupled from upgrades to other SMG components such as SCSs. This however requires that the proxy be able to convert between the two versions of the interface transparently, so the Framework continues to operate as it usually does. Note that this approach is limited in application to only those situations where the delta between the versions of the interface does not require too many changes to the statefulness of the associated transactions.

Figure 10.8 shows an example of an inter-version Proxy Architecture, used for upgrades. The figure shows a proxy for the Framework. Similar Proxy Architectures for proxies to the SCS can also be used.

This may also be used as a very simple, but rather ugly technique to resolve implementation differences between SCSs and Frameworks from different vendors that are deployed into a given service provider network to ensure interoperability and compatibility between the various products. In the authors' opinion, such differences of implementation should be resolved between the various equipment vendors involved through bake-offs, implementation changes, etc., rather than resort to such proxy implementations.

10-6.7 Summary

The mixed mode, multi-step upgrade process is more involved, but it is the preferred implementation where an SMG upgrade cannot be done by bringing down the entire SMG and then bringing up particular nodes with the upgraded SCSs. The service provider may want to notify enterprise domains (that administer applications) of the time and duration of these upgrade and maintenance windows. The proxy architecture may seem very simple but makes future upgrades more and more difficult, as more significant changes are rolled in.

---

[21] It must be pointed out however that adding more steps to an upgrade only increases its complexity, and that is to be avoided as far as possible.

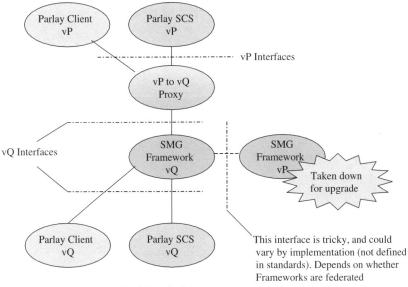

Two versions: vP (old), and vQ (new)

Some SCSs and applications have been upgraded, others have not.

Figure 10.8 Proxy Architecture for upgrades

10.2   Chapter 10 Summary

In this chapter and all its sub-chapters, we have tried to give the reader some flavor for the various kinds of real-world issues involved in building products for telecommunications networks. No claims are made that the considerations covered here are exhaustive, merely that they are indicative of the kinds of things that need to be addressed as carrier-grade products are designed and architected. The intent here was mainly to stimulate thought and provoke discussion on the kinds of questions that engineers should ask as they try to design, build and deploy systems; the specifics of one's solution architecture, 'surround' technical choices, environment, and to some extent, corporate culture, would dictate the solutions that one chooses to resolve issues in each of the above areas.

Part IV

Realizing Parlay

So you've built a Parlay product and are evaluating applications to host atop your gateway. Or you are negotiating with application vendors to determine if you want to partner with them. Or you have developed a new and exciting Parlay application that you now want to use in different network contexts. Or you want to deploy a newly acquired gateway into your network but not risk your existing revenue or infrastructure. How to go about this?

These are the questions covered in this section. The focus here is on deploying Parlay gateways and applications, application testing, usability of Parlay components in different network contexts, etc. Technical readers, especially those working as application developers or in telecom equipment provider or service provider companies, will appreciate and relate to this technical content more easily.

11

Deploying Parlay Gateways

11.1 Introduction

So far, we have discussed the benefits of Parlay technologies to end-users, service providers and application developers, Parlay standards, and issues relating to Parlay gateway product implementation. But what about actual deployments in real-world networks? Do implementations really meet the goals that the standards set out to achieve? How? Is the promise and potential realized? These and other such questions are considered in this chapter.

Furthermore, in the concluding sections of this chapter, we shall present a simple mathematical model that attempts to 'prove' the value proposition of Service Mediation Gateways in service provider networks, and show why the investment in these gateways can be easily recouped while simultaneously enabling such a deployment to tap into new revenue streams.

11.2 Parlez-vous Parlay?

By now, the reader should have a fairly good understanding of what Parlay is, how Parlay-based systems work, and what the technology promises to do. However, building systems conformant to the standards-defined specifications is only one part of the puzzle. The proof of the pudding, they say, is in the eating. Let us therefore look at how systems, once built, can be deployed in networks, what deployment considerations this entails, and how this is simpler than doing so with traditional systems. In later sections of this chapter we present a theoretical proof of the Parlay gateway value proposition. Here, we try to ground those arguments in more concrete reality.

Speaking loosely, there are two types of Parlay deployments – one, where the network does not possess a well-defined services infrastructure, and is 'starting out' with Parlay as the technology of choice (also called a 'green field' deployment[1]), and the second, more prevalent situation, where the network already has a well-defined, widely used services layer employing legacy components that provide end-user services. As Parlay deployments get more widespread, there will also be a third type of deployment, namely that wherein a new Parlay gateway or other components are deployed in a network that already has another Parlay gateway or components in place, possibly from another vendor. Each of these presents its own challenges from an integration perspective.

---

[1] There aren't very many of these kinds of networks extant today. Virtually every telecommunications operator network in existence has an IN implementation. However, as new access technologies become more prevalent, and other kinds of networks besides the telephone networks are used to carry voice traffic, there may be opportunities to evolve such 'green field' networks to support additional services through Parlay and OSA technologies.

Parlay/OSA: From Standards to Reality Musa Unmehopa, Kumar Vemuri, Andy Bennett
Copyright © 2006 Lucent Technologies Inc. All Rights Reserved

Where Parlay is used to complement, then supplement, and finally displace legacy services infrastructure, unless services are carefully designed and deployed to ensure that there is no conflict between the legacy and new components put in place, trigger contention and feature interaction or feature interference issues might arise in such networks. The feature interaction and trigger contention problems merit closer study, and are discussed separately in Chapter 12, which is dedicated to those issues. Here, we consider what needs to happen from a network perspective for Parlay deployments to be possible, even if such contention were avoided from a services standpoint.

For example, if a service mediation gateway were deployed with User Location and Charging and Account Management SCSs and the Framework only, but without any call control related capability, then the potential for trigger contention with existing SCP infrastructure elements could be avoided. We consider situations such as these in this chapter. In a sense, there is some commonality between 'green field' deployments and these kinds of deployments, though there may be some integration considerations that apply to the latter, though not the former.

11.3   Growing the Parlay Network Footprint

Service providers may start out small, with a small subset of offered or available Parlay SCSs deployed on their gateways, and then grow these as demand for these technologies picks up or as competition heats up to offer more services more quickly and at lower prices (thereby mobilizing the large pool of Internet-based developer talent), and the struggle to offer feature parity becomes an issue. As we have seen in previous chapters, several alternative candidate architectures could be used to achieve the same from a gateway software architecture perspective.

Some SCF APIs are also more tightly coupled to the underlying network technology than others. The former may require more integration effort than the latter. For example, Call Control (CC) SCFs require a lot more integration work than the GMS SCF APIs. CC APIs utilize service control protocols and the SCF needs to interface with core network infrastructure that is tied to switching and related critical functions, while the GMS SCF merely uses IP-based infrastructure with less real-time sensitive service characteristics. A different variant of the CC SCS may need to be developed to interface to switches in the CDMA, GSM, SIP and other environments, whereas the same GMS SCS could be deployed independent of core network technology in use (since it depends only on there being support for SMTP or IMAPv4 or equivalent protocols over IP in the network, and is independent of core network technology). These differences also factor into costs associated with deployments. But, as is the case with all other things, you get what you pay for – the more tightly coupled systems tend to be the ones that deliver greater value. At any rate, it is likely that service providers will start with one or two SCSs that they perceive to be of greatest value to their networks, and then grow from there.

11.4   Simplifying the Labors of Hercules

We have discussed in previous chapters how Parlay simplifies the integration problem by reducing the one-off integration required in networks today where each application needs to be separately interfaced and connected with the underlying network component, before it is able to provide any value to end-users in terms of end-to-end flows. And also how, with the gateway being deployed, this integration is done only once, but this time with the SCS hosted by the gateway, which later interfaces to other northbound client applications. Thus, there is a simplification, and once things work with one clone of the service, the process can be made transparent to other clones that are then deployed. Complexity is reduced, but it is not eliminated. Rather than hand-hold multiple independent application developers, the service provider now has to help the gateway vendor hook into the core network infrastructure.

Green-field networks offer the gateway vendor greater flexibility in which provisioning, billing/charging, or administration infrastructure may be used. Legacy networks do not. In the latter case, in addition to the integration with core network components to support end-to-end call flows, the SCS may also have to interface with deployed element management, OA&M and billing and charging systems. This will require additional work and perhaps some amount of customization on the part of the gateway vendor. But let us look at the advantages here – with Parlay, the gateway vendor has to perform this added layer of integration; the application developer is completely exempt from it[2].

Again, this is an N-way simplification, for the same gateway is used by multiple applications. Once the integration is done, the vendor that has carried out the task is in a good position to repeat the process with other new SCS elements the service provider may be interested in acquiring, especially where support for these interfaces transcends the functional characteristics of the SCS in question. (For example, both the GMS SCS and the CC SCS may be required to interface to the same element management system. In such cases, if the vendor has already deployed the GMS SCS in the network and completed the integration with the element management system he will be in a better position to manage the integration of the CC SCS more quickly than otherwise[3]). Similar arguments may be made for customization, if any, required for call detail records (CDRs), service detail records (SDRs) or event detail records (EDRs) that are generated by the service.

Parlay manages to make the core network evolution completely transparent to the application domain. Core network infrastructure may change, new network elements may be deployed, etc., but the application view can be kept reasonably constant[4] if so desired. SCS mappings from the northbound API to the southbound protocol may change quite drastically, but the applications remain unaffected. SCS changes incur costs, but each of these costs is masked behind an N-way savings, since it need not be done once per deployed application. These points are further illustrated by means of the following case study.

Case Study: Utopia goes Parlay!

Jim Chase, senior VP at Utopia, was aware of subscriber perceptions that they trailed their competitor, Freedom Wireless, rather badly when it came to services. It was universally acknowledged that their coverage was much better than their competition, but coverage without services was starting to cost them. They were losing subscribers, and it was beginning to worry the company.

'We are at par with Freedom Wireless as far as traditional IN services go,' he thought, 'Where Freedom really differentiate themselves is in the area of "surround" capabilities. We too have WAP, SMS, and other infrastructure in place, but haven't yet bothered to integrate them into neat service bundles and offer value added capabilities around them, and that is what Freedom seems to be doing so well.' He subscribed to Freedom Wireless to experience first-hand the services they offered, and liked what he saw.

---

[2] Admittedly, the application developer has to perform some integration with the gateway itself, but this process is much simpler than having to perform said integration with the network elements. This is so because gateway elements built to the same standards-defined interfaces, though they may have some quirks, would be relatively very similar to each other. Network elements on the other hand could vary widely from one network to another, and some may even rely on the use of proprietary protocols. Also, as indicated in previous chapters, some gateway vendors provide SDKs with simulators that give application developers easy access to a test environment that closely parallels the real SCS they may encounter in the actual deployment.

[3] Note how the same argument does not immediately hold for multi-vendor environments. Some hand-holding is needed for each new vendor whose gateway integrated into the system, but since the number of gateway vendors whose components are integrated is still likely to be far less than the number of supported client applications, there is a proportionate reduction in the scale of the problem, making it still a viable solution.

[4] We say 'reasonably constant' because, in some cases, if old network elements are replaced with new ones this is not network evolution, this is more like node upgrade for capacity purposes, and as additional capabilities become available, the service provider may choose to have the SCSs upgraded to make those new capabilities accessible to applications via proprietary API extensions. In such cases, greater flexibility and control of the underlying elements may be achieved, but at the expense of seamlessness and transparency.

At a recent executive meeting, they decided to give Parlay technology a try, investing cautiously at first, perhaps running concurrent three month trials with multiple vendor gateways, and a small subset of subscribers interested in newer applications and willing to take the chance with beta-grade services on a trial basis at no charge, and then, if all went well, explore acquisition, integration, administration and maintenance costs for a full-blown deployment with a selected vendor. They had looked at alternative technology possibilities, but could find no other way to achieve their goals at anything approaching a reasonable cost.

Truth be told, this was a change in paradigm for the company. Utopia was known to be very cautious when it came to accepting new technologies. But they did not seem to have any other choice. The integration costs for new applications developed separately were way too high, and the concept to deployment cycle took much longer than they had planned for. They had to deploy services within the next six months to remain competitive. They had to act NOW!

They liked the gateway manufactured by Luminant. This vendor also had partnerships with several application developers that had built client applications interfacing to SCSs that Luminant developed, and one of these was an address book, content push and end-user alerting service very similar to the one that drew so many subscribers to Freedom Wireless. The three-month trial got rave reviews from the participating test groups (subscribers trying out the beta-grade services). And Luminant promised more services through their partners at very competitive rates, at very short intervals. Jim liked what he saw very much.

Next, Utopia had them integrate the trial system with their legacy billing, charging, OA&M and element management infrastructure. Here, they faced the usual problems they saw each time a new application was deployed in their network. Long hours for Luminant engineers, and some hand-holding by Utopia employees to ensure all the integration was carried out properly. But then, once the integration was done, the end-to-end call flows ran flawlessly. To prove their point Luminant offered to trial another application that used the same SCSs, already deployed in the trial setting. Users could perform a limited form of collaborative browsing with their buddies over the WAP infrastructure. Jim was ecstatic. He was completely sold on this technology, though he still advised that more thorough testing of the new application take place before it was offered officially.

Utopia decided to go with a medium capacity deployment with only three SCSs – PAM for explicit user presence sharing across subscribers, a WAP Push SCS built to leverage the UI SCF for non-call related near real-time content push, and a Framework for access control. Luminant assured them that new SCSs could be added at any time should Utopia want to provide API access to new or other services for applications, or open up other network elements via programmatic APIs. The marketing blitz started as Utopia was going through the final stages of their integration and end-to-end testing. Subscribers that were about to leave Utopia for Freedom Wireless decided to give them another chance. And good news through word of mouth did more for Utopia over the next few months than all the advertising in the world.

Freedom Wireless tried to keep pace, but started falling behind. Suddenly Utopia started offering new applications very, very frequently. Jim Chase had saved his company. 'Until the next disruptive technology comes along,' he thought, with satisfaction.

11.5 The Value Proposition for the Service Mediation Gateway in Service Provider Networks

For any new technology, there is always interest in evaluating its value proposition to ensure that the promise and potential of the technology can in fact be realized in actual real-world deployments. So far, in this book we have looked at the value of Parlay technologies to end-users, application developers and service providers. We have studied how Parlay gateways can be built, and how applications can leverage these exposed standards-defined APIs to enhance end-user experiences. We have studied examples.

This section tries to 'prove' the value of such gateways in today's networks. This is not meant to be a business case statement, nor is it input to the business case for Parlay/OSA. This is merely a somewhat rigorous treatment of the properties associated with the SMG in real-world network deployments, and is intended to demonstrate how the SMG adds value to customer networks from both qualitative, as well as quantitative, perspectives.

The intent here is to present an engineering view, not an accountant or a mathematician's viewpoint. Simple mathematical models based on elementary arithmetic are used to drive points home. We advise the reader not to be intimidated by the complex looking mathematical formulations. The word 'proof' is used somewhat loosely. However, the arguments are expected to be convincing and to stand on their own merit.

This is intended to make a convincing case for every single technical argument made in favor of Parlay/OSA Service Mediation Gateways in general (in our discussions from previous chapters, for example), for service provider network deployments. The various propositions and their associated proofs are covered as granular components to a) isolate errors, and b) use proved propositions as building blocks for other proofs. Also, some parts of the 'proof' are more rigorous than others. We acknowledge that perhaps more cogent, complete arguments for the model presented herein, or perhaps even altogether different models may be constructed to drive home the same, or similar points, more forcefully. This is 'a' view, and a good way to look at things, even if not the only way.

11.6   Propositions and Proofs

Proposition 1.   New applications serve as new generators of revenue for Service Providers.

Proof:
Assume a Service Provider network that has a set of applications $A = \{a1, a2, a3, \ldots, aN\}$. Let us further assume that these services are of two types: a) subscribed applications (SA) where subscribers can sign up for them and get billed for them independent of usage characteristics; and b) pay-per-use applications ($PPUA$), where the subscriber is billed for each use when a usage statement is issued at the end of each billing cycle. True, other charging and discounting models can be supported, but we restrict our considerations to these two kinds of applications from a modeling perspective, further stating for simplicity that no given application simultaneously falls into both the subscribed and pay-per-use categories (if this assumption does not hold, simply treat that application as two distinct applications, one in each category).

$$A = SA \ \cup \ PPUA \tag{11.1}$$

$$SA \ \cup \ PPUA = \emptyset \tag{11.2}$$

$$\text{Revenue from Subscribed Applications} = \sum_{i=1}^{k} C(Ai) * Si \tag{11.3}$$

$$\text{Revenue from Pay-per-Use Applications} = \sum_{i=k+1}^{N} Tr(Ai) * F(Si) \tag{11.4}$$

With these assumptions, the following statements hold.

Equations (11.1) and (11.2) are self-evident and do not require any further explanation. (11.3) indicates that the revenue from subscribed applications (k of the total N applications in number) is the sum of the products of the individual subscription fee per application ($C(Ai)$) and the number of subscribers for each application Si. (11.4) indicates that the revenue that is obtained from the remaining ($N - k$) Pay-per-Use applications is the sum of the products of the transaction processing fee per application Ai ($Tr(Ai)$), and the number of transactions each application Ai sees, which are a function of the number of subscribers that can use Ai (indicated by $F(Si)$, potentially the total subscriber pool if unrestricted access is permitted to particular services).

The total revenue obtained from deployed applications is given by the sum of the terms in the RHS of Equations (11.3) and (11.4). If more applications are added, assuming there is at least one user in the billing cycle for each new application, there are more product terms in at least one of the component equations, and therefore there is an increase in the associated realized revenue.   **QED.**

Note: The above analysis looks at revenue from applications only, not at the expenses involved in deploying and maintaining deployed applications, and associated costs. The costs may over-shadow the revenue and the applications may not bring in sufficient money even to break-even. That is, however, not something the current proposition is concerned with.

Proposition 2.   'Context Aware' services improve the end-user experience and further accelerate revenue recovery.

Proof:

In order to assert the verity of this self-evident statement, we must first define what we mean by the 'context' of a service or an application in a given network context.

A Wireless Service Provider (WSP) subscriber has the following kinds of information associated with her 'identity' as known to the network:

a) at least one agent, and potentially multiple agents, with their associated information;
b) (relatively static) profile information and preferences associated with the user identity, each of the associated agents, billing IDs, and policies that govern user interaction within the network context;
c) (relatively dynamic) information tied to the immediate 'user environment' as the user operates within the network context – this includes elements such as location, presence, availability, terminal status, etc.

We define the union of these three elements to form what we call the user's 'self-context'. Metcalfe's law states that 'the usefulness, or utility, of a network equals the square of the number of users'. A network of users is, in effect, a community, and both a given user's own self-context and a sub-set of elements from those of the user's buddies together provide information that could be leveraged by applications to provide significantly greater value in enhancing the end-user experience.

Some of this value is an 'intangible' adder to the end-user experience in that it cannot be exactly quantified. However, let us attempt to assign some metrics to this in order to be able to illustrate that context aware services do in fact provide value above and beyond that which is normally perceived by end-users.

Every application provides the end-user with an output – some kind of information in response to an immediate or pre-established request (e.g. in the case of Push services). In every case, useful output is a result of some input. Or, to state this another way, the information received (the output) is a function (Ψ) of one or more inputs provided.

$$\text{Output(O)} = \Psi(I_1, I_2, \ldots, I_k) \qquad (11.5)$$

Ideally, the user would like to get this same output while explicitly providing as little input as possible.

Now, if some subset of these parameters $\{I_1, I_2, \ldots, I_k\}$ is part of either the user's own or her buddies' contexts in the network environment, this subset need not be provided by the user as input, if the application can derive this information from the network. Thus, we have

$$\text{Output(O)} = \Psi(I_1, I_2, \ldots, I_p, \{\text{network context information}\}) \qquad (11.6)$$

where, $p < k$. The application is now able to perform the same set of computations as before, but with less explicit user input. This argument, though convincing, is not yet complete.

We also consider new services that can be provided given the ability to leverage network context information that could not have been provided earlier. In (11.6), for example, there could be elements within the network context information of which either the user is unaware, or does not know, and so cannot provide this information to applications that could use this to provide a better user experience. If these applications are authorized to access this information however, with the user having control over who is able to get at the information at any given point in time, this enables WSPs to support whole new classes of 'context aware' applications capable of leveraging user-related dynamic context data, that would otherwise not have been possible. This adds further value to the end-user experience through support for more compelling service scenarios.

Last, but not least, the more user context information stored in the network, and the greater the facility afforded to authorized applications to leverage these data in providing end-users service, the harder it would be for users to switch carriers, thereby retaining existing subscribers (customer loyalty) who may end up spending more per billing cycle as the number of compelling applications increase. This also accelerates revenue recovery.   **QED.**

Proposition 3.   The SMG enables a Service Provider to build applications more cheaply.

Proposition 4.   The SMG serves as a catalyst to the deployment of new applications.

Proposition 5.   The SMG facilitates quicker development and deployment lifecycles, and faster time to market for new applications.

Proof:
Since these three propositions are closely related, and tied to the SMG, we try to prove all three of these together. Before we start, however, we must first define what we mean by a 'catalyst'. Here is the definition we will use: 'a substance that enables a chemical reaction to proceed at a usually faster rate or under different conditions (as at a lower temperature) than otherwise possible; an agent that provokes or speeds significant change or action'.

Also, since we are now talking specifically about a Parlay/OSA-based service mediation environment, the words 'service' and 'application' are no longer as freely interchangeable as in the previous context, and we shall use them in the strictest and most correct sense from this point on.

We also use two other terms in what follows: PMO (Present Mode of Operation), which refers to the mechanism of deploying point solutions that are tightly coupled with network elements that they leverage to extract user context information used in the processing of their transactions, and FMO (Future Mode of Operation), which refers to the support for a common mediation platform such as the SMG that is used by multiple applications.

Let us assume it takes k_i man-months of effort to support the integration of protocol P_i into a given application. Proposition 2 already proves how the effective use of multiple components from a user's current network context may be used to provide a more satisfying and compelling user experience.

Since different components of the user context are normally distributed across or 'known to' different network elements, each of which could potentially require a separate specialized protocol to support application queries, it is likely that applications which provide a more compelling user experience would need to interface to more network elements than those that do not.

In the PMO scenario, the total WSP effort required to support the integration of a single application into the network in the context of a point solution is given by

$$\text{PMO Integration Effort per Application } (E) = \sum_{i=1}^{N} P_i * k_i \qquad (11.7)$$

where N is the number of protocols that the application utilizes in providing a user experience.

As more applications are deployed that use some of the same interfaces, it is apparent from (11.7) that some of the same product terms appear in the integration efforts of each of these applications. The picture is slightly different in the case of FMO scenarios. Here, the following holds instead

$$\text{FMO Integration Effort per Protocol } (E) = P_i * e_i \qquad (11.8)$$

where e_i indicates the costs of integrating a protocol stack with the service mediation gateway. Technical details dictate that e_i is not very significantly different from k_i for any protocol where the API provides a good degree of semantic correlation between the functional characteristics supported by each. Note that the integration effort in (11.8) is amortized across a set of applications that utilize the protocol, and **the cost per application** is thereby reduced.

$$\text{FMO Integration Effort per Application } (E) = \sum_{i=1}^{N} \frac{(P_i * e_i)}{A_i} \qquad (11.9)$$

where A_i is the number of applications that utilize the protocol across the entire set of WSP deployed applications, and N is the number of protocols needed for the application in question to function (as in (11.7)). The RHS of (11.9) is typically less than that of (11.7), and gets smaller each time a new application that utilizes a particular protocol binding is deployed. Thus, as more applications are deployed, the savings accumulate, reducing the break-even point of each application lower. This proves Proposition 3.   **QED.**
    Notes:

1. One may be able to extend the case to cover even the first deployed application that utilizes a given protocol, but this case is harder to make, since this varies based on the particulars of the protocol and the characteristics of the API on a case by case basis. Typically, the ratio (e_i / k_i) is expected to be larger than one, but less than two, and depending on how much it is fractionally greater than one, it would be possible to make judgments per protocol.
2. This proof only considers application development and deployment costs. A more complete analysis should also factor in capital and operating expenses to make the case more strongly.

Proposition 1 indicates that new applications are new sources of revenue for WSPs. Proposition 2 shows how context aware services provide a more compelling user experience, generate customer loyalty, help increase revenue (in terms of subscriptions and transaction processing fees), and attract more subscribers. Proposition 3 proves that new, context aware applications can be developed and deployed most cheaply when a service mediation gateway is supported. The three statements taken together prove Proposition 4 that the SMG serves as a catalyst to the deployment of new applications since all WSPs are continually interested in increasing their revenue streams.   **QED.**
    Assuming that the WSP has S engineers available to support the integration effort, the total time-line to support the integration of each application is given by *PMO(E)/S*, assuming an equal division of work across the entire support staff (an assumption that does not generally hold for software development – 'Adding more people to an already late software project makes it *later*' [Brooks 1995]). This severely limits the WSPs ability to deploy applications more rapidly.
    In contrast, since all the integration in FMO scenarios is carried out once per protocol, and the applications themselves are all built to standardized API interfaces, the need for hand-holding or one on one partnering with each application provider is obviated. There may be small incremental costs associated with certifying that applications work as intended, but this is typically a function supported by the vendor of the service mediation gateway and the costs involved are not that significant to begin with. There are no limits on the number of applications that can be deployed since this is only constrained by the number of developers that can build these applications, and is in no way constrained by the WSP support staff.

In other words:

$$\text{PMO timeline per Application } (T) = \sum_{i=1}^{N} \frac{(P_i * k_i)}{S} \tag{11.10}$$

$$\text{FMO timeline per Application } (T) = \vartheta(D) \tag{11.11}$$

$$\text{PMO } (T)\alpha \left(\frac{1}{S}\right) \tag{11.12}$$

$$\text{FMO } (T)\alpha \left(\frac{1}{D}\right) \tag{11.13}$$

$$D \gg S \tag{11.14}$$

The FMO development and deployment timeline is a function of the number of developers D trained in technologies such as XML, Java and C++ that can build to the standards-defined API interfaces. (11.12) and (11.13) merely reiterate that the greater the number of developers available, the smaller the development and deployment intervals. Given that the number of developers D is significantly greater than the WSP support staff S, it stands to reason that the time intervals involved in developing and deploying applications in the FMO case is much smaller than those involved in the PMO case. This proves Proposition 5.   **QED.**

11.7 Conclusion

The Service Mediation Gateway provides real, quantifiable value to a Wireless Service Provider network. The best time to deploy service mediation gateways into a network is during a 'moment of change' – the time immediately preceding the rapid addition of new applications to the network – so that costs incurred in deploying new components such as service mediation gateways are defrayed by their effective distribution across a larger number of application elements. The current time, when many Wireless Service Providers are transitioning to support new core network technologies such as CDMA 1XEVDO, etc. provides an excellent opportunity to drive through service mediation gateway sales. Parlay is crossing the technical chasm at this time.

It is inevitable that as service providers come to rely more and more heavily on Parlay technologies a gradual 'Parlay-ification' of their network will occur. Once a tipping point is reached and the new application floodgates open, all major new service infrastructure additions will be made primarily in the Parlay arena and legacy equipment will be gradually subsumed and deprecated in its favor. As of this writing, there are a number of Parlay deployments in service provider networks across the world, and there seems to be a rather strong demand not just for the SCSs based on standards-defined functional APIs, but also for new SCSs built to conform specifically to operator requirements.

12

Parlay and Legacy Systems – Handling Feature Interactions

12.1 Introduction

In previous parts of this book, we have discussed some of the hopes, fears and frustrations of engineers building and deploying services and applications, and of consumers who want access to more ubiquitous communications capabilities and are willing to pay for them. In the last few chapters, we have studied some of the more practical issues relating to the architecture, design and development of both service mediation gateway elements and of the client applications themselves, with a view to better understanding how the more theoretical aspects of the solution as defined in the standards can be implemented in real world systems.

In this chapter, we focus on yet another practical aspect as it relates to deployments of Parlay technology, looking at things first more from a network perspective, and then from a services angle. The focus here is on how Parlay-based systems can be made operational in network contexts and how interactions with other services and with existing legacy network hosted elements can be managed.

12.2 Out with the Old, In with the New? Not quite

As we have seen already in Chapter 11, generally speaking, there are two kinds of deployment scenarios extant today. One is what is referred to as a 'green field' network environment, where the components being introduced have no parallel or peer in the network, or the network itself is being built from the ground up, and there is little that needs to be done from the perspective of getting the new equipment to interoperate with legacy elements already deployed. These kinds of environments are easier to handle from a new services/applications/network elements deployment point of view, but these kinds of situations are also very rare.

More often than not, however, engineers – both those working for the service provider and those that build the gateway while working for the telecommunications equipment provider – have to contend with issues relating to how the new solution component can be 'plugged into' their existing network, and made to work without disrupting the existing services and applications.

In earlier chapters we have seen how Parlay/OSA technologies improve the overall concept to completion cycles for new services. However, for the gains to be immediately realized, and for this

Parlay/OSA: From Standards to Reality   Musa Unmehopa, Kumar Vemuri, Andy Bennett
Copyright © 2006 Lucent Technologies Inc. All Rights Reserved

new technology to be feasible from a practical perspective, it is rather important that the gateway, when deployed, operates seamlessly within the network context and does not break existing services. It is also unacceptable if deploying the gateway entails the re-deployment of all legacy services and applications within the context of the new paradigm.

Thus, we need to address issues pertaining to the simultaneous support for Service Mediation Gateways (SMGs) and elements such as Service Control Points (SCPs) from traditional IN in the telecommunications networks of today.

Service mediation gateways are manifestations of new technology that is unlikely to be deployed in a 'direct cut-over' mode whereby legacy technology elements are replaced all together or all at once. An SMG is not a consumer gadget, where you need to have the latest and sexiest model to secure your standing with the in-crowd, and older models are instantly written off ('That model is so last year'). It is rather more likely that these SMG elements and legacy entities such as SCPs will co-exist in networks for some time to come, with a gradual migration of services from the older equipment onto the newer infrastructure components.

There needs to be a strong, non-technical impetus that drives the acceptance of new technologies. After all, it is unreasonable to expect service providers to invest large sums of money in upgrading infrastructure for the 'technical elegance' that the new solution will afford them. Deployment of IN into telecommunications networks was an early 'moment of change'. This technology let service providers build and deploy new services more cheaply than previous models would allow. The driving force in the case of Parlay and OSA comes from the ability to build and deploy services even more quickly and cheaply than in the IN context, in large numbers (leveraging the large Internet application development talent pool, perhaps even buying some off the shelf), and then experience a sharp intake of new revenue while growing one's subscriber base.

12.3   Parlay and Legacy IN Co-existence

It is not only probable, but also very likely, that SMGs will be deployed in networks that already host a number of elements that play the role of an SCF (Service Control Function) from the Distributed Functional Plane in the generic Intelligent Network architecture. These elements, introduced in Chapter 1, receive requests from Service Switching Points or other call-processing elements in the network and return appropriate responses back to them. In traditional wireline networks or AIN[1], these messages are INAP encoded. In other networks, different protocols may be used to support analogous functions.

Since the SMG provides functions in a similar capacity, though however, not restricted to call control, it becomes important to specify how the two components – namely the SMG and the more traditional (or legacy) peer entity – may communicate with each other and/or otherwise interact.

In legacy (e.g. IN) systems, whenever a trigger on a switching element fires, a message is formulated in the appropriate protocol and addressed to the destination element capable of processing this message. This destination element is identified by a pre-configured point code. Different destinations may be specified on a per-trigger, per-user basis. Sometimes, multiple features need to be invoked in response to the firing of a single trigger (this is commonly referred to as 'trigger contention' between the services involved).

To an extent, switch-based functions are standardized, and therefore a priority can be defined for the inter-relationship between these in the standards. But with a plethora of third party applications, without knowledge of one another, dynamically provisioned and brought into and taken out of service, an a priori priority scheme can never be defined. This is why the feature interaction

---

[1] AIN or Advanced Intelligent Networks is a variant of IN developed and marketed by Telcordia (earlier Bell Core) and deployed primarily in the NAR (North American Region). There are some differences between AIN and other IN flavors such as those defined by ETSI used in Europe, though we shall treat these as equivalent from the standpoint of this book.

problem exists (even in traditional IN contexts). It is also why solutions are designed to try to minimize, mitigate or eliminate it at some expense (given that these SCP-hosted services provide that much more value to the network in terms of new and exciting applications for subscribers) and why we keep it despite its downsides.

In such cases where trigger contention occurs, some logic has to be specified that indicates which services are to be executed, in what order, and how their responses are to be processed. Several solutions are possible, and as is always the case, each solution has its own advantages and demerits. The key point to note, however, is that trigger contention is not a new problem that is caused by the introduction of Parlay solutions into the network; this is a potential issue even in networks that rely only on IN technology, and has to do more with how the triggers are used, what services are deployed, and more importantly, which services need which triggers and how these trigger messages are processed.

Sidebar

The trigger contention problem is widely known, has been well studied, and a lot of literature exists that documents both the desirable and un desirable effects it has on user experience [FIW 2003]. Features deployed in telecommunications networks need to be tested rigorously to ensure that all undesirable 'feature interactions', also sometimes called 'feature interference' situations, are somehow managed, or even eliminated. For instance, if an end-user subscribes to both 'call forward on busy' and the 'voice mail' feature, unless some priority order is defined between these two services for that user profile, unexpected behavior or call treatment may occur, leading to end-user dissatisfaction.

Appropriate treatment logic needs to be specified somewhere in the network to ensure that such situations are dealt with properly. It must be emphasized that the feature interaction problem has no generic solution, and all potential interactions for deployed services must be examined carefully and addressed either through provisioning logic (Alice cannot subscribe to both services A and B at the same time), or some kind of implicit priority ordering of selected conflicting capabilities (Service A has a 'gold' SLA, and hence takes precedence over 'silver' service B, or, the good old First Come First Serve mechanism).

In some contexts, the feature interaction problem is addressed by selling pre-packaged sets of features together. These feature sets contain features that are already prioritized within them, with the possible interactions or interference between elements of the set eliminated or carefully designed out of the call flow. Here, elements of the technical solution (focused on 'avoidance') are percolated through via marketing, to ensure that subscriber satisfaction is not impacted.

Note that this chapter is focused primarily on the SS7 domain (where most of these kinds of interactions between legacy and new equipment occur), so SS7 concepts useful to the development of the SIM (Service Interaction Manager) idea are examined in greater detail. However, by no means is trigger contention restricted to the SS7 domain. As more VoIP networks are built and supporting services deployed, mechanisms to address trigger contention in these other environments will also get more attention.

12.4 Managing Trigger Contention

Besides the 'avoidance' approach described in the previous section, several other technical options exist and are used in practice today to address trigger contention related issues. In this section, we look at three of these. Other techniques also exist, and the three described here are by no means the only possibilities. However, an understanding of the techniques presented here will give the reader a flavor for the kinds of solutions in use. Each of these utilizes some mechanism to prioritize service access to trigger information where contention occurs, tries to route the messages to these services in the sequence that makes the most sense, then collate responses from there and generate a single response to the underlying switching element that made the service request. Let us look at these techniques in greater detail.

12.4.1 Supporting a 'Gateway' Service

When this technique is used, if services A and B both need the same trigger to be able to provide appropriate call processing or service related treatment information to the switching element, one of the two services A or B is designated the 'gateway' service, and is given priority. This service (say B) then implements additional logic to feed the request on to the other service (in this case A), the message from the switching element, appearing to the other service as a switch, while appearing to the switch as some combination of both services (A and B). When the other service returns a response, the gateway service receives it, parses it, factors it into its own response to the service requestor, generates a consolidated response message, and sends it back to the requestor.

Thus, the switch sees only one service, namely the gateway service, and assumes that single service provides both the individual capabilities required. The gateway service acts as a smart proxy for the incoming request and performs not only its local service processing but also the collation of the responses from one or more other services, and generates a single response sent back out to the switching element.

Sounds simple in principle, but there are some issues with this:

1. Handling a large fan-out of secondary services is more difficult, as is the associated collation of returned responses.
2. This solution only works in networks where the gateway service is deployed. If there are other network deployments where trigger contention is still an issue but the gateway service is not deployed, then the 'gateway' or proxy logic has to be ported onto another service in that network so trigger contention resolution can take place as required.
3. The performance of the gateway service suffers because it is doing additional work besides just what it is required to do to provide service. Thus, the service provider may need to deploy additional elements that can provide the functional capabilities associated with the gateway service.

Therefore, this solution, although used in many networks, can perhaps be improved on.

12.4.2 Service Granules and Intra-service Routing

Another approach that is sometimes used is one where new service logic is implemented within an existing service (let us call each piece of service functionality a 'granule') and an intra-granule router is also built into the ingress of the service logic. Here, any request that is destined for this service is first processed by the intra-granule router. This module decides which granules need to be invoked, and in what order, and whether in sequence or in parallel. Once all this work has been completed, and responses from these various service granules are received, the router then stitches together a consolidated response and generates the final message that is sent back to the service requestor. The astute reader will note that this is very similar to the pattern used in the 'gateway' service example above. The main difference here is that the gateway service and the other services are integrated into one logical service component, and the router is shared across them.

It may not be practical to implement services and applications in this way all the time, however, since there may be reasons why particular services need to be deployed separately (provided by a different vendor, on a different platform, etc.) and in such cases, this mechanism may not work.

12.4.3 Service Interaction Manager or Service Combination Manager

A network element called the SIM (Service Interaction Manager) or SCM (Service Combination Manager) may be used to provide capabilities for feature interaction management, essentially providing the same generic capability as that of the 'gateway' service, but completely decoupled from the service logic of any of the required functional capabilities so as to mitigate some of the drawbacks we listed in such scenarios.

Where a SIM is used, it can be specified as the destination for such requests where trigger contention is an issue. The SIM would then mediate between multiple SCPs to get the request processed and responses from these different SCPs are then collated and combined at the SIM into a single response that is then forwarded back to the originating entity[2].

Since the fan-out is handled at the SIM and not a 'gateway' service, service degradation from a functional standpoint for any one service (that would otherwise have to act itself as a gateway) is not observed. Also, the SIM could be deployed into any network where there is feature interaction (obviously with some customization or re-development done in each case to handle the specifics of each interaction to be addressed), thereby becoming a re-usable component. The SIM is an entity dedicated to resolving trigger contention, so costs associated with it will have to be borne. However, a good SIM would be extensible, and when new services are deployed, if additional contention results, the SIM could perhaps be modified to handle those cases as well at lesser cost.

12.4.3.1 SIM Deployment Configurations

Figure 12.1 depicts four service flows. Part (a) illustrates the simplest case – namely that of direct SSP-SCP interactions. Part (b) depicts the role of the SIM in mediating service request interactions between multiple SCP elements. The SMG may be viewed as an SCP peer, or may itself be used to support SIM functions as well – Part (c) depicts the SMG as an SCP peer while Part (d) indicates how the SMG provides not just SCP-like functionality, but also the SIM capability, and incorporates logic that indicates how the SCP returned responses are to be factored into the single response that is sent back to the original requesting entity.

In theory, any element playing the role of an SCF from the IN DFP (e.g. the SCP or its peer the SMG) can act as a SIM. However, the SMG can effectively take on the SIM role rather more easily, since it is a network element that not only supports policy-based processing of requests, but also supports a whole series of external interfaces thereby enabling it to interact with other network elements as required.

Does this mean supporting SIM capabilities is an essential requirement of all Parlay/OSA Service Mediation Gateways? Definitely not. In fact, there may be advantages in some network environments to having a separate SIM entity completely disjoint from the SMG. As remarked earlier, feature interactions and trigger contention issues exist in networks today. Solutions already in place (which may include one or more of those discussed in this chapter) should not be disturbed as an SMG is deployed. Telecommunications equipment vendors would do well to factor in the needs of individual networks as they design or configure their products to operate in particular environments.

Thus far, we have discussed how the message routing to and from the services needs to happen. Where a SIM is used, request message forking and response collation may be done by specifying a routing key for the SIM service (hosted on the SMG or to the separate network element hosting the

---

[2] Here, we introduce two terms – SIM and SCM. In sections that follow, we use the two terms interchangeably to mean the same thing.

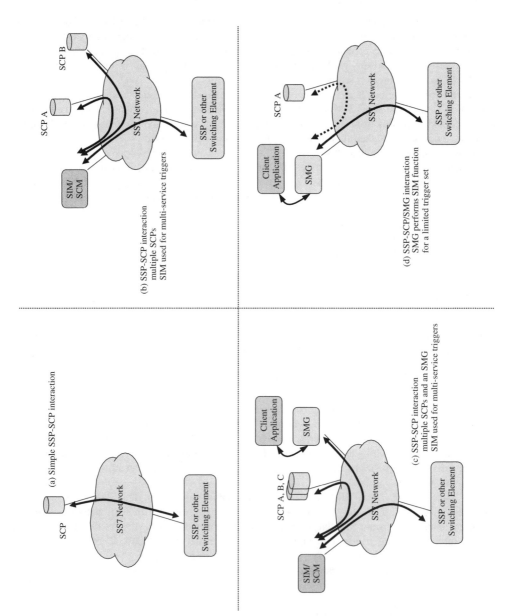

Figure 12.1 SIM deployment configurations

SIM capability). Switching elements that require support for multi-service triggers are provisioned to use that routing key and the SIM destination point code.

12.4.3.2 Support for Global Title Translation (GTT)

In order for a network element effectively to play the role of a SIM, it needs to be able to act as a proxy between an SSP and an SCP. It needs to do so in a manner that is transparent to that SSP and SCP (recall that the SSP and SCP are legacy elements). This objective can be achieved by making use of the Global Title translation mechanisms.

Global Title Translation (GTT) is a procedure whereby the destination point code (identifying the signaling point), and subsystem number are not explicitly specified by the entity originating the request message, but are derived based on the digits and other information encoded as message parameters and translation tables.

The Signaling Transfer Points (STPs) in the network of SS7 nodes provide this Global Title Translation (GTT) function. Therefore the originating entity need not necessarily know the specific destination node to which a particular request message is to be routed. Only the STPs need to maintain tables that associate particular services with specific point codes and subsystem numbers of possible destinations.

GTT mechanisms may also therefore be used to route messages that originate at various switching elements in the network to the SIM. These mechanisms may also be advantageously used when the SIM forwards messages on to other IN SCF (Service Control Function from the IN DFP, not Service Capability Feature from Parlay) entities hosted by the network. Where the SIM is the SMG, it can also mediate between IN SCFs and Parlay SCFs or Applications, the former through SS7 mechanisms, the latter through implementation differentiators built into the SMG product.

The details of the inter-application interference from a Parlay application perspective are addressed in Chapter 14 where we introduce the idea of the Parlay Proxy Manager (PPM). However, feature interference between Parlay SCFs merits some discussion, and we look at this in a later section.

12.4.3.3 Complicating Factors

In the preceding sections, we have discussed some of the issues with feature interactions and trigger contention, and have looked at some possible solution techniques to mitigate, and resolve the problem, or to avoid it completely. From the foregoing, it appears the SIM solution is perhaps the most flexible and versatile, but we would be lax if we were not also to point out some of the other issues related to SIM development. Knowledge of these issues does not take away from the niceness of the SIM solution, it only serves to highlight some of the difficulties with handling feature interference, and why a general solution to this problem is not yet prevalent. The devil is, after all, in the details.

As we have seen in Chapter 1, different service control protocols are used in different networks. These are typically some variants of INAP over TCAP for SS7. However, these protocols vary in some important respects based on network context (CDMA, GSM or other network type). In some network contexts, each exchange between the switch and the SCP is a separate transaction, while in others a single transaction is supported whose lifetime coincides with the duration of the call. Since trigger contention can occur each time a trigger fires at the switch element, depending on the services supported in each network context, these considerations of transaction duration need to be factored into the design of the SIM.

For instance, in networks where the transaction is long lived, the SIM may have to proxy only some messages associated with the transaction (for triggers where there is contention) to other network elements, while not others. Or it may have to proxy forward all messages associated with the transaction to all services, but only selectively process responses from these services based on where they are best able to provide instructions that can affect further call processing. This may depend on factors such as network context, the service control protocol in use, how the services themselves are implemented, etc.

Last but not least, there may be cases where service providers wish to make services developed for one network, accessible to switching elements in another, thereby employing the SIM as some kind of protocol converter proxy. These kinds of situations pose their own unique problems and the relative characteristics of the two network environments need to be factored in carefully. The reader is referred to [Vemuri 2000] for a more in-depth treatment of this subject.

Note: The IMS standards (recall mention of the IP Multimedia Subsystem from Chapter 1?) also refer to a component similar in scope to the SIM discussed earlier in the chapter, called a SCIM (Service Capability Interaction Manager) that mediates event distribution and service control between Application Servers in the IMS harmonized network architecture. A description of this component and related details are discussed in Chapter 1, which describes the important aspects of IMS.

12.5   Service Level Feature Interactions

Thus far, in this chapter, we have studied how feature interactions occur and are handled from a network perspective. Let us now look at some related considerations from a services angle, and how these may be addressed.

Service level feature interactions can occur in service mediation gateway scenarios from two perspectives – at the Inter-Service level (between two services), and the Intra-Service level (between multiple instances of the same service). These, and their associated considerations for management, are addressed in the sections that follow.

12.5.1 Inter-Service (Parlay) Contention

Someone once said, 'In theory, there is no difference between theory and practice, in practice, there is'. There may be situations in real-world deployments where there are overlaps between different Parlay SCFs where multiple distinct services need to be notified of the same network event. Let us illustrate this by means of an example.

The User Status (US) interface of the Mobility SCF could be implemented as a separate US SCS, as we have seen in Chapter 9. The US SCS may be required, in a certain deployment, to report the changes in user status based on when a subscriber goes off-hook and makes a call, and again when he goes on hook or disconnects. Such call-trigger tracking may be useful, for example, in reporting subscriber presence information.

Now, the Call Control (CC) SCS, or even legacy call control applications may want the same network events in order to provide call processing related support. In many cases, the switching element or other network component can only issue one message each time a network event occurs. The question then becomes, which element should get this message? Does US have priority or does CC/Legacy IN?

Typically, such situations can be cleanly addressed by noting that only one element can have control of the call at any one time. CC/Legacy IN applications are normally[3] required to provide some kind of response back to the switch on such messages, while US merely needs to be notified of this event. So CC/Legacy IN takes precedence and the SIM can be programmed appropriately

---

[3] Please note that IN triggers are of different types – there are R or request triggers, and N or notification type triggers. The former require the SCP to issue a response back, and call processing on the switch halts till an SCP-generated response is received. The latter are simply SCP notifications, and call processing proceeds while the SCP is notified of the event from within the call-processing context. For more, the interested reader is referred to [Faynberg 1996].

to route the messages and notifications of this network event to the various parties in keeping with this knowledge. Of course, there may be cases where multiple applications tied to the CC SCS may want access to the same trigger. Issues of this latter nature are addressed in Section 12.5.2 on 'Intra-Service (Parlay) Feature Interactions'.

12.5.1.1 Managing Service Clones

In Chapter 9, we discuss architectures that implement multiple cloned SCSs. A question might arise here: how does one handle event distribution across clones? If a network event is received at clone #1 of the service, and clone #3 is the only one that has an interested client application attached, what is one to do? This is indeed an interesting question and needs to be handled by means of communication between the clones.

The quality of an SMG implementation depends on many factors – the degree of coverage of the API, standards-compliant behavior, ease of interoperability with different application types, support for useful, efficient and compatible API mappings with underlying network protocols and so on. A key element that defines quality in multiple cloned SCS implementations is the degree of seamlessness the gateway offers to applications from a services perspective. The more the clones are able to communicate and appear as a cohesive whole that provides a service, rather than as fragmented servers of functionality, the more useful the gateway would appear to be from an application perspective.

While saying this, we do note that some aspects simply cannot be made seamless. Service sessions are to individual clones in such architectures and not to the cohesive set of all SCSs of that service type. However, looking from the underlying network up towards the service layer, it is highly advisable that a single network event sent to one of the clones should be made available to an interested Parlay application connected to another clone of the same type.

12.5.2 Intra-Service (Parlay) Feature Interactions

In the previous section, we have studied how feature interactions can occur between different services or SCSs supported by the SMG component deployed in a service provider network. Here, we shall examine how, if at all, there could be interference between different elements at the services layer and above. For this, we need to look at two things:

1. Inter-Application Interference, and
2. Multi-clone Architectures.

12.5.2.1 Inter-Application Interference

Events that are reported to client applications may be classified into two types just like triggers in call models can be of two types (please see Appendix A for details pertaining to 'N' and 'R' type detection points [Parlay@Wiley]). The type of the trigger defines the exclusivity characteristics of the associated event. Let us illustrate by means of an example.

Applications A and B both use, among other SCSs, the CC SCS for call control operations (call control is the best example since one can most easily draw parallels between this and the discussion of traditional call models). Per the API specifications, A and B can define a set of criteria for events to be reported to them from call contexts. This reporting can be done in one of two modes, called 'monitor modes' – 'INTERRUPT' and 'NOTIFY'. These are explained in Chapter 6, in the section on Call Control, but let us look at them again briefly.

In the former ('INTERRUPT'), the client application requests that it be notified of the event and that it retain control of the call context as it may wish to alter the outcome. Call processing on the switch is put on hold till the application responds. At any point in the life-cycle of the call, the application can relinquish control by simply invoking the deassignCall() method.

In the latter ('NOTIFY'), the client application merely requests to be notified of the occurrence of the event even though it does not have, or indeed require, any control or influence over the further processing of the call.

It will be immediately apparent to the astute reader that the event being reported has different exclusivity characteristics in the two cases. There can only be one entity that controls a call at any given time to eliminate contention. The standards specification today explicitly forbids more than one application from subscribing to a given event in the INTERRUPT mode, though potentially many applications could subscribe to the same event in the NOTIFY mode at the same time[4]. This is called 'overlapped criteria handling'.

Thus, event contention at the intra-service (within the CC service in our example) is eliminated by defining rules that do not permit the condition to exist in the first place. Stated another way, overlapped criteria conditions are eliminated by the standards by a careful definition of what is permitted by the specifications.

12.5.2.2 Multi-clone Architectures

We have studied in Chapter 9 how the SMG could be constructed of multiple SCS clones each providing the same type of service. Each clone contributes towards the total required capacity. Each clone independently registers and announces itself, can be upgraded independently, and can be put into and taken out of service independently of its peers. Each clone supports its own set of service sessions to client applications directly connected to it.

The question then naturally arises, if the clones behave independently of each other, but provide the same (say CC) service, how is the event exclusivity model implemented across the set of clones as a whole? After all, if Applications A, B and C are connected to clone X, and Applications D and E to clone Y, what prevents A and D from subscribing to the same event with the same criteria? How is the criteria overlap avoided here?

Obviously, there needs to be some kind of solution, for otherwise, we have the same problem the standards have so carefully eliminated from a conceptual perspective. We need to find a means to support the standards-defined solution from an implementation viewpoint. This is quite simply achieved through a sharing of event subscription, the events themselves, and associated reporting responsibilities across the various clones within the set. Some kind of inter-clone communication mechanism needs to be defined for this, along with techniques to manage race conditions, failure cases and other such scenarios.

Let us look at a couple of examples of the kinds of issues a solution here needs to address:

1. *Race conditions*: Applications A and B both subscribe to an event X in INTERRUPT mode at nearly the same time, A at clone P and B at clone Q. The clones need to communicate as each INTERRUPT mode subscription request comes in, to ensure that only one of A or B's subscription requests is accepted, and the other rejected. The behavior across the set of clones needs to be deterministic and repeatable. There must be a mechanism to 'break ties' even if two applications were to make their requests at the exact same time. And it would appear that since time considerations factor into the equation, some kind of clock synchronization mechanism (e.g. NTP – the IETF defined Network Time Protocol [RFC 1305]) is put in place between the various clones providing the service.

2. *Failure cases*: Different kinds of failure cases may occur. For example, a clone may die suddenly, all subscriptions associated with clients connected to it may be lost, and need be recovered (no point locking out applications from accessing those events when there are no other conflicting

---

[4] In fact, a combination of the two conditions is also possible, so long as there is only one application subscribed to the event of interest in INTERRUPT mode at any given time. Note that although call control is the example used here, other SCSs like DSC (Data Session Control) also support the same type of event exclusivity model.

applications that can receive them anyway). In fact, the applications that get disconnected when a clone terminates abruptly might want to discover, select, and connect to another clone and re-establish subscriptions for the same set of events.

Another situation that may occur is that somehow the inter-clone messaging mechanism is broken or disabled – how does one get the clones to share information on INTERRUPT mode subscriptions now? This is commonly referred to as the 'cluster-partitioning problem'.

It must be noted that these are merely examples of two kinds of issues that merit consideration. Developers will need to address these and more, and the attention each issue merits may vary significantly from one implementation context to another based on some of the design decisions.

The details of solutions used vary from one vendor's implementation to another and form the basis for (you guessed it) product differentiation. These are often closely guarded secrets, though the effectiveness of each solution can be compared and contrasted by service providers as they test different vendors' products with a test suite of different failure cases.

12.5.2.3 Advanced Parlay Scenarios

Some vendors are not satisfied with the 'elimination' techniques defined in the standards for 'overlapped criteria handling' and strive to provide greater value-add in their solutions. They achieve this by adding more sophisticated logic to the event distribution mechanism across the set of clones. This can be done by maintaining lists of interested subscribers for events in the INTERRUPT mode just like these lists are maintained for NOTIFY mode event reporting, and then assigning a priority-based ordering for events based on some criteria.

Recall that in Chapter 6, we talked about the use of the Policy Management SCS to policy-enable various applications and services. Policies could also be used to design conveniently a priority-based event distribution and communications mechanism between the various component parts of a solution.

12.6 Summary

In this chapter, we have studied a challenge pertaining to the deployment of a Service Mediation Gateway into service provider networks, namely that of trigger contention resolution and feature interactions among deployed services, and potential solutions for the same. The objective here was merely to provide the reader with a good understanding of the issues involved, not necessarily to present all the details of associated solutions. Such solutions could vary widely based on the particular needs of, or constraints imposed by, individual service providers and their respective networks.

13

Application Implementation Perspectives

13.1 Introduction

Parlay is designed around the principle that applications can be written once to utilize particular network capabilities in a completely network and protocol independent manner, and could then be deployed across different network types (technologies and generations), different gateways, different service providers, etc. with minimal or no changes. Also, as described in previous chapters, Parlay based development could allow for more rapid development and deployment of services (since the integration aspects of service deployments are centralized in the gateway, and reused over and over again, thereby saving on costs).

These factors could have great impact on the extensibility and testing of applications. Parlay also provides a rapid feedback environment for one to build, test and deploy applications, and get a relatively quick sense of whether a particular application is successful (has adequate take-rate to justify its continued existence in the network or not). Sometimes, depending on the life span of the application in question, or as a result of customer demand, it may need to be extended with added capabilities, and this is more easily accomplished in some cases than in others.

Does this technology really live up to its promises? Yes, but it does assume that although the application developers themselves need no longer possess a detailed knowledge of telecommunications networks and protocols, they do design their applications defensively to ensure that the end-user experience is invariant across core network protocol changes, Parlay gateway peculiarities, and idiosyncrasies associated with particular service provider deployments.

Applications that are not built defensively will:

a) require more 'custom' support per gateway;
b) will have a hard time interoperating with different gateways;
c) will be less 'adaptive' to missing or unsupported functionality in particular gateway implementations.

This also inherently implies that such applications will reduce the overall value or cost-savings that result from the use of Parlay technologies.

In this chapter, we study the art of defensive application design[1]. We pick up the example from Chapter 7 once more, and look at it in greater detail and examine not just the business logic

---

[1] Defensive application design and implementation, like defensive driving, is an acquired skill and something of an art. It requires that the developer build more resiliency into his implementation so that unforeseen behavior from the other end of the interface is handled 'appropriately'.

Parlay/OSA: From Standards to Reality   Musa Unmehopa, Kumar Vemuri, Andy Bennett
Copyright © 2006 Lucent Technologies Inc. All Rights Reserved

aspects, but also some 'surround' capabilities dealing with High Availability, scalability, application deployment considerations, etc. Finally, we look at some considerations for minimizing the costs of the testing of Parlay applications while trying to maximize the effectiveness of the testing.

13.2   The Theory of Defensive Applications Design

How would one go about designing an application? What are the design goals? Application developers want to be able to sell their application as widely as possible. They want to build it once and run it on any Parlay gateway across any network type within limits of reason. What ideas merit consideration as we proceed?

a) We identify a basic set of application transactions and identify the needed set of methods associated with these: Let's say $\{a.a1(), a.a3(), c.c4(), d.d2()\}$ where a, c and d are interfaces within the standards, in either one or distinct SCF APIs, and the element after the '.' is a method from within that interface.

b) Next, we construct simple call flows for the scenarios we want to support using these method calls and ensure we have the complete set that we need. We compose examples around them.

c) Once the sunny day scenarios are dealt with, we work on failure legs for those flows – failures may be either those that result from error legs during application logic execution, or as a result of exceptions thrown in the Parlay context due to unsupported or poorly supported methods in the gateway[2] or an SCS not being included in a given service provider deployment of a mediation gateway.

d) Good application design philosophy requires that we now build alternative flows that enable the application to provide an invariant user experience (or as invariant as possible, given the circumstances), even with a degraded gateway, and try hard to hold to this invariance as much as possible when the network and/or gateway do not co-operate with the application intent in specific cases. The designer therefore now considers an expanded method set to account for some of this. Updated method set: $\{a.a1(), a.a3(), a.a5(), c.c4(), c.c5(), d.d2()\}$.

e) Since invariance cannot always be guaranteed, and there are limits to how far one can plan for these cases, the application designer should also consider cases that result in degraded user experience, 'stubbing off' some aspects of operation that may not be emulated easily across different network types.

Thus, application developers would use Internet technologies when building Parlay applications, but they are still bound by telecommunication users' and their service providers' expectations of truly carrier-grade service experiences. After all, Parlay claims to alleviate the need for telecom specific technical expertise, not to get rid of requirements that are implicitly demanded of applications by every phone user.

Here is a simple model that brings out this idea more clearly with a little more rigor in the presentation.

Let us denote an application's use of the network by AU.

$$AU = fn \ (GW, Pr, Z) \tag{13.1}$$

---

[2] This happens, for example, when the Parlay gateway vendor does not implement the method and a P_METHOD_NOT_SUPPORTED exception is thrown when the application invokes it, or, the method is supported as best it can be within a network context that does not provide the required capability. As an example of the latter case, consider support for 3rd Party Call Control (3PCC) in the ANSI™ context. The ANSI service control protocol does not support this implicitly, so a work-around technique using service nodes may need to be built instead.

where GW denotes gateway capabilities, Pr denotes network protocol capabilities, and Z denotes all other variable aspects that impact the applications usage of the network – including service provider deployment peculiarities.

Now, we can define a User Experience Index or UEx as a function of AU.

$$UEx = g(AU) \tag{13.2}$$

If an application is robust (as previously explained), UEx will remain invariant regardless of how AU changes. In other words

$$UEx = g(AU) = \text{CONSTANT, or}$$

$$UEx = g\ (fn\ (GW, Pr, Z)) = \text{CONSTANT} \tag{13.3}$$

As can be seen, $fn\ (GW, Pr, Z)$ may vary widely from one application deployment to another, but the overall UEx index should remain the same because users will be unsympathetic to why the application functions differently, they will **demand** parity. Good applications will seek to minimize user visible disparities across network deployments.

In reality, UEx will vary slightly from one deployment to another for even the best-designed applications, but one can use this metric to measure application quality. The smaller the range in UEx, the better the application. Of course, if an application was written specifically for a particular service provider deployment, all bets are off if it has to be later reused in a completely different network context.

Now that we have looked at the theory behind the principles of good application design, let us look at an example.

13.3   Example Scenario Revisited

Let us go back to the example we used to explain the operation of Parlay applications in Chapter 7. There, Alfie, our end-user, visits Chicago and takes a walking tour downtown with the assistance of an application called Tourism Genie (TG) which communicates with him over the various interfaces supported by his cell-phone or mobile. In that chapter, we studied how the application may be built from a purely functional perspective – how the various SCF APIs available from the standards could be used, how the protocol mappings may be leveraged transparently for the end-to-end call flows to work, etc. Here, we look at other aspects surrounding the call flow: aspects such as availability, reliability, performance, etc. that we discussed from a service mediation gateway perspective in previous chapters can also be examined from the perspective of application design.

Let's say the Acme Computer Applications Company built the TG application for Utopia! Wireless but retained the right to resell it to other service providers per the business agreement. Now, if Acme tried to deploy this application into a different network type where 3rd Party Call Control (3PCC) primitives were not supported by the service control protocol, either custom integration into this network would be required or only a degraded end-user experience could be supported. Let us assume that the new service provider customer did not want to invest in custom integration just yet. In such a situation, Acme could still reuse the same TG software, but configure it such that when the 3PCC method call failed, it would simply present the user with the phone number of the restaurant to be called, and let the user make the call herself.

Various other shortcomings or differences from one deployment to another could be dealt with in similar ways. Even if the same set of required SCSs were deployed in different networks, there are no guarantees that the different deployments have the same reliability characteristics. SCS components in one network may fail more frequently than another – good applications will be able to deal with these failures by providing degraded user experiences, even as they try to access backup or alternative SCS clones from within the network that are capable of providing the same service. If a particular SCS is very critical to its operation, an application may go so far as to set up multiple service sessions – each one to a separate clone of an SCS of a given type – to ensure that there is adequate coverage to address overload and reliability considerations. This way, individual SCS failures and overloads are kept hidden from the end-user – there is no visible impact to their perceived quality of service.

Thus, the application can use knowledge of the service provider service mediation gateway deployment configuration, even very high-level information, to good effect. In the preceding paragraph, Acme engineers are able to utilize the knowledge that the service mediation gateway supports an 'N + K' deployment architecture to set up multiple simultaneous service sessions – each to a separate clone – to achieve the required capacity[3] and High Availability (HA) requirements from a client application perspective. Of course, if this kind of setup is permitted by the gateway implementation, the service provider will want to ensure that the right kind of service agreements are put in place for that application's use of network hosted capabilities. Note also that Acme does not necessarily have to know about the HA constructs used in the gateway implementation, but application engineers may be able to put this information to good use if they did know it.

Of all these aspects, HA merits particularly careful consideration. Different service mediation gateways may support different mechanisms for high availability in their implementations. For example, some may use fault-tolerant CORBA[4] or other fault-tolerant middleware capabilities. Others may use yet other mechanisms – does this mean Acme will have to re-write the application each time it interfaces with a different service mediation gateway in a service provider network?

Not necessarily. If the gateways themselves are standards-compliant, standards-defined HA mechanisms should be transparently available to client applications for their use. Standards documents provide the ability for a single application to provide multiple callback references so SCSs can report asynchronous responses to them through alternative means if the primary callback interface registered by the application were to fail or become unreachable for some reason. These and related concepts from the API could be used by applications to their advantage to provide the desired HA characteristics. The interested reader is referred back to sub-chapter 10-3 on Scalability and Reliability for more details.

Does this discussion imply that this absolves the server of providing a high-availability carrier-grade implementation? Absolutely not. As we said before, the intent behind building applications defensively is to account for unexpected situations that might arise in the interactions between clients and servers (perhaps even in different networks), and to provide deterministic reactions to such situations. Each end bears some responsibility, and here we are simply focusing on what a client could do to make the overall user experience more pleasant and reliable.

Let's continue on with our example. When TG is deployed in a third network, Acme engineers realize that no prepaid system was ever deployed by that service provider. All transactions are postpaid, and this customer tells Acme that although they very much like the TG application, they are not comfortable letting people make all kinds of purchases and then charge them to their telephone bill for each monthly cycle. Acme chooses to utilize the credit card recharge module (already in place to support the recharging of depleted prepaid accounts) to enable customers from this service provider networks to pay for ticket purchases and the like. Since the service provider does not support the required underlying infrastructure, the application uses an alternative means to try to provide similar capabilities.

---

[3] Application throughput in terms of the number of transactions it may be able to process per given time duration is a function of both the network and the application itself. Although Parlay/OSA enables transactions to be designed independently in the application and services domains, as is evident from our discussions on performance in earlier chapters, optimal call flows will require that optimizations be made in both components – optimized network and application transactions are best, always better than those that optimize only one but not the other.

[4] CORBA, or the Common Object Request Broker Architecture, is an object-oriented technology that is widely used in the industry today to support remote API invocations, location transparency and object access. This technology is particularly relevant to Parlay/OSA implementations. The interested reader is referred to [Henning 1999,Schmidt 2004] for more details.

We also note that some capabilities are central to the operation of a given application – the TG application, for instance, cannot operate without some form of location input to it. Loss of location capability severely reduces the value of the application and Acme may choose not to deploy this application in networks where location capabilities are not available. There are no workarounds in some situations.

13.4   Where to Deploy?

Someone once remarked insightfully, 'There is no such thing as bad weather, only inappropriate clothing'. True, Parlay gives you many options regarding the deployment of applications, but one must ensure that wise choices are made to ensure that maximum benefit accrues in each case. As described in the previous section some choices need to be made as applications are developed – these include things how HA, reliability, performance and related characteristics are met. Other decisions need to be made when an application is being deployed, and sometimes, assumptions relating to these aspects needs to be factored into application design as well. And yes, sometimes, it is the act of choosing that matters most, not the choice itself – there are cases where either of the possible choices may be equally appropriate, but it is crucial that the choices be made in time so that the application is developed, tested and ready to deploy in a timeframe that meets the ever important market window. Miss the market window and sometimes even the best application may not succeed.

Let's go back and look at some deployment time considerations. Parlay applications can be deployed either within the secure confines of the service provider network boundaries (i.e. within the service provider network) or outside, in enterprise domains with some kind of secure transport between the two domains. The choices here could be important.

Applications deployed in service provider networks may offer better round-trip time (RTT) behavior, seeing as how the applications are closer to the services they leverage and given that some security restrictions can be relaxed slightly (resulting in less overhead for service session setup). In addition, the service provider is itself responsible for the maintenance, administration and operations aspects of the application, which means that the desired HA and reliability characteristics can be engineered into the deployment following some of the same principles outlined in Chapter 10 for services supported by the SMG.

On the other hand, service providers may decide not to host applications (to save on Operating Expenses – commonly referred to as OPEX), and instead either let the application vendor themselves host it, and provide the required 'carrier-grade' aspects for the application, or farm out the application management aspects to a completely different third party charged solely with this responsibility. A business relationship can be established between the two parties based on some kind of revenue-sharing agreement per transaction processed, or based on a flat-fee. This last model of outsourced application management is typically termed a 'service bureau'.

13.5   Building the Application: Designing High-Level Logic

In Chapter 7, we studied an application scenario and then the call flows to understand how it really works under the covers. The last few sections of this chapter have discussed some considerations for application design. Here, we look at how one might go about putting this knowledge into practice, engineering in the 'surround' aspects of an application, factoring in some ideas we picked up in Chapter 10. In this section, we first start out with a simple view of what the application might look like, and see how that view morphs into one that is more robust, as other criteria associated with factors not directly related to business logic are considered one by one, to make the application more 'deployment ready'.

Some readers may consider what follows to be 'pseudo-code'. But there is pseudo-code, and then there is pseudo-code. Sometimes, pseudo-code comes very close to real code that is simply language agnostic. Readers expecting that level of detail here are likely to be disappointed. Hence the section title makes no pretence of actually providing the pseudo-code itself, but guidelines on how to go about building the application.

13.5.1 A First Cut View

Begin Application

A. Application Initialization Routine
   1. Perform variable initialization
   2. Establish connectivity to various local systems – application-based subscriber information database, databases for maps, configuration tables etc.
   3. Indicate readiness to administrator if required
B. Connect to the Service Mediation Gateway(s)
   1. Note: This section depends on deployment characteristics.
   2. If application is hosted in the service provider network, the SMG to connect to may simply be the one in that network. If it is hosted in an enterprise domain, and is shared across multiple service provider networks, it may have to connect to multiple SMGs, one from each network.
   3. Application implements code to authenticate and authorize itself with each SMG, to discover the services it needs – in the TG example from Chapter 7 this would include SCSs for Mobility (User Location), Charging, Presence and Availability Management, and Multi-Party Call control.
   4. Code to sign service agreements digitally with each SMG to which it connects for the service, to initiate service sessions with SCSs in each service provider network.
   5. Code to report errors to the application administrator per gateway, per service provider network – what services it was unable to connect to, in which environments, and whether degraded call flows can be supported in those cases. In some cases, the degraded mode of operation may have been necessitated by the SCSs deployed in that particular network, in which case the configuration file in A.2 should provide guidance on the exact flow the application should use in processing requests – this is known prior to deployment, and the application design should make allowances for it.

In other cases, the 'inability to acquire an SCS' may be a result of SCS failure in the network, and a well-designed application will tell the administrator whether it is possible for it to provide service in that network context, if so, what call flow it will use, and if not, what error it will return to end-users who make requests of it.

   6. Logic to indicate it is 'ready to go' to the administrator at the console if it is, based on checks performed thus far.
C. Listen for end-user requests and process them
   1. This is where the revenue comes in, with end-user requests. End-users may be authenticated by the application, and then authorized[5] to ensure only requests of the type they are permitted to make are serviced. This requires appropriate service logic to perform local database dips as necessary.
   2. A threading model – thread pool or something similar[6] – may be put in place to support a concurrent server for efficient processing of incoming transactions from a multitude of users.

---

[5] One could argue that the end-user be provided only with interfaces that expose only those features they are permitted to invoke, per their profiles. This ties in to usability engineering for the handset-side components or of the web-based graphical user interface the user can use. These considerations are not treated in any detail in this book.

[6] Quite a few threading libraries (e.g. POSIX) are available today, depending on the particulars of the platform and operating system (and other environment details) of the system being used for application development and hosting. Various patterns or models relating to multi-threaded servers are also extant. The reader is referred to [Hyde 1999, Robbins 2003] for more details.

This is often very convenient since Parlay, as we have learnt in Chapters 5 and 6, supports a number of asynchronous operations (some of which may be long-lived, for example, periodic or triggered location reporting requests), and it is generally more efficient not to have code that blocks frequently expected responses.

3. Leverage the threading model to process new requests. In the case of the TG application, this would involve, for each user:

   a) accessing end-user preferences from the application database and those stored in the network (common to multiple applications including TG) – the application stored preferences may be used to authenticate and authorize the user, and determine which of the subtended network contexts to use to process the requests in that user's session;

   b) setting up a periodic user location query with the network;

   c) keeping a loop open for user input via voice or dialed digits on the phone;

   d) consulting the application map database to access and download to the user handset content that may be relevant, while playing the correct voice content describing the same, offering options, factoring them into further business logic related processing;

   e) accessing the user's prepaid account and making debits to it for tickets or other purchases the user may make during his session along with charges for using the service, if any;

   f) performing lookups of phone directories in the area for restaurants etc. in the application space;

   g) utilizing the network provided Call Control capabilities to set up a call to the party or parties (e.g. restaurants) with which the user wants to communicate, and associated call management.

Some of these transactions (a), b), e), g)) are Parlay based and involve communication with the underlying gateway. Others (a), c), d), f)) are purely application business logic related: a) has aspects of both. In some networks, the call flow may differ or be degraded, as described in B.5, and may miss some steps, or add others. But the design should be clear in what flows are supported, and how the application logic will handle user interactions in each subtended network context.

13.5.2 Enhancing the First Cut

So the above is a simple logical flow of the application. But this just shows the absolute bare bones requirements for the application to be practical (admittedly, we have embellished the basic structure with some nice capabilities in step B.5 and C.3 to make it more appealing to the reader). Application developers may want to enhance this structure further to provide additional capabilities or improve the user experience through 'surround' features. These could be wholly in the application space, or also factor in some aspects of the communication with the gateway. For example:

1. Step A.3 could be modified to support different kinds of authentication handshakes depending on network type, supporting P_OSA_AUTHENTICATION or P_AUTHENTICATION and the use of different support ciphers as required, to make the application more usable across network contexts[7].

2. For service level HA, step B.4 could be enhanced to support multiple service sessions (to different clones in a multi-clone (see Chapter 9) SCS SMG deployment) between the client application and a given type of SCS in a particular service provider network.

3. Step C.3 could be modified to use more advanced primitives exposed by the API where they are available. For instance, e) could utilize the reservations capability if the prepaid system supported it, thereby holding some money from the user account for the duration of the session, enabling him to make multiple purchases of small amounts (or use this to charge for content

---

[7] If something like this is done, it must necessarily be done very carefully so that weaknesses in an authentication handshake being used with one service provider cannot be exploited by malicious interlopers to compromise other interactions. As they say 'A chain is only as strong as its weakest link'.

downloaded, depending on the billing model) and then refunding the remainder at the end of the session.

4. A whole new series of steps could be added as 'Section D' in the application logic, which provide support for application replicas for application level HA. The mechanisms in the so-called Section D, could, among other things, support check-pointing of user session state, the sharing of callback references with other replicas of the service, the ability to pick up sessions in the middle and support user requests within, in case of sudden replica failure, inter-replica communication, and other such capabilities making the application truly more robust.

5. Other optional Parlay features such as those for Integrity Management, as described in Chapter 5, including, say, heartbeat management, fault management and load management could also be added further, to a 'Section E' of the application logic, thereby enabling each application replica to monitor services, communicate with the Framework, and be monitored by the Framework (and perhaps other replicas locally).

As can be seen, once the basic application is built, enhancements can be made not just to the business logic, but also to surround capabilities from a Parlay perspective to make it more attractive to service providers that might want to deploy them. Well-designed applications strive to strike a balance between just doing things right, and doing things well... and they do this without over-whelming the user with the number of options or configuration aspects that may need to be programmed.

Applications are designed and built by engineers, but engineers would be well advised to be mindful of the fact that for an application to be successful, the logical flow of the interaction must be very intuitive to even the most naïve of users. An application that requires the user to have a PhD in order for him or her to derive maximum benefit is one that will likely fail from a market acceptance (and even marketability) perspective. Parlay gives developers great power to do things well from an application standpoint, but, as they say, 'with great power comes great responsibility'.

Some enhancements can be added over time, as the application evolves, and market needs change. Others, however, need to be factored in from the beginning or the application will be difficult to modify, or there may be a significant amount of 'throw-away work'. Sometimes, the old adage 'measure twice, cut once' does indeed hold. Thus, good application design, just like good gateway design, and all other things worthwhile, is an art form that has its beginnings in some sound engineering principles that we have tried to cover in Chapter 10 and in the earlier sections of this chapter.

Sidebar: Application Extensibility

Elevators are seldom stand-alone. Typically, they are deployed in banks. An elevator bank generally offers a single control panel where the user can press the up or the down button, and in response to that input one of the elevators eventually stops at the floor and the user can get on. Elevators are programmed to react to requests coming in for rides from various floors, and from requests coming in locally from within each car for where users want to get off. (Elevator algorithms are interesting in their own right, some computer operating systems use algorithms derived from elevator operations to handle disk seeks in computers.)

A teacher in a programming class challenged her students to write a simple algorithm for elevator operation (with a very restricted feature set), which they did. After they turned in their assignments, she asked them to think if their solutions would continue to work if their elevator was one of a bank. Of course they would not. But some solutions would adapt to support that late requirement more easily than others. It would appear that tracking some problem context information and proactively allowing for some 'feature creep' might be one way of dealing with change (or evolution) that may be unavoidable. As they say 'change is the only constant'. But design for change is only beneficial for applications that have a relatively long lifetime.

13.6   On Lower-Cost Testing of Applications across API Interfaces

(The reader is encouraged to note the title carefully – one cannot guarantee 'least cost' in all cases)

Now that we have briefly looked at client application design considerations, let us also briefly look at software testing, certification, and related issues. There are some patterns or concepts in the area of application testing across API interfaces that could perhaps be applied to reduce the time and effort involved with a view to ensuring interoperability across implementations.

Let us consider an SCS X that implements a single SCF Y (recall from Chapter 5 that an SCS is the physical manifestation of one or more SCFs or APIs defined in the standards). For the sake of simplicity in explanation, let us assume that Y supports only a single interface I, with k methods, marked $\{M1, M2, \ldots, Mk\}$. Similar arguments as those presented below apply to SCSs that implement more than one SCF each, but the focus here is on trying to uncover the patterns of interest, so some complicating factors are ignored.

Each application that uses this SCS X (i.e. invokes methods on the implemented SCF Y) would invoke a certain method or set of methods in some sequence (or in parallel, though if this is the case, due to idempotence and state management considerations, a serial schedule can be produced per transaction with the invoked methods cataloged into one or more possible strings of method invocations).

The serial schedule per invoked method would vary per transaction type (for example, at the simplest level, the User Location SCF supports three transaction types – single-shot queries, periodic queries, and triggered location queries) that was supported by the application. It could also potentially vary based on the application's reactions to responses or input received either from the end-user (EU), or from other SCSs (OS), or from other sources not known to the service mediation gateway (UkS).

So a transaction's serial schedule as seen by a service mediation gateway SCS is:

$$SCS\_X\_transaction\_view\ (for\ transaction\ x) = Fn\ (EU, OS, UkS) \qquad (13.4)$$

(we abbreviate this as $SCS\_X\_TxV(x)$)

However, one can compute the 'closure' of a transaction view by first generating the sunny day scenario serial schedule, and then branching out from that scenario at appropriate points to handle the various special cases.

For example, let us say that partner application A that uses $SCS\ X$ (among other SCSs), supports three types of application transactions, a, b, and c.

For each of these transactions, one can generate a serial schedule for the sunny day scenario, for $SCF\ Y$, but in a manner that is totally independent of the underlying protocol bindings for the SCS.

This results in something like the following:

$$SCF\_Y\_TxV(a) = [M1, M3, M5] \qquad (13.5)$$
$$SCF\_Y\_TxV(b) = [M2, M4]$$
$$SCF\_Y\_TxV(c) = [M1, M2, M6]$$

Note that these were generated from what the application expects to invoke on the SCS as it processes each type of transaction. Now, one can compute the 'closure' of each transaction sequence by determining the branching out of the 'sunny day sequence' for exception cases. For example, this might lead to something like

$$(c3)$$

$$SCF\_Y\_TxV(a) = [M1, M3, M5] \qquad (13.6)$$

$$M2, M3$$
$$(c2)$$
$$M4$$
$$(c1)$$

($c1$, $c2$ and $c3$ indicate various exception conditions that occur as the methods from the sequence are invoked. Note that there may be multiple branches from each state, though only a single one is depicted for simplicity – an example of this is where the same method could return different results based on the kinds of input it gets, the underlying protocol bindings, etc.)

From this, one can extract the 'complete' set of test cases (this assertion needs to be taken with a grain of salt – not all exception cases can be planned for in all cases, since the graph could get too complex, and sometimes the cost of testing involved – including the cost of setting up a suitable test environment and then designing and executing tests that simulate the desired behavior – does not justify the benefits; one needs to make judicious choices here):

Sunny day scenario : $[M1, M3, M5]$

Other cases : $[M1, M4]$, $[M1, M3, M2, M3]$, $[M1, M3, M5, M3, M5]$,   (13.7)

and the loop$[M1, M3, (M5, M3)^*, M5]$

In the last case, one needs to check there is a terminating condition so the loop does not cycle forever, but that is an application requirement; the SCF will continue processing each transaction it sees as an independent one so long as it does not conflict with the transaction state it sees.

The SCS itself may only implement a subset of the set of methods defined in the API for SCF Y (it may not implement $M2$ and $M5$, for example). In such a case, Equations (13.5–13.7) could be re-written with 'SCF_Y' replaced with 'SCS_X', and the sequences in (13.7) properly adjusted to factor in the missing methods. This may not necessarily be as simple as deleting the unavailable Ms from the sequence, since the application designer needs to factor in application behavior as the unsupported methods are invoked and result in errors.

If the closure in (13.6) were computed accurately, a resulting simplified graph extracted from it could be easily built to account for the missing methods from an implementation. Note that this is the step where different protocol bindings for SCF Y can be accounted for in the test schedule.

For example, if two SCSs (say from different vendors), X and X' implement the same SCF Y, but build support for different subsets of methods (due to underlying protocol capabilities), the two closures can be computed on paper per application transaction tr, as $SCS\_X\_TxV(tr)$ and $SCS\_X'\_TxV(tr)$.

The two can then be contrasted and tests run a single time for application A against a single SCS of the type that implements SCF Y (either X or X'). One can be confident that the application behavior would remain the same for all other SCSs of the same type (for SCF Y, independent of southbound protocol mappings) that implement that subset of methods, with potentially incremental testing only required in cases where additional methods were also supported, or for those methods sequences that could not be tested in their entirety from earlier SCS implementations.

For this to work, one needs to be able to configure the SCS (or equivalent simulator) to provide the set of responses that each of the real SCS implementations is likely to return. Some of these may in fact be made possible by changing the input set of variables to a method, though perhaps some other cases cannot be handled as easily that way. For cases where input variable changes can subsume other aspects of the behavior, the more complete the SCS implementation used in the test, the more confident one can be with the results across SCS tests for that SCF. Ideally, the application should be modifiable (where the SCS is not) to encode it with the equivalent of 'method Mx invocation returned a "method not supported" exception', or 'method Mx invocation returned a CORBA exception' as the case may be.

The same scheme could be applied across application transactions as a whole, subsuming the operation of multiple SCFs, through a simple extension of the concepts discussed above. However, soon one hits the problem of combinatorial explosion with unwieldy graphs, if the transactions are not reasonably simple.

We conclude that it would be advantageous if, in testing service mediation gateway compatible applications, the testing could be done once against an SCS of a given type, with reasonable assurances that it would exhibit expected behavior when deployed with another implementations of the same type of SCS (same SCF, different protocol binding) in a network. A framework, such as that described above, could help reduce costs for such interoperability testing between applications and Service Mediation Gateway implementations.

The more rigorous treatment of this testing aspect in the last section is along the lines of what was covered in the previous section, relating to how the TG application may be tested, modified, or reconfigured to adapt itself to provide the best possible end-user experience in different networks.

13.7 Summary

Anything worth doing is worth doing well. In Chapter 7, we had studied an example of a Parlay application with a view to understanding how the application works, what happens beneath the covers, and how the call flow holds together. In this chapter, we took a deeper look, building on the knowledge of the last few chapters, to understand some of the finer aspects relating to application engineering and design. Merely building functioning Parlay applications is not enough, one needs to test them to ensure they can function and be reused in network contexts that may vary in terms of their relative sophistication of support for functional capabilities. Thus, we have also explored how some of these issues could be factored in as new applications are designed. Now that we have studied how service mediation gateways and Parlay applications may be built, we cover more advanced topics in the Parlay/OSA domain in Part V of the book.

Part V

Advanced Topics and their Implementation

As we have indicated before, in common with most standards, Parlay doesn't mandate a particular deployment architecture. This leaves implementers free to exercise their creativity in building systems that meet a particular set of requirements. These may put a priority on lowest cost and highest performance, or some other set of conflicting criteria. Each set of criteria can lead the implementer towards a particular implementation architecture. Part V illustrates this aspect by describing some of these architectures.

We start out with a discussion of a new Parlay architecture entity called the Parlay Proxy Manager. Then we study federation and multi-network operator scenarios. Chapter 16 focuses on Parlay and related Web Services technologies. An additional chapter is included on the web site [Parlay@Wiley] as advanced reading, and covers future evolution paths for the standards.

This part of the book will be of greatest interest to technical readers (such as architects and designers) who want to learn more not just about the technical aspects of today's new solutions but also about their possible evolution into the future. The discussion of federation and multi-network scenarios will also be of interest to business and marketing audiences.

Parlay/OSA: From Standards to Reality   Musa Unmehopa, Kumar Vemuri, Andy Bennett
Copyright © 2006 Lucent Technologies Inc. All Rights Reserved

Part V

Advanced Logics and their Implementation

14

The Parlay Proxy Manager

14.1 Introduction

Chapter 9 introduced the concept of the Parlay Proxy Manager[1]. As we have seen it is an entity that looks – to the 'outside world' – as if it is a single Parlay SCS. This simple, single appearance hides a complex underbelly, however, as we shall now explain.

If a Client Application needs a highly reliable and flexible location information provided to it by a Parlay Gateway, it may be necessary for the gateway to provide a number of individual Location SCSs. The Client Application can then distribute its requests across these SCSs and switch to using the remaining SCSs if one of them fails. Figure 14.1 illustrates this scenario.

Clearly the Client Application bears a great deal of responsibility for managing its relationships with the individual SCSs and is required to perform the load balancing itself. In an ideal world it would not need to do this. Step forward the Parlay Proxy Manager.

The concept of the Parlay Proxy Manager is fairly simple, as illustrated in Figure 14.2. From the point of view of Client Applications it seems to be a Parlay SCS like any other. The Client App can discover it and start sessions with it as it does with any other SCS. The Client App doesn't see, however, that the PPM doesn't typically provide any SCS functionality directly but instead is in communication with a set of SCSs of the same type. This communication takes the form of standard Parlay interactions between a client application (the PPM) and a set of SCSs. In other words the PPM has two roles; it is both an SCS and a client application.

For example, the PPM may appear to be a Location SCS. Although it has no intrinsic Location SCS functionality of its own (it doesn't interface directly to a network element such as a GMLC) it is able to behave as a Location SCS by proxying the functionality provided by other Location SCSs.

So why bother? After all, this introduces an extra layer of software (and hardware to run it on) and thus introduces something else that can go wrong and increases the delay in processing messages from, or to, a Client Application. The answer is complexity, or rather the hiding of complexity. A key tenet of Parlay is that it should be possible to make life as simple as possible for Client Applications whilst still delivering the functionality or performance they require. As we shall see the PPM concept is one way of keeping life simple for the Application while at the same time improving the performance and reliability of the functionality delivered to it.

---

[1] Note that the Parlay Proxy Manager is not an entity defined by Parlay.

Parlay/OSA: From Standards to Reality   Musa Unmehopa, Kumar Vemuri, Andy Bennett

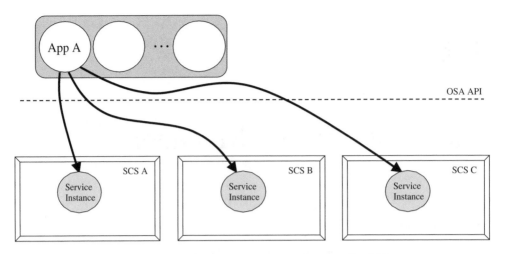

Figure 14.1 Client Application accessing three Location SCSs

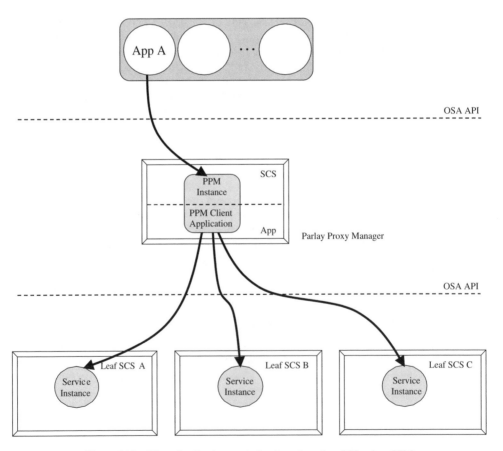

Figure 14.2 Client Application accessing three Location SCSs via a PPM

14.2   Prising Open the Parlay Proxy Manager

Cloned SCSs were introduced in Chapter 9. They are SCSs of the same type and can be used to provide load-balancing functionality, for example. They are a useful solution but have a number of drawbacks, principal among these is that the Client Application is responsible for managing all of its relationships with these multiple SCSs. The result is a complex Client Application. Chapter 9 went on to introduce the idea that this problem of Client Application complexity can be solved by hiding the SCSs behind a single SCS (the PPM).

Assume we have a set of SCSs of a given type distributed across a set of nodes within a Parlay Gateway. We call these the leaf SCSs. The leaf SCSs do not necessarily have to be homogeneous, i.e. the set can consist of any combination of monolithic and distributed SCSs, as long as they are of the same type: the leaf SCSs must have the same service type and the same set of service properties (though we see later that there can be some exceptions to this). For example, the leaf SCSs might all be MultiParty Call Control SCSs supporting a maximum of three call legs within a single call. The Parlay Proxy Manager SCS is identical in this respect to any of the leaf SCSs, though one additional service property denotes it to be a PPM SCS. For now we will assume that the PPM SCS and all leaf SCSs are trusted entities from the point of view of the Framework to avoid over-complicating the discussion.

At system initialization time let us assume that all SCSs, including the PPM SCS and the leaf SCSs, register with the Framework. Again, this keeps our initial description simple but we will examine more dynamic situations later. In addition to the standard set-up (registration) the PPM SCS enters into a service discovery sequence with each of the leaf SCSs. The sequence includes service discovery, service selection, and the creation of service sessions between the PPM SCS and the discovered leaf SCSs, i.e. service instances are created on each of the leaf SCSs. The PPM SCS now knows of a population of suitable leaf SCSs and has a relationship (service session) with each of them.

As we noted earlier the PPM SCS behaves as a Parlay Client Application towards the leaf SCSs. The Framework will grant the service sessions, provided all policies can be successfully applied. The fact that this particular Parlay Client Application is in fact an SCS as well, engaged in a service session of its own with the original Parlay Client Application, is completely transparent to the leaf SCSs. At this stage, service sessions are in place between the Parlay Proxy Manager SCS and each of the subtended leaf SCSs. Figure 14.2 shows these service sessions as lines between the PPM and the Service Instances it is using (one per leaf SCS).

The Parlay Client Application (the one offering a service to the end-user, not the one that the PPM SCS is performing towards the leaf SCSs) then sets up its service session with the PPM SCS, using the reference to the service factory of the PPM SCS it has obtained from the Framework[2]. The PPM SCS can now, based on some local policies or load-sharing algorithm, farm out service requests to the various subtended leaf SCS nodes with which it has previously set up service sessions. The fact that it is the leaf SCS that is executing the service request towards the network entities is again completely transparent to the Parlay Client Application. The Parlay Proxy Manager SCS effectively functions as a 'man in the middle'.

The Multiple Clone SCS configuration, introduced in Chapter 9, presents a simple and elegant architectural solution. It achieves both the distribution of SCSs as well as Integrity Management capabilities at the granularity of individual service requests. Through application of the proxy pattern, the configuration is not only powerful but also provides a seamless extension of Parlay compliant implementations. Due to the achieved transparency, no modifications to the standards-defined interfaces are required in order to support multiple clone SCSs.

---

[2] It can be a Framework policy only to expose Parlay Proxy Manager SCSs during the service discovery phase and thus only return service factory references of Parlay Proxy Manager SCSs to client applications. Alternatively, the Framework may expose PPM SCSs and leaf SCSs alike. The case where a client application selects a leaf node for the establishment of a service session, rather than a PPM SCS, is the degenerate case; i.e. it defaults to the standard Parlay scenario as introduced in Chapter 5.

14.2.1 Discussing the Merits of the Parlay Proxy Manager

As always, there are of course some drawbacks as well. In what follows we will describe some of the more important ones.

14.2.1.1 A Single Point of Failure?

The first question that would spring to mind is what would happen if the primary node[3] or proxy node fails? It is a fair question. We claim increased service availability through distribution and redundancy, but at the same time seem to introduce a single point of failure.

In the case of distributed SCS configuration, presented in Chapter 9, where only the primary node hosts a service factory, we have a problem. Ongoing service sessions can continue, as the primary node is not involved anymore once the service session is in place between the Parlay Client Application and the secondary SCS. New requests for service session establishment can no longer be honored though, as the only node hosting a service factory is down. This still provides a performance improvement over the monolithic SCS configuration, where failure of the monolithic SCS means that the service session(s) with the application are lost and all bets are off.

In the multiple cloned SCS configuration both the PPM SCS as well as the leaf SCSs host service factories. Therefore, failure of the PPM SCS does not impact the ability of the application to engage in new service sessions. Ongoing service sessions though are lost, of course, as PPM SCS failure implies that the service session between application and PPM SCS, as well as the service sessions between PPM SCS and leaf SCSs, are dropped. The Framework needs to supply the application with references to service factories on the leaf SCSs in order for new service sessions to be set up. If at service discovery time the Framework had only exposed PPM SCSs towards the application, the recovery scenario gets a bit more involved. In that case the entire Parlay handshake may be required to engage in service level agreements before the Framework will provide the service factory references.

14.2.1.2 What's Your Policy?

As we have seen before, all SCSs have a set of policies (whether implicit or explicit) that guide their behavior depending on the Client Application using them. This is true of each of the leaf SCSs and, for appropriate policies to be applied, the SCS needs to be able to differentiate between the Client Applications interacting with it. Remember though that when the PPM is being used it is the PPM that is the Client Application as far as the leaf SCS is concerned.

This could present a problem. In the scenario that we described earlier, the PPM discovers and starts sessions with each of the leaf SCSs at system initialization. Clearly at that point it cannot (necessarily) know which Client Applications will come along at a later point and so the PPM supplies its own AppID when starting these service sessions. Effectively all interactions from Client Applications will be treated exactly the same by the leaf SCSs since there is only one Application (the PPM) interacting with them.

As an illustration (Figure 14.3), consider two Client Applications: A and B. They both want to use the same leaf SCS. At system start the PPM will have started a session with the leaf SCS and so when the two Client Applications start using the PPM they are in general forced to abide by the same policies and Service Level Agreements as each other since only one session with the leaf SCS is in existence.

Note that there was an 'in general' inserted in the last sentence. That is because it isn't entirely true that they have to be bound by exactly the same policies and SLAs. It would be possible to construct a PPM such that it would perform policy enforcement based on whether, in our example,

---

[3] This term has been introduced in Chapter 9.

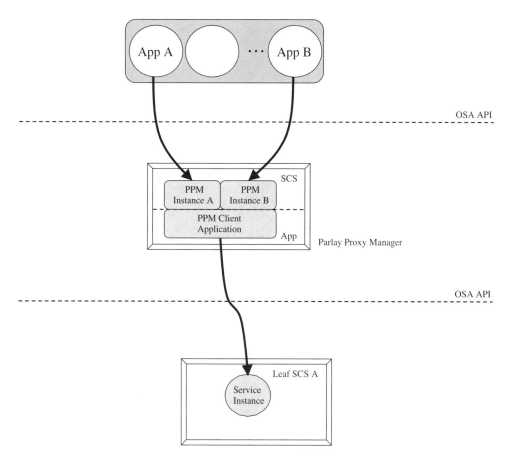

Figure 14.3   Two Client Applications; PPM acts as a single Client Application

Client Application A or B is making a request. For example, if the PPM has an agreement with the leaf SCS that it can perform up to 50 transactions per second, it could establish policies with the Client Applications such that A can perform up to 30 transactions per second and B can perform up to 20 transactions per second. There is one big drawback to this approach, however. Since the leaf SCS doesn't know that the PPM has performed policy enforcement already, it will perform it, wastefully, for the second time.

To avoid either the restriction that identical policies are to be applied or the performance (and complexity) disadvantage that the PPM has to perform policy enforcement, another approach needs to be found. Fortunately this isn't too hard to find. All that needs to happen is that the PPM can pretend to be more than one Client Application. One way for it to do this, as depicted in Figure 14.4, is to assume the identities of each of the Client Applications (literally, since it uses the AppIDs) when interacting with the leaf SCSs. This solution does however require that the procedures followed on system initialization be changed.

Rather than the PPM starting service sessions with each of the leaf SCSs on system initialization, it will wait until a Client Application interacts with it to start a service session. At this point it knows the AppID and uses this when starting a service session with one of the leaf SCSs.

In fact the PPM can start sessions with the leaf SCSs regardless of whether the PPM is going to use multiple AppIDs or not. This is described in more detail in Section 14.4.1, 'Dynamic Leaves'.

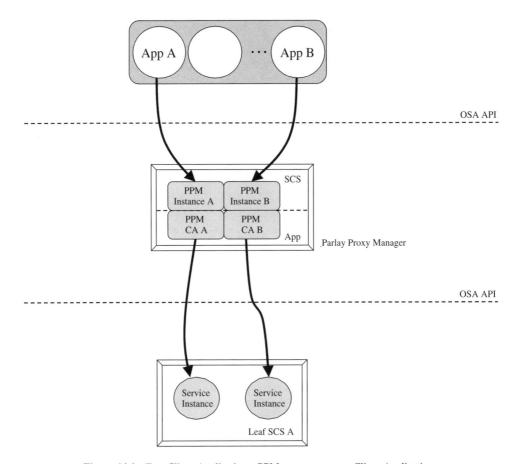

Figure 14.4   Two Client Applications; PPM acts as separate Client Applications

14.2.1.3 Lack of Integrity?

The preceding discussions have avoided discussing the important issue of Integrity Management. In the simple PPM scenario the PPM will start using the Integrity Management interfaces when it begins a session with the leaf SCS. Load and Fault information will be reported by the leaf SCS and will reach the PPM via the Framework. Similarly, Load and Fault information will be reported by the PPM and will reach the leaf SCS via the Framework. When the Client Applications (A and B, from above) start to use the PPM they will also start using the Integrity Management interfaces, but this time the flow will be from PPM to Framework to Client Application (and vice versa).

So what happens when the leaf SCS goes into overload? It may be that this information will be sent to the PPM and the PPM will send it to the Client Applications. However, that assumes that there is only one leaf SCS, or that only one leaf SCS can be used by the PPM to fulfill requests from the Client Applications. Such an arrangement wouldn't be a particularly good use of the architecture. It is more likely that a number of leaf SCSs can be used and that the report of a particular leaf SCS going into overload wouldn't necessarily result in the PPM having to report that it has gone into overload.

In general, therefore, a Client Application's interactions with the PPM will be distributed to one or more of the leaf SCSs. The load or fault information from one of these leaf SCSs will be reported to the PPM and may be used by the PPM when making decisions but in general won't translate directly to load or fault information reported to the Client Application.

So does this also apply to the dynamic mode of operation? As we have seen there are a number of ways in which the dynamic mode can be used. In the example above the PPM assumed the identities of the Client Applications. As far as the leaf SCS is concerned the PPM is the Client Application. When the leaf SCS reports load and fault information this will be passed to the Framework as usual. Now the report from the leaf SCS doesn't indicate which Client Application this information is destined for. It is up to the Framework to make the association, and it does this by keeping track of which Application started the service session. In our case it is the PPM that started the service session and not the Client Applications so it is the PPM that receives the Integrity Management information.

We can extend this argument to the use of the dynamic mode in general. Since the PPM is the entity that started the service sessions with the leaf SCSs, it is the PPM that receives the Integrity Management information, regardless of what AppID it uses to identify itself to the leaf SCSs.

14.3 Applications of the Parlay Proxy Manager

In this section, we discuss some examples of how the Parlay Proxy Manager could be used in real world deployments. Some of the more advanced ideas merit their own chapter and are covered in Chapter 15, so here we present only the highest level conceptual description with a forward pointer to the later chapter where the details are presented.

14.3.1 Crossing Continents

There are a number of network operators (particularly mobile network operators) that own networks in more than one country. This situation lends itself to economies of scale and the potential for reusing successful ideas across the different networks that an operator owns.

Parlay Gateways can be used in each of these networks to allow new Services to be rolled out in each. If one of these Services becomes particularly popular there is the possibility to deploy it quickly and easily on one or more of the other networks by having it installed on the Parlay systems in those networks. This is the basic advantage of Parlay as technology. The Parlay Proxy Manager (PPM) concept takes this a step further. If the operator group wants to deploy a Service on more than one network, it can do so very easily by deploying a PPM that sits between the Service (Client Application) and the SCSs that it wants to use in each of the networks, as illustrated in Figure 14.5.

14.3.2 Premium Blend

Extending the scenario from the previous section even further, in some cases, multi-network deployments may be such that certain SCSs are run in some networks, but the operator wants to make these available to other network operators (for a fee, of course, or in keeping with reciprocal agreements), so that those operators can in turn expose these services to their own client applications, thereby enabling every party to become part of the end-to-end value chain (and thereby make money). The Parlay Proxy Manager is indeed quite well suited to this kind of situation. An example of this kind of deployment – 'Proxy-regulated Window-to-the-World Model' – is illustrated in Section 15.4 of Chapter 15.

Alternatively, in some cases, a Mobile Virtual Network Operator (MVNO) (recall the discussion of these in Chapter 2?) may be permitted to access the service capabilities exposed via the PPM by the service provider hosting them, thereby supporting new business models for service usage, selling services 'wholesale' to parties that may broker them further to their own clients. This scenario is a variant of the one previously referenced in this section, and is also described in Chapter 15.

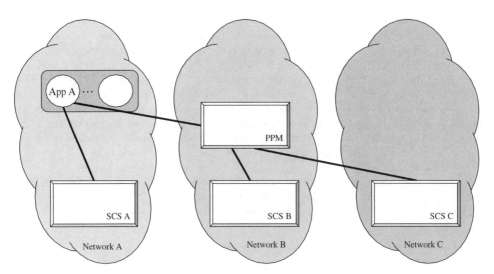

Figure 14.5 Using PPM to deploy a Client Application on other networks

14.4 Taking the Proxy Model Even Further

Naturally the proxy design pattern can be applied to other elements of the Parlay architecture. Some of these are explored in the sections that follow. First though we return to the PPM concept that we met earlier and describe an enhancement that makes it even more flexible.

14.4.1 Dynamic Leaves

The simple PPM-based Parlay implementation described earlier involves the PPM discovering and starting service sessions with leaf SCSs when the system is initialized. In effect the configuration and behavior (policies to be applied) of such a system is fixed at initialization.

This fixed configuration makes it impossible to gain full advantage from a PPM-based architecture. In the real world we would like to be able to replace leaf SCSs that have failed or have been withdrawn. We would also like to be able to increase or decrease the overall capacity of the system, increasing or decreasing leaf SCSs as needed. This is all possible with some alterations in the behavior (and if necessary the design) of the PPM.

As well as starting sessions at system initialization, the PPM starts sessions with new leaf SCSs during its lifetime. This process may be triggered by certain conditions (the failure of an existing leaf SCS perhaps, or the announcement of the existence of a new leaf SCS) or may be a regular activity running at some appropriate interval. Existing Parlay mechanisms (Framework Event Notifications for example) can provide all of these triggers and functionality.

If the analogy can be pushed a bit further, the picture that emerges is of a system resembling a coniferous tree. Leaves (needles) are constantly lost and replaced and extra leaves added when times are good. The tree itself always looks much the same despite this constant growth and loss and it adapts to the prevailing conditions. (It should be noted that designing a system modeled after an oak or beech is probably less of a good idea.)

Having explored the nature and usefulness of the proxy model as applied to SCSs, it is only natural to consider whether this model can be applied to the other key Parlay entities. The next two sections do just that, starting with the Framework.

14.4.2 Framework Proxy

We have explored how applying the proxy model to the Parlay SCS can bring benefits in terms of reliability, scalability and Client Application simplicity. Could we also gain similar benefits in

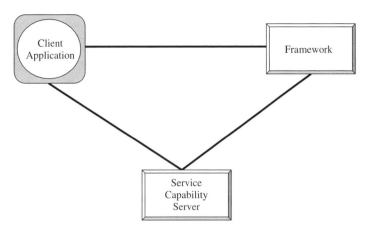

Figure 14.6 The Parlay triangle

proxying the Framework? To determine this we first need to understand what a Framework Proxy is. Consider the Parlay triangle of entities: Client Application, SCS and Framework (Figure 14.6).

Now when we proxy one of these roles, the proxy must implement two roles. We have seen that the PPM (an SCS proxy) takes on the role of both an SCS and a Client Application. Similarly a Framework Proxy must take on the role of both a Framework and of one of the other two entities. In other words it must also be an SCS or Client Application. Figure 14.7 shows a Framework Proxy that is a Framework and an Application.

Figure 14.8 shows a Framework Proxy that is a Framework and an SCS.

Note that by extension the last two diagrams could be combined as shown in Figure 14.9. Note that in each case the service session always runs between the Client Application and the SCS.

So much for the theory. Do any of these configurations make sense? Are they useful for anything?

The purpose of a proxy of any kind is to 'hide' some complexity from the entity using it. The interfaces the Framework Proxy provides can only be standard Framework interfaces. We therefore need to examine the functionality provided by a Framework and then consider how the life of an Application or SCS might be made easier by using a Framework Proxy.

Let's consider things from a Client Application's perspective first. Beginning at the start, a Client Application needs to find a Framework and authenticate with it. If the owner of the Client

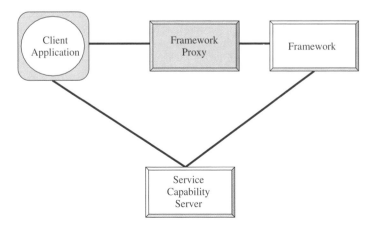

Figure 14.7 Framework Proxy as Framework and Client Application

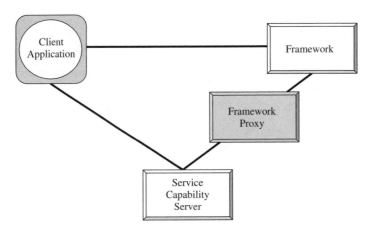

Figure 14.8   Framework Proxy as Framework and SCS

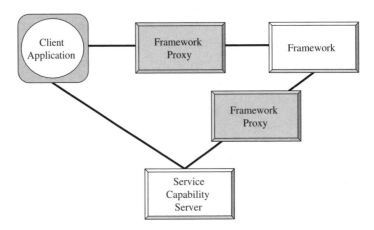

Figure 14.9   Framework Proxies towards both Client Application as well as SCS

Application provides a Framework Proxy then it is this proxy that is found (and in fact may well be provisioned into the Application software). A real Framework is still required and so the Framework Proxy is responsible for finding one, using any of the possible discovery methods. Not a great benefit to the Client Application on its own, so we need to look at what else might be provided.

The next thing the Client Application needs to do is authenticate. If the Framework Proxy is in the same security domain as the Client Application this authentication can be quite simplistic and so the Client Application isn't obliged to implement some of the more complex algorithms. Instead the Framework Proxy implements any security algorithms required by the real Framework(s). This provides a clear benefit to the Client Application.

The remaining functionality of the Framework can be broadly split into finding a suitable SCS, starting a service session with it and maintaining this service session using the Integrity Management interfaces. Let's consider discovery of SCSs first.

There isn't much in the Discovery interfaces in themselves that can usefully be simplified by a proxy. However, let's think about the purpose of Discovery: to find SCSs. The wider the available set of SCSs the more likely it is that a good match can be found. A Framework Proxy could expand the available set by making contact and setting up an Access Session with more than one other Framework. The Client Application is of course blissfully unaware of them, and the Framework Proxy takes each of the discovery requests and sends them to the other Frameworks. This is depicted in Figure 14.10.

In summary then, a Framework Proxy can be used to hide functional complexities such as Framework 'discovery' methods and security algorithms or it can be used to hide topographical complexities such as the existence of multiple Frameworks.

14.4.3 Application Proxy

By concentrating on the SCS and Framework entities, our exploration of the proxy model has to a certain extent ignored the needs of the owner(s) of the Client Applications. Of course, we have seen that using a PPM can allow the complexities of a resilient and flexible Gateway to be hidden from the Client Applications, but there other issues that need to be addressed.

One barrier to any new technology, such as Parlay, is that there are many existing applications implemented using older technologies that have already been deployed and are operating success-fully, generating revenue. Typically there is a financial imperative to be able to continue to operate (and profit from) them and rewriting them is impractical.

One solution is to provide translation between the existing interfaces implemented by an appli-cation and the Parlay interfaces. In other words, requests by a non-Parlay application are routed to a function that performs a translation into a Parlay request and forwards it onto the Parlay Gate-way. Any responses from the Gateway are of course translated into the non-Parlay format. Such a translation function can be termed an Application Proxy.

For example, an enterprise operator may have a set of applications that don't support Parlay but they would like them to gain access to network functionality via a Parlay Gateway. Re-writing the applications would be prohibitively expensive so the development of an Application Proxy that can translate between the two technologies is a reasonable investment.

The success of such an Application Proxy depends on a number of factors. Clearly the Parlay and non-Parlay interfaces must be somewhat similar in nature. The closer they are in terms of semantics the easier it is to implement a translation.

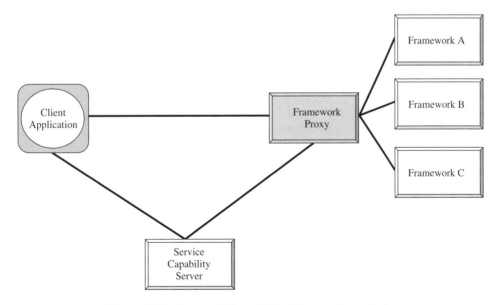

Figure 14.10 Framework Proxy 'hiding' Framework complexity

Of course, if the existing applications have been implemented in a well-partitioned way it may be feasible to remove the existing interface implementation and replace it with a Parlay-compliant one. This has clear benefits in terms of performance since it avoids the additional layer of translation.

The preceding paragraphs have included an implicit assumption that it is the owner of the non-Parlay applications that implements and operates the Application Proxy. This is certainly a valid approach and is particularly suitable when the existing interfaces are very proprietary in nature. If, however, the existing interfaces are in accordance to a popular standard or are a widely used proprietary technology (a de facto standard) then there may be alternative solutions. In this case the same Application Proxy may be usable with a number of applications owned by a number of Enterprise Operators. It may make sense for the network operator (or the owner of the Parlay Gateway, if different) to implement (or buy into) the Application Proxy and deploy it in its own domain.

Such an approach does have implications for security. Where the Application Proxy is owned and deployed by the Enterprise Operator it looks, to the Network Operator, as if it is a standard Parlay Client Application and all of the Parlay rules and solutions apply. If the Application Proxy is part of the Network Operator's domain then clearly a security solution is needed for operating the non-Parlay interfaces across the domains.

14.4.3.1 Application Proxy 2, or is it an SCS Proxy?

We have seen previously how there can be requirements on SCSs to be reliable and flexible in terms of capacity. Overall reliability and capacity for a system is determined by all of the elements in that system, and limited by the weakest link. Since the Client Application is part of the overall system, it may be necessary to design it to match the SCS in reliability and capacity terms.

One last illustration of the proxy pattern is to use an Application proxy of a somewhat different nature to those examined in the previous section. The proxy in this case appears to be a standard Parlay Client Application as far as the SCS is concerned. In reality, however, it is an SCS as far as another set of Client Applications is concerned. This set of Client Applications provides the reliability and flexibility that is desired by distributing requests and responses across them as appropriate.

In effect what appears to be a highly reliable Application as far as the SCS in the Network Operator's network is concerned is in fact another Parlay system. This system consists of a Framework, an SCS (the Application Proxy) and a number of Client Applications.

This may appear to be stretching the proxy pattern a little, but it is intended to be an illustration to show how Parlay can be applied to many different situations.

14.5 Summary

We have seen in this chapter that applying the proxy concept to Parlay opens up many powerful options to the designer of a Parlay system and we have examined some of them in detail. These options (and many more that we are sure the creative reader will be able to devise) enable a range of business and deployment models to be supported. At the same time they keep the life of the Client Application implementer as simple as possible, which for Parlay is what it's all about.

15

Multi-Network Deployment Scenarios

15.1 Introduction

As more and more Parlay gateways are deployed, more applications will be built in support of this model. This leads to more revenue being generated as subscriber experiences improve and new subscribers are attracted and retained through the availability of newer and more exciting services. Aspects of federation, data sharing and interaction between networks also rise in importance as this happens.

This is not unusual. Recall Metcalfe's law from Chapter 1: the interconnectedness of networks contributes directly to their value. An application connected to a network of networks is accessible to more users than one tied to a single lonely isolated one. Groups of users in an isolated setting are still somewhat alone, though not necessarily lonely. Community is key. Enter the federation.

Also, many of the larger service providers typically tend to own multiple properties. Some of these may be geographically dispersed, while others may be geographically close but built to support different technologies. In the former case, the service provider might want to support more efficiently the roaming of their subscribers from one of their networks to another. The objective is to provide to the user in the visited network an experience as close as possible to the one the user would get at home, within the limits of reason, by sharing whatever data are necessary between their subsidiaries to make this feasible. In the latter case, they may want to hide network protocol peculiarities from the application or services domains to enable a single application to operate seamlessly across network types, and permit subscribers using differing underlying technologies to have the same end-user experiences. As we have seen before, Parlay definitely supports that goal. In fact, one might even argue that Parlay was conceived especially to meet these needs.

In this chapter, we study these and related issues. What makes Parlay a technology that can bridge networks? How can Parlay gateways and applications straddle network boundaries and support the notion of federated networking? How can end-users benefit, and how will this translate to benefits for the service providers?

15.2 Some examples

We will try and illustrate the concept of seamless service operation by means of two examples that it is hoped will help towards an improved appreciation of the value that federation, data sharing, and interaction between networks can bring to subscribers.

Parlay/OSA: From Standards to Reality   Musa Unmehopa, Kumar Vemuri, Andy Bennett
Copyright © 2006 Lucent Technologies Inc. All Rights Reserved

15.2.1 Example 1

Alexandra Darden traveled a lot on business. What she really liked about her service with Freedom Wireless was that the services she subscribed to worked the same way when she was in France as they did at home in England. She would connect to the WeatherHere application and it would give her the local forecast. She would look up the nearest Italian restaurant or ATM machine or directions to a local destination and things would always work seamlessly without a hitch. She could even track the presence of her buddies in England and could chat with them via instant messaging sessions with no trouble at all. She was truly impressed with how well the business agreements between Freedom Wireless and their French service provider partner were working out. They had a reciprocity agreement where the tariffs were distributed based on the average number of subscribers transiting networks in each direction (computed over the past year), and so the charges were almost the same as she had to pay for the service when she was in her home network.

15.2.2 Example 2

Jimmy Buffett enjoyed shopping these days. He would buy something from one online store and get coupons that could be used in other stores (online or otherwise) that belonged to the same federation. He wondered when the idea really took off. It seemed so natural it was hard to imagine the practice was not as prevalent just a couple of years ago.

It had all started with inter-working agreements between network operators and a technology called Parlay that everyone was talking about but no one seemed to understand very well. But it did work wonders from an 'average user on the road' viewpoint. Service providers were in an all-out war initially, trying to best each other to grab subscribers and revenue. But once things settled down a bit, there was a notion of sharing, open, boundaryless, unrestrained freedom where the subscriber could roam across regions and get their favorite services unhindered ... and billions of dollars of business was being done in the mobile commerce arena these days.

Groups of firms selling diverse products got together and formed federations with one or more service providers, and together these parties all leveraged the wonders of Parlay, the open application development environment, to sell to the vast addressable market of mobile cell-phone subscribers.

15.3 Federation: What is it? Why is it a Good Thing?

As we said before, the value of a network increases with the number of users and the number of nodes it connects. A federation overlays business relationships that drive revenue atop a core network and makes a large customer pool available to merchants (such as Internet-based vendors like Amazon.com, and even 'brick-and-mortar' stores that provide an Internet portal); and it makes these merchants accessible to end-users who want convenient, secure access to their products in a seamless way. Services and applications can be easily and most naturally infused into this setting to make the user experience very natural, very intuitive. And with this kind of added value, everybody wins.

However, for such a model to be successful, several requirements need to be met:

1. Businesses involved must agree to a common set of privacy and security rules and not be anti-competitive, for that would stifle co-operation and collaboration and fragment the nature of the federation.
2. Data sharing, wherever possible, is encouraged, to promote the seamlessness of the end-user experience while reducing 'ownership' related issues (admittedly, this would be difficult to do, at least in the initial stages, for although a lot of data really belong to an end-user, not many businesses would see it that way)
3. Business agreements that enable synergistic relationships to develop across the federation as a whole are set up so growth in some areas fuels growth in others and so on.

4. The network, which provides the life-blood of the dynamic aspects of end-user interactions, is actively leveraged, with the network operators being adequately recompensed for their contributions towards making this vision a reality.

These are but some of the constraints that need to be satisfied for a truly beneficial federation to exist. Now, although end-users may provide information to individual businesses as they go about their transactions, centralization of critical information would ensure that Single Sign On (SSO) and similar capabilities could be provided to make their experiences as effortless as possible without compromising security or privacy constraints that are so important to the end-user view.

SSO allows an end-user to register and authenticate (i.e. log on) only once while entering the federation of service providers and gaining seamless access to the applications of all service providers belonging to the federation, as opposed to having to provide their credentials to each and every individual application time and again.

For the remainder of this chapter 'federation' is defined as a multilateral affiliation of various administrative domains (i.e. service provider domains) to create a community all trying to achieve a common goal. Contracts or business agreements need to be in place between the domains, defining the terms of the relationship.

As an example of federation and how it supports SSO, let us look at Vanessa, a Freedom Wireless subscriber. Vanessa is a user of services from a federation to which Freedom Wireless belongs. Other members of this federation include Renter's World Rent-A-Car, Galaxy Coffee, and the Walton Group of hotels. Now, all her accounts (where she has them) across these various distinct corporate entities are linked together, along with her profiles and preferences with each vendor. There is careful sharing of her information based on previously established business agreements between the members of the federation. Vanessa benefits by having to sign into just one service in the federation and then seamlessly accessing services from multiple vendors. Also, now she no longer has to remember separate passwords for each vendor's site. So, while logged on to her Freedom Wireless service, Vanessa could access the online reservation system of the Walton in Chicago to book a stay over the weekend and arrange for a Rent-A-Car convertible to be delivered to the hotel for her to pick up, all while having provided her authentication credentials only once. And a truly great federation would access Vanessa's preference profile at Galaxy, and make sure her favorite blend was available next to her in-room coffee machine.

Parlay promotes federation – consider deployments where other parties in a federation could leverage telecommunication network hosted capabilities to provide value added services, for instance. But the true power of federation is only realized when Parlay is coupled with other standards such as those defined by the Liberty Alliance [Liberty] in this arena. In what follows, we focus on how Parlay technology promotes federation at various levels. Details pertaining to Liberty Alliance and other such standards are outside the scope of this text.

15.4 Models for Multi-Network Deployments of Parlay Gateways

Service providers may see the need to support multi-network deployments of Parlay gateways for a variety of reasons. As also previously mentioned, a single corporation may own wireline and wireless properties and may want to interconnect the services infrastructures for the two...Or they may own exclusively wireless properties, each of which operates on a different technology but where it appears service efficiencies may be derived through support for a common services infrastructure... Or multiple different corporations may make agreements to share hosted service capabilities in a carefully regulated way.

In each of these cases, Parlay provides a means for how this 'sharing' can happen and how it may be controlled. The reader should note that Parlay merely functions as a facilitator for these kinds of architectures and that some amount of roaming support is already built into many of the underlying protocols that operate in the core telecommunications network behind the Parlay APIs.

For instance, you can call any phone in the world from any other phone in the world – you are not bound by technological or network boundaries. The networks themselves are connected 'at lower levels' to make this possible. Parlay support for federated architectures merely serves to provide parallels from a services perspective.

Several distinct models exist for multi-network and multi-operator deployments of Parlay gateways and include the following:

1. Federation of Frameworks: Frameworks from different networks each accept registration and announcement requests from service capability servers (SCSs) in their local networks. The registration databases are then shared across networks that have business agreements in place between them. Thus the Frameworks from different networks are federated, share knowledge of registered and announced services, and can make these discoverable to client applications that connect with them regardless of network boundaries. Thus, a Framework in Operator A's network is able to broker accesses to SCSs registered and announced with Operator B's Framework.

   This is depicted in Figure 15.1. Service providers may define policies on which services are shared with which networks and under what conditions (i.e. access may be closely regulated, perhaps using policies).

2. Cross-Boundary Service Registration: In this model, Frameworks do not share data, but selected services (meant to be shared across service provider network boundaries) are configured with the addresses of Frameworks in other networks, and are permitted to register and announce themselves with them. This way, these selected services are now visible to client applications in other networks during discovery and may be selected for use in service sessions if the supported capabilities match the selection criteria. Here too, policy enforcement may be used to regulate the visibility of particular services at the network where discovery operations are performed.

   This is depicted in Figure 15.2, where applications register both with local and remote Frameworks. Subscriber data, as in the Federation of Frameworks scenario, may still be shared across service provider boundaries.

   In this model, SCS B_service in Operator B's network is provided with the addresses for Frameworks in both Operator A and Operator B networks, and registers and announces itself with both of them. Now, client applications talking to Frameworks in either network can discover and use the service in keeping with the service agreements in place.

3. Application-level Federation: Applications connect to different Frameworks in two or more totally disjoint networks and utilize service capabilities exposed by each network in a manner that is totally transparent to the end-users.

   This is best illustrated by means of an example. Alice is a Freedom Wireless subscriber, while Bob is a customer of Utopia. However, both Alice and Bob use IMNow! for instant messaging. The application seamlessly connects to the Frameworks in both the Freedom Wireless and Utopia

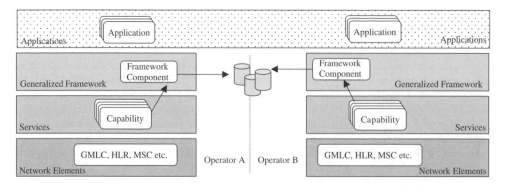

Figure 15.1 Federation of Frameworks

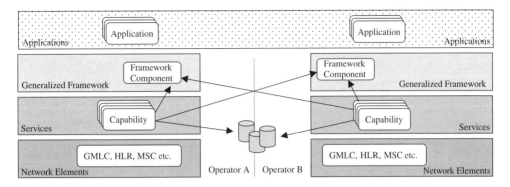

Figure 15.2   Cross-boundary Service Registration

networks (which may not have any business arrangements with each other) and is able to provide
Alice and Bob with the presence and availability information they need about each other as IM
buddies. We call this the 'buddies across boundaries' scenario.

Figure 15.3 illustrates Application-level Federation, a federated architecture through data sharing at the application level. The databases in Figure 15.3 could represent data shared across
multiple replicas or instances of a single application (implemented that way for high availability
and redundancy reasons) or between multiple applications that share dynamic network context
information within a federation so this can be reused in different application contexts to provide
value-added services to end-users. Any of the other federation schemes (Federation of Frameworks, Cross-boundary service registration, and the Proxy Manager model described below) can
be supported in conjunction with Application-level Federation.

4. Proxy-regulated 'Window to the World' Model: In Chapter 14 we discussed the Parlay Proxy
   Manager (PPM) component that, among other things, enabled gateway vendors to provide higher
   availability, transaction-level load balancing, and service proxy capabilities using a PPM component that appeared to be an SCS to client applications and an application to the subtended
   SCSs. This PPM element can also be used to open up selectively services to other networks
   while limiting the exposure of individual leaf SCSs to external Frameworks. Hence the term
   'Window to the World', since only the proxy is registered and announced with external Frameworks and the leaf SCSs themselves are not. This limits exposure or dissemination of network
   service capabilities and topology related information outside the network, while providing all
   the flexibility naturally inherent in the PPM model.

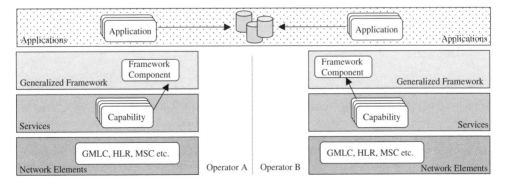

Figure 15.3   Application-level Federation

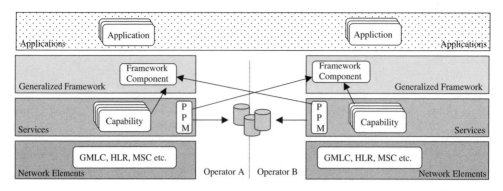

Figure 15.4 Proxy Manager Federation

Figure 15.4 presents the Proxy Manager model for federation, where the PPM serves to expose selectively service capabilities to external Operator networks and applications.

It must be emphasized that these are but examples of models that can be implemented to support multi-operator Parlay gateway deployments. In some cases, multiple models may be applied concurrently, for different services, or for inter-connection with different service provider networks, or altogether different feasible models may be designed to meet most optimally the needs of the business agreement between two collaborating networks.

The models described above have far-reaching implications for gateway providers, service providers, application providers and subscribers or end-users:

- Gateway providers need to realize that the Parlay series of standards do not constrain an implementation to within the strict boundaries of a service provider network and that there are opportunities to build standards-compliant solutions that promote and support federation across operator networks.
- Service providers have several network architecture options to choose from, can leverage Parlay-based solutions to their advantage, and can further increase revenues by leveraging federation capabilities.
- Application providers can, through support for models such as those providing for federation support in the application layer (option 3 above), drive up their own revenues while providing better end-user experiences.
- End-users probably have the most to gain: seamless access to services from within the federation; support for single or reduced sign-on flows; greater security since less personal information is needed to transit the network; perhaps cheaper services overall since the reuse of services data and infrastructure across the federation drives costs down for everybody, etc. Through this process, the cell-phone or mobile device also grows to a greater importance, also in keeping with Metcalfe's law.

15.5 Mobile Virtual Network Operator Scenarios

In Chapter 2, a passing reference was made to Mobile Virtual Network Operators or MVNOs, and how they were different from the more traditional service providers. MVNOs function as resellers of services hosted by service providers and thus effectively increase the subscriber base that might end up utilizing the network capabilities and infrastructure from any given service provider network. MVNOs typically do not host any network infrastructure other than perhaps a portal through which the services are made available. MVNO architectures fit in nicely with the Parlay model and in this section we will look at these in more detail.

The Parlay Framework is the entity that authenticates client applications and 'secures the perimeter' so to speak for access to network hosted service enablers or SCSs. It may informally be referred to also as a 'traffic cop' since it regulates service session distribution among clones of a given service.

A reseller (or MVNO) could host just the Framework and broker access to services hosted by the service provider if the SCSs from within the service provider network were registered with the Framework in the MVNO domain per business agreements between the MVNO and the service provider. In this way, the MVNO hosts a minimal amount of service infrastructure and is able to resell access to the service provider network capabilities (if the MVNO hosts both the Framework and SCSs in the MVNO Parlay gateway) or just the service provider SCSs (if the MVNO Parlay gateway only hosts the Parlay Framework but no SCSs) to client applications from third parties.

Client applications can now negotiate with the MVNO (that resells services) for use of particular services and the MVNO could then permit these clients to discover and select services for SCSs (from one or more subtended underlying service provider networks). Once service selection is complete, and the service agreement signed by the client and MVNO Framework, the MVNO Framework could invoke the createServiceManager() method on the SILM of the selected SCS located within the service provider domain. Technically speaking, this appears exactly the same as option 2 from the previous section (Cross-Boundary Service Registration), where the service provider hosting the MVNO Framework does not support any services or SCSs within its own network.

In such a scenario, the access session between the client and the Framework would be to the MVNO domain, while the actual service session itself would be directly established between the client application and the SCS in the service provider domain. Thus, 'the Parlay triangle' as described in Chapter 5, which normally transits just two domains, now potentially covers three, but the overall operational model itself remains virtually unchanged.

15.6 Revenue Settlement between Federated Entities

All businesses (except perhaps non-profit organizations) are founded with the objective of making money and delivering value to their customers and shareholders (if the company is publicly held). Federation, we have seen from previous sections, does help significantly to drive up business and sales for members of the collective. However, what does the service provider gain in all this? After all, dynamic user data may be queried by one member of the federation and then shared with others. Does this not result in a loss of service provider revenue? And what about MVNO situations?

Appropriate business arrangements should be put in place to address this scenario. More value can be obtained per service mediation gateway transaction in a federation scenario than in other cases with stand-alone applications. The data sharing arrangements must factor in not just the number of transactions but also the value of the data to the collective, to ensure that service providers grow their revenue in proportion to the growth of business across the federated collective.

Similarly, reciprocal agreements between multiple service providers that share services using either the Federated Framework or Cross-Boundary Service Registration or Proxy Manager models must be supported so that as business grows, and service access becomes truly seamless between the various interconnected networks, it translates to a win-win scenario for all the concerned parties (including the end-users).

15.7 Summary

As Parlay technology finds greater market penetration and more widespread acceptance and deployment, service providers who have already individually deployed Parlay gateways and supporting infrastructure will want to find ways of exploiting partnership agreements with each other or partnering with businesses to accelerate further revenue generation. Federation acts as a simple but powerful enabler to support such scenarios. In this chapter, we have explored architectures in support of federated deployment of Parlay gateways.

16

Parlay/OSA and XML-based Technologies

16.1 Introduction

One of the objectives of Parlay is to enable large communities of practitioners to design and build value-added business solutions in a complex telecommunication network. The idea here is to attract developers with a knack for creating innovative applications, without necessarily being privy to the often arcane and privileged information on complicated call models and extensive signaling protocols. Even though thus far in the book we have shown that Parlay has very successfully lowered the threshold for application development and deployment, through abstraction and mediation, we owe it to ourselves to recognize that there may still be parts of the developer community who find the remaining complexity too daunting. There are two ways to combat this and pull this remaining contingent of developers onto the Parlay bandwagon; even higher levels of abstraction and realizing the Parlay technology using the tools of their trade.

This chapter deals with Parlay and the Web Services technology realization of the Parlay interfaces. Two efforts can be distinguished, each with their own objective, but related in their use of XML-based technologies. These two efforts are called Parlay WSDL and Parlay X. Both are aimed at realizing the Parlay concepts in the tools and technologies more familiar to the score of Internet savvy application developers. We shall explain each in detail and point out how they relate and where they differ.

In order to be able to distinguish between these newer Parlay efforts and the Parlay API specifications as covered thus far in this book, we will refer to the latter as the existing Parlay interfaces, or simply Parlay, whereas the newer initiatives will always be identified by the additional qualification, i.e. Parlay X and Parlay WSDL.

As with any technology area, Web Services comes with its own vocabulary of acronyms and this acronym soup gets real thick real soon. So before we get to the good stuff, we need to introduce a number of terms and technologies.

16.2 The Acronym Soup

This section is intended to provide the reader with sufficient understanding of the Web Services lexicon required to fully appreciate the discussion of Parlay WSDL and Parlay X. It is specifically not the goal of this section to be a tutorial in this vast and exiting new technology area. Interested readers are referred to the various available excellent textbooks for a more complete introduction and coverage of these individual topics. [Newcomer 2002] is provided here as a good example.

Parlay/OSA: From Standards to Reality   Musa Unmehopa, Kumar Vemuri, Andy Bennett
Copyright © 2006 Lucent Technologies Inc. All Rights Reserved

16.2.1 A Brief Recap

Some of the terms and acronyms have already been introduced, mainly in Chapter 4, but will be reiterated here briefly for the benefit of the reader.

- *UML* – The Unified Modeling Language serves as the formal analysis model used to define the Parlay APIs. UML consists of a collection of modeling techniques including class diagrams, sequence diagrams, and state transition diagrams. The UML model for Parlay is defined in a technology independent way such that the various technology realizations can all be generated and derived from this common source.
- *IDL* – The Interface Definition Language is used to specify the CORBA realization of the Parlay APIs and is automatically generated from the Parlay UML model, using standardized language mapping rules. IDL is often prefixed with the name of the body that standardized it, i.e. OMG IDL.
- *OMG* – The Object Management Group is an industry consortium, which has published, among other technologies, the specifications for UML, CORBA and IDL.
- *CORBA* – The Common Object Request Broker Architecture is an Object-Oriented middleware technology for distributed computing. CORBA uses IDL to define the interfaces between remote objects.

16.2.2 The X-files

The aim of this section is to aid the reader in navigating through the new and vast vocabulary in use in the Web Services field.

- *XML* – The eXtentable Markup Language is a textual syntax, or markup language, designed to describe data, using tags you can define yourself. In order to describe the data, XML makes use of either a Document Type Definition (DTD) or XML Schema Definition (XSD).
- *DTD* – A Document Type Definition is a template describing the structure of XML documents. Since an XML document is just a text file, a DTD can be used to validate whether an XML document is constructed correctly, made up of allowed elements and using the defined tags.
- *XSD* – The XML Schema Definition defines XML Schema, which is an XML based alternative to Document Type Definitions (DTD), and describes the structure of an XML document. XSD is in itself an XML document.
- *XMI™* – XML Metadata Interchange defines rules for the generation of XML DTD representation from formal models.
- *SOAP* – The Simple Object Access Protocol is an XML-based communication protocol for exchanging messages between applications.
- *UDDI* – Registry where Web Service descriptions and definitions can be registered by service providers and discovered by businesses or end-users.
- *WSDL* – The Web Services Definition Language is an XSD-based language specifically designed to describe Web Services.
- *Web Service* – A piece of software, its interfaces described in WSDL, which can be accessed over the Internet, using SOAP.

Now that we have recapitulated some of the acronyms introduced earlier in this book, as well as providing an initial understanding of some of the Web Services lingo, we are all geared up to sink our teeth into the main subject of this chapter, i.e. Parlay and XML-based technologies.

16.3 Parlay WSDL

Parlay WSDL is the term that we choose to assign to the WSDL technology realization of the Parlay interfaces. The reader will recall that this is the third Parlay technology realization, next to the CORBA technology realization (using OMG IDL) and the Java technology realization (using J2EE and J2SE).

The Parlay WSDL technology realization was proposed in September 2001 in recognition of the rising popularity of SOAP and XML as an alternative distribution mechanism to the CORBA/IDL approach. This effort can be seen in the support of the high level objective of Parlay to be independent of the transport technology or distribution mechanism. The approach, as introduced earlier in Chapter 4, can be summarized as defining an XML definition of the Parlay methods and data types, to allow for interoperability between elements in the Parlay architecture that are using an XML-based RPC protocol such as SOAP or XML-RPC as transport mechanism.

Since this body of work is based on UML-to-WSDL language mappings, we speak of WSDL technology realization, rather than Web Services-enabling Parlay. This is an important distinction with the Parlay X initiative, as described in the next section. As with all technology realizations supported by Parlay, the WSDL technology realization is functionally equivalent with its IDL and Java (J2EE) counterparts. This includes the support of for instance the asynchronous communication patterns and the associated callbacks, and the use of (object- and interface-) references. We shall come back to this later, when assessing the values of Parlay WSDL and discussing the most recent developments in standards.

16.3.1 The UML-to-WSDL Language Mapping

This section provides some more information on the UML-to-WSDL language mapping rules developed by Parlay and used for the automatic generation of the Parlay WSDL definitions.

The Parlay UML-to-WSDL language mapping rules are based on the OMG XMI (XML Metadata Interchange) specification [OMG 2002a], which defines rules for the generation of XML DTD representation from formal models, including analysis models like UML. XMI defines rules for the generation of XML DTDs from a formal object model and rules for the generation of XML documents from the objects themselves (i.e. the instances of that formal object model). The OMG has defined an extension to XMI [OMG 2003a], which adds rules for the generation of XML Schema (XSD). Parlay in turn has extended these rules to generate WSDL, which is based on XSD.

In general, data type definitions, like 'constant' or 'enumeration' are mapped to XSD constructs, whereas API components such as 'interface', 'method', or 'exception' are mapped to WSDL constructs. For example, a UML Structure is mapped to an XSD sequence element, whereas a UML interface class is mapped to a WSDL portType.

Formalizing these UML-to-WSDL language mapping rules allows the automatic generation of the WSDL definition files for every Parlay interface once a new version of the standards specification is available and the UML model is updated. Semantic and functional equivalence between the technology realizations, which are all generated automatically from the UML model based on formalized language mappings, is thereby guaranteed.

The UML-to-WSDL language mapping rules are contained in the Parlay WSDL Style Guide [Parlay 2002a]. The WSDL that is produced using the best practices, guidelines, and mapping rules from this Parlay WSDL Style Guide is conformant to the Basic Profile published by the Web Services Interoperability organization [WSI 2004]. This ensures that the Parlay X Web Services are consistent with the tools available on the market and can be deployed on most commercial Web Services platforms.

The WSDL interface definition files are added as a single separate archive file to the archive for each part of the 3G TS 29.198 series. For instance, the WSDL technology realization for the Multiparty Call Control SCF in 3GPP Release 5, dated September 2004 [3GPP 2004i], can be found as archive 2919804-03V580WSDL.zip in the archive for the specification itself (28198-04-3-580.zip). The WSDL archive in turn contains a separate WSDL file for the data definitions (mpcc_data.wsdl) and one for the interfaces themselves (mpcc_interfaces.wsdl).

16.3.2 Parlay WSDL in Relation to Parlay Web Services

As mentioned before, Parlay WSDL is not entirely the same as Web Services-enabling Parlay. The Parlay WSDL technology realizations merely provide a realization of Parlay in a definition language, which is used in Web Services deployments. This is where the Parlay Web Services working group comes in. The Parlay Web Services working group does not publish any specifications but rather has produced a number of white papers outlining how Parlay Web Services interfaces, such as Parlay WSDL, are deployed in a Web Services environment. Specifically the Application Deployment Infrastructure white paper from Parlay covers a number of deployment models for the registration, discovery, and use of Parlay X Web Services [Parlay 2002b].

Any Parlay Web Services environment requires the presence of a Parlay Web Services Gateway. This is a new entity, implementing a Web Services interface, for instance Parlay WSDL[1]. The Parlay Web Services Gateway may be realized either as an integrated part of the Parlay Gateway (Figure 16.1), or as a separate proxy element (Figure 16.2).

When acting as a proxy element, the Parlay Web Services gateway will translate Parlay WSDL requests to the Parlay technology realization in use by the Parlay Gateway and hence act as Parlay application towards the Parlay Gateway. As explained before in Chapter 9 this is completely transparent to the Parlay WSDL Client Application (as well as to the Parlay Gateway), and hence for reasons of brevity here we will only cover the integrated deployment

The first deployment scenario covered here in Figure 16.3 is deployable today with published Parlay and Web Services technologies to date and is applicable in both trusted as well as untrusted domains. What follows is a brief description of an end-to-end scenario, from the publish phase until the service usage, involving a Parlay Web Services Gateway.

As a first step the Parlay Web Services Gateway publishes itself in the UDDI registry to make itself available for discovery by a Parlay application: that is, it ensures it can be found. Subsequently,

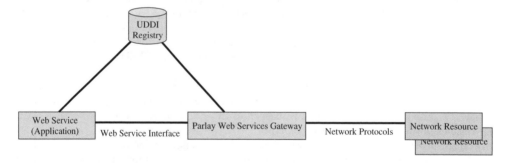

Figure 16.1   Parlay Web Services Gateway – Integrated deployment

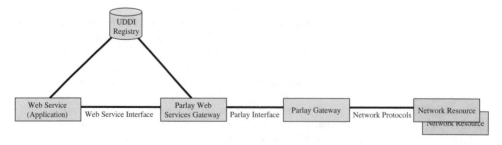

Figure 16.2   Parlay Web Services Gateway – Proxy deployment

---

[1] In later sections we will see that Parlay X Web Services are applicable here as well.

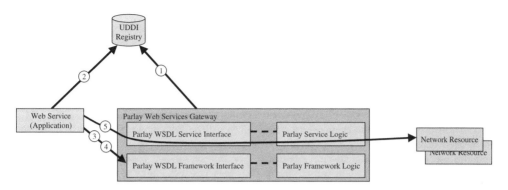

Figure 16.3   Parlay Web Services Gateway – Combined deployment

in step 2, the Parlay Web Services application consults the UDDI registry and finds one or more service providers. The application then selects a specific service provider that supports a Parlay Web Services Gateway deployment. The application will then enter the 'bind' step (step 3), where the application will perform the Parlay Framework handshake, through its Parlay WSDL realization, for authentication, authorization and the signing of a Service Level Agreement. Once a contract is in place between the application and the service provider operating the Parlay Web Services Gateway, the application can in step 4 perform the processes of service discovery and selection. The Parlay WSDL service interfaces implemented on the Parlay Web Services Gateway are available for this process. Step 5 can now take place, which is the usage of the Parlay service capabilities by the Web Service application for service execution.

In the deployment scenario above, the Gateway publishes itself, and subsequently the authentication with the service provider as well as discovery of available services takes place through the Parlay Framework mechanisms. Hence the deployment scenario above is applicable for untrusted, third party access. The following deployment scenario, depicted in Figure 16.4, makes use of Web Services technologies only, and hence the controlled access to network capabilities provided through the Parlay Framework mechanisms is no longer available.

The first step (step 0) in this untrusted deployment scenario is the publication of the Parlay WSDL technology realizations in the UDDI registry within the Parlay namespace. In step 1 the service provider now not only publishes the availability of its Parlay Web Services Gateway but also all its Parlay WSDL Services, which comply to the WSDL realizations published before in step 0. The application can now simultaneously find in step 2 both the Parlay Web Services Gateway as well as all Parlay WSDL Service interfaces it supports. At this stage, the application is all set to commence using the Parlay network capabilities in step 4. In order to still provide some form of

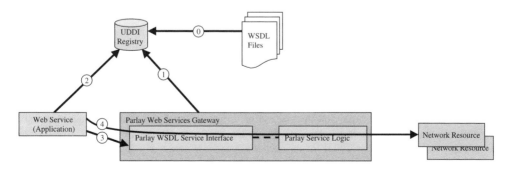

Figure 16.4   Parlay Web Services Gateway – Web Services deployment

secure communication between the application and the Parlay Web Services Gateway (recall that no Parlay Framework handshake has taken place), a simple web log-on or some more involved means of authentication may be required.

16.3.3 Assessment of Parlay WSDL, and Recent Developments

We have seen that Parlay WSDL is a true technology realization of the Parlay APIs, as defined in the Parlay UML model. Any design construct or communication pattern, as laid down in the UML, is ported as is to the specific language used for a given realization. Two examples we have already mentioned are the support for the asynchronous communication patterns and associated callbacks, and the use of object and interface references. For this reason we have called Parlay WSDL a WSDL technology realization rather than a Web Services technology realization. Chapter 17, which is included as advanced reading at [Parlay@Wiley], will provide more insight in the Web Services paradigm and its applicability in the telecommunications domain. Here, we suffice by observing that a realization of a given technology using only the language of another technology may possibly yield less than optimal results. After all, a communication paradigm is more than just its interface definition language. For a more detailed analysis of the issues with the WSDL realization of Parlay, the reader is referred to [Lagerberg 2002].

The Parlay WSDL realization is an elegant and non-intrusive means to deploy already supported Parlay capabilities in a Web Services infrastructure, as depicted for instance in Figure 16.3. Operators wishing to expose their already deployed Parlay assets as Web Services, or wishing to integrate their Parlay assets with a larger Web Services based service environment, may successfully utilize the Parlay WSDL technology realization.

Since the introduction of Parlay WSDL, however, significant changes have occurred in Web Services technologies as that technology matured and gained a foothold in the mainstream of service paradigms. As a result, the understanding of how best to apply Web Services as a realization of Parlay has changed accordingly. For this reason, the Parlay WSDL realization is not continued in later versions of the Parlay Release 5 specification set. As the technology further matures, future standards activities may provide a replacement for the Parlay WSDL realization, reflective of the improved understanding and increased experience with the Web Services technology and its expected usage.

16.4 Parlay X

The Parlay X Working Group was created in September of 2001. The requirement for Parlay X was born out of the desire to define even simpler interfaces (or indeed a more natural mapping to Web Services technology), targeted towards application developers without even a basic understanding of network signaling protocols, network state machines, etc. These include for instance applications from the financial domain (e.g. banking, insurance), content creation domain (e.g. music, motion pictures), retail (e.g. online stores), etc. We have seen that, even though for instance the Parlay Call Control interfaces make it simpler for application developers to build call related applications without specific signaling protocol expertise, some state behavior is still involved. Also, for instance, the asynchronous subscribe-notify pattern requires state and is a communications pattern, though common in telecommunications, that is not intuitive to most application developers. Applications in which the communications component is only a small part merely require primitives like 'make a call' or 'give me a location'. For such applications the existing Parlay Call Control API and the Parlay User Location API may be too heavyweight in terms of functionality and complexity. The Parlay X initiative in part builds on the success of scripting language based interfaces to service capabilities. Some examples include SIP CPL [RFC 3880], and VoiceXML [W3C 2004]. The reader is referred to [Bakker 2002] for an overview of the evolution of scripting and XML-based technologies in service creation.

The objective of Parlay X is to define a set of powerful and imaginative building blocks, defined at a chosen level of simplicity and abstraction, so that developers and the IT community can generate new, innovative applications without the long learning curve typically involved with 'old-school' telecommunications protocols. The basic idea is to define each Parlay X building block as an abstraction of the service capabilities exposed through the existing Parlay interfaces. The Parlay X building blocks are specifically aimed to fuel the development of innovative third party applications, not necessarily by developers skilled in the area of telecommunications, but more by the heretofore untapped vast pool of developers in the IT community. Simplicity of the interfaces, coupled to the use of tools and technologies familiar to this particular audience, is of key significance to the success of the Parlay X initiative. The motivation for Parlay X is also described in [Lofthouse 2004].

16.4.1 Parlay X in Relation to Parlay

The Parlay X Web Services are designed from the ground up as Web Services, rather than a Web Services realization of a technology neutral interface specification.

The Parlay X effort builds on some of the design principles of the existing Parlay interfaces but differs in other areas. This section outlines how the Parlay X solution is differentiated from the Parlay solution. Whereas the existing Parlay interfaces, defined for the various SCFs, are homogeneous in terms of their capabilities (e.g. Presence is separate from Location), the Parlay X building blocks may be heterogeneous if so desired (e.g. Presence may be combined with Location). Another desired feature of the Parlay X interfaces is to design each interaction as a simple synchronous message exchange, e.g. the request-response pattern, and not use asynchronous exchange patterns, or triggered exchange patterns. The underlying thought is to adhere to the KISS-principle, i.e. keep it simple stupid. Other Parlay design principles, for instance the principle that the SCFs are defined application independent, are maintained. In fact, the Parlay X building blocks will attempt to address a wider application range, rather than trying to achieve feature richness and high functionality. This is referred to as the '80/20' rule, i.e. compared to the existing Parlay interfaces the Parlay X building blocks are designed to address 80% of the application space while defining only 20% of the interface functionality. The challenge here is to minimize the complexity of the API while not needlessly reducing the addressable application space. Note that the '80/20' rule serves more as a guiding principle than as a strict, true rule.

Although it is the strong desire to define each Parlay X building block as an abstraction of the service capabilities exposed through the existing Parlay interfaces, in order to allow for evolutionary solutions as well as preventing parallel competing approaches, some exceptions are present in the Parlay X specifications. In those cases where there is a compelling use case to define extended capabilities, Parlay X has allowed such instances to occur (Figure 16.5).

16.4.2 Parlay X in Relation to Parlay Web Services

Parlay X defines Web Services, that is, service capabilities deployable in a Web Services environment. Parlay X does not define that environment or specify the means by which to deploy these service capabilities in such an environment. Issues like security for Parlay X Web Services, registration and discovery of Parlay X Web Services are addressed by the Parlay Web Services activity.

The deployment infrastructure for Web Services technologies only (i.e. not making use of the Parlay Framework capabilities) as outlined above in the description of the relation between Parlay WSDL and Parlay Web Services is equally applicable to Parlay X, where the Web Services interface implemented on the Parlay Web Services Gateway is the Parlay X interface, rather than the Parlay WSDL interface. This deployment scenario is described in detail in [Parlay 2002c] and summarized diagrammatically below.

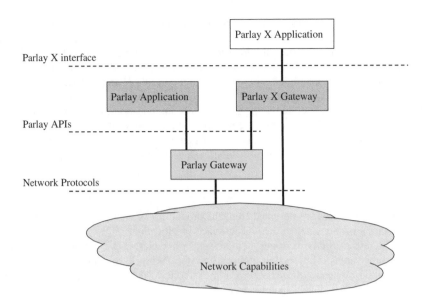

Figure 16.5 Parlay X in relation to Parlay

16.4.3 The Parlay X Building Blocks

The first public Parlay X Web Services specification, v1.0, was published as part of Parlay version 4.0, in May 2003 [Parlay 2003]. This specification contains the interface definition of eight Parlay X Web Services, which are briefly introduced below.

16.4.3.1 Terminal Location

The Terminal Location Parlay X Web Service provides applications with the means to request the location for a given end-user's terminal device. The Web Service operation getLocation returns the location in terms of a latitude-longitude pair, a time stamp, and an indication of location accuracy[2].

The Terminal Location Parlay X Web Service is a good example of how the 80/20 design principle has been applied successfully. Whereas the User Location API's support for instance triggered reports, and periodic reports, as well as location to be returned in all sorts of formats[3], the Parlay X Web Service interface simply supports a synchronous request for the lat-long coordinates for a given user. This serves the purpose of most non-telecommunication applications requiring location information as part of the overall service they provide.

16.4.3.2 User Status

The User Status Parlay X Web Service is intended for application scenarios that require information on the status of an end-user terminal device, e.g. 'busy' or 'offline'. A single operation is supported on this Web Service interface, getUserStatus. There are not many differences when it comes to

---

[2] The accuracy can be requested and is returned in one of three values Low, Medium, High. The Parlay X specification does not specify exactly, for instance, Medium accuracy. It is therefore required to make proprietary arrangements between an application provider and a Parlay X Gateway operator to agree upon the value ranges for these accuracy indications.

[3] The formats include network location information (e.g. cell ID, VLR number), geodetic position (e.g. a slice of an elliptic sector), and geographical coordinates (e.g. latitude-longitude pair).

the equivalent functionality in the User Status API, other than the fact that the User Status API provides an asynchronous means for obtaining the status, and that the status request can be issued for a set of users rather than just a single user.

However the User Status API provides more functionality above and beyond the single status request, e.g. User Status Parlay X Web Service does not support triggered status requests. Again, the 80/20 design principle was put to good use for this Parlay X Web Service.

16.4.3.3 Third Party Call

The Third Party Call Parlay X Web Service is used to support application-initiated calls from a Web Service environment. Using the Parlay Call Control API to support an application-initiated call would require the application to request the call manager object on the gateway to instantiate a call object and request routes (or call legs) to be set up to both parties in the call. Each request for a route to be set up can be accompanied with a variety of additional requests, such as requesting reports on the call's progress, including for instance 'alerting'. This pretty much reflects how such functionality would be realized in basic call processing in the network.

The Third Party Call Parlay X Web Service however simply provides a single request for a call to be set up between two identified parties, makeACall. In addition, operations are provided to end a call (endCall), or to cancel a previous request for a call to be set up (cancelCallRequest).

A fourth operation is supported by the Third Party Call Parlay X Web Service to allow an application to retrieve information regarding an application-initiated call set up previously on its behalf, getCallInformation. The operation returns information on the state of the call in progress (e.g. 'connected' or 'terminated') as well as a call termination cause, when applicable. One could argue that simple use cases for application-initiated calls would settle for knowing whether a call ended normally, or abnormally, or even just for whether it ended or not. However, the termination causes supported for this Parlay X Web Service include 'no answer', 'busy', and 'not reachable' for both parties, as well as 'hang-up' and 'aborted'. The use cases for this operation do require more than just superficial knowledge of basic call processing in telephony networks, though not as much as when using the Parlay Call Control APIs.

16.4.3.4 Network-Initiated Third Party Call

The Parlay X Web Service for Network-Initiated Third Party Call provides Web Service applications with the means to control and influence the progress of calls initiated by subscribers in the network. Of course if you wish to exercise control over a call and its progress in the network, you would have to have some idea of the stages a call typically goes through during its lifetime. Some knowledge of state behavior and service control protocols seems essential to know where you can interrupt in the flow and what you are allowed to do when the flow is interrupted. This makes Network-Initiated Third Party Call both an interesting as well as a challenging candidate for realization as Parlay X Web Service.

Service control in communication networks is typically managed by using event driven state machines (see also Appendix A on Call Models [Parlay@Wiley]). Certain events are only valid in certain stages of the call and some events may cause a transition from one state to the next. The Call Control APIs provide applications with the means to control and influence the progress of a call in the network by providing an application view of these state machines. In order to be notified of certain events, the application would have to register for them.

The Parlay X effort has addressed the issue of state and asynchronous communication by introducing atomic stateless operations that provide a synchronous response to events from the network. For instance, when a subscriber places a call to another subscriber who is already engaged in a call, the Parlay X gateway will ask the application what to do using the operation handleBusy. The action recommended by the application is provided in the synchronous return of this operation.

Similar operations are available for the 'not reachable', 'no answer', and 'off hook' network events, and for calls initiated to a specific destination number.

The model described above does present Web Service applications with an easy means to perform service control on calls originated in the network, without requiring an in-depth understanding of the service control mechanisms. Of course, application subscription to these network events still need to take place, however, this is left out of the scope of Parlay X, to be dealt with through offline means. So in a sense the asynchronous nature of this service control mechanism is made to fit a synchronous model by reducing the scope to the synchronous requests for action.

16.4.3.5 Payment

The Payment Parlay X Web Service is spread over four packages and supports Web Service interfaces for both direct as well as reserved amount charging, and both direct as well as reserved volume charging.

The Amount Charging package provides operations for crediting and debiting an account associated with an end-user. The Volume Charging package provides similar functionality but for volumes rather than amounts[4]. The Reserved Amount Charging package provides operations for amount charging where a reservation is required, including reserving an amount, charging the reservation, and releasing the reservation. The Reserved Volume Charging provides similar functionality, but again for volumes rather than amounts. Both volume-based packages support one additional operation to convert volumes to amounts.

The functionality supported by the Payment Parlay X Web Service is rather involved and may be perceived as unbalanced when compared to the other Parlay X Web Services. We believe the statement that the 80/20 design principle was applied to Payment with less rigor to be a valid point of critique. As a result of this the Payment Parlay X Web Service package approaches the functionality provided by the Web Services technology realization of the Parlay Content Based Charging API to the extent that differentiation between the two is less apparent. On the other hand one could argue that any service architecture that does not provide a mature and fully functional charging solution will be a less likely candidate for commercial success and large market uptake. In addition, charging may be considered less telecommunication specific than, for instance, call control, and hence an understanding of its principles and functionality is more widespread.

Compared to the Parlay Content Based Charging API, the Payment Parlay X Web Service has dropped support for sequence management using request numbering and omitted the reservation lifetime management. Also, there is no support for unit charging in the Payment Parlay X Web Service.

16.4.3.6 Account Management

The Account Management Parlay X Web Service provides an application with the ability to manage the account of a pre-paid subscriber. Although the Account Management Parlay X Web Service and the Parlay Account management API share the same name, there is a difference in approach. Whereas the existing Parlay feature supports solely management functionality for an end-user account, the Parlay X version provides application access to the account in addition to the management capabilities. Furthermore, the management functionality of the existing Parlay API is intended for use by the network operator whereas the Parlay X Web Service addresses the third party domain. For this reason, the concept of an end-user PIN is added to the Parlay X Web Service.

Three operations are supported in the management category of functionality, i.e. the ability to request the current balance of an end-user's account (getBalance), the ability to request the expiry

---

[4] The Parlay X specification does not specify what types of volumes are supported (e.g. number of bytes transmitted), and hence these need to be negotiated between the application provider and the Parlay X Gateway operator via some offline means.

date of an end-user's account (getCreditExpiryDate), and the ability to request the transaction history of an end-user's account (getHistory).

The direct operations supported for the end-user's account in the Account Management Parlay X Web Service are the ability to update the balance using a voucher (voucherUpdate) and the ability to top up the balance directly (balanceUpdate).

A part of the simplification with respect to the existing Parlay API for Account Management is obtained through a reduction of the complexity of the data types involved.

16.4.3.7 SMS

The SMS Parlay X Web Service provides the application with the means to send and receive SMS messages. Three packages are supported: the Send SMS API, the SMS Notification API, and the Receive SMS API. Whereas in the existing Parlay APIs, generalized messaging and user interaction mechanisms can be used to achieve the same functionality, the SMS Parlay X Web Service is specifically designed for the SMS service. Dedicated operations are supported for the sending of an SMS message (sendSms), the sending of an SMS ring tone (sendSmsRingtone) and the sending of an SMS logo (sendSmsLogo). Each of these three operations implicitly requests a delivery status for the sent message. In the absence of support for asynchronous communication patterns, an application has to poll explicitly for the SMS delivery status. These operations are supported in the Send SMS package.

SMS messages that are delivered to the message store for a given subscriber can be retrieved from that store using the operation getReceivedSms. This operation is supported in the Receive SMS API package.

The application is notified of the availability of a received message for a given subscriber at the message store by the operation notifySmsReception. As this is a synchronous operation, this Web Service needs to be implemented at the application in order for the Parlay X Gateway to be able to invoke this operation. Supporting Web Services at the application has heretofore been avoided in Parlay X as it introduces a whole suite of complexities for the application developer. The omission of asynchronous communication patterns so far is partly a result of attempting to avoid this additional complexity. The inclusion of the SMS Notification API can be interpreted as straying from the design principles for Parlay X.

So for SMS we see that the details of the underlying messaging technology, which is a store-and-forward mechanism, cannot be shielded from the application. The delivery status report for a sent message and the availability notification of a received message are typical examples of asynchronous events inherent to the store-and-forward mechanism for SMS.

16.4.3.8 Multimedia Message

Whereas the SMS Parlay X Web Service was specifically designed for a single messaging technology, i.e. SMS, the Multimedia Message Parlay X Web Services is intended to be applicable to a multitude of messaging technologies. Examples include MMS, EMS, and also SMS.

The Multimedia Messaging Parlay X Web Service also consists of three packages: the Send Message API, the Message Notification API, and the Receive Message API.

As opposed to the Send package in the SMS Parlay X Web Service, here only a single sendMessage operation is supported. The content, that is the actual message, is sent as a SOAP-Attachment [W3C 2000], encoded as either MIME [RFC 2045] or DIME [Nielsen 2002]. The sendMessage operation of the Multimedia Parlay X Web Service can also be used to send an SMS message, and hence we here have two ways of achieving the same functionality.

The Receive Message API consists of three operations. The application has the ability to retrieve messages from the message store individually (getMessage), in a bulk (getReceivedMessages), or using URIs that point to specific message parts of a multipart message (getMessageURIs).

Also, similar to SMS, the MultimediaMessage Parlay X Web Service supports a Message Notification package, containing a notifyMessageReception operation. Again, in order to support this package, a Web Service needs to be implemented at the application side.

16.4.4 Assessment of Parlay X, and Recent Developments

The Parlay X effort has been initiated to address the requirements of a developer community more familiar and at home with IT toolkits and development environments. In conjunction with this, such a developer community typically sports less affinity with telecommunications skills and expertise and hence the resulting requirement for an even higher level of abstraction than is provided through the Parlay APIs. This is a good development as it furthers the pooling of application development talent and creativity with the feature rich service capabilities from the telecommunications domain, potentially leading to more innovative and compelling applications.

One of the lessons learned though is that intrinsically difficult matters are not easily simplified through abstraction. Also, not every dynamic behavioral model can be force-fitted into a stateless, synchronous mold. Having said that, the set of Parlay X Web Services do offer more simplified building blocks for the controlled access of network capabilities.

One possible scenario is that application developers from an IT background get acquainted with the vast potential of incorporating network service capabilities in their application suite through the use of Parlay X Web Services. As they get more familiar with those capabilities they might get compelled to tap even further into the broad range of network service capabilities to enrich further their applications through the use of the full-fledged suite of Parlay APIs, be it in their WSDL realization or some other deployment choice.

In September 2004, the specifications for the Parlay X building blocks were submitted and approved as 3GPP technical specifications. Together, they comprise the 3G TS 29.199 series. In addition, the initial set of Parlay X building blocks was extended to a total of 13 Parlay X Web Services. Some existing Parlay X Web Services were extended, although a number of completely new Parlay X Web Services were introduced to augment the Parlay X suite. The most notable additions are Multimedia conference, Address list management, and Presence. The interested reader is referred to the set of Parlay X Web Services specifications in 3GPP Release 6 [3GPP 2005d, 3GPP 2005e, 3GPP 2005f, 3GPP 2005g, 3GPP 2005h, 3GPP 2005i, 3GPP 2005j, 3GPP 2005k, 3GPP 2005l, 3GPP 2005m, 3GPP 2005n, 3GPP 2005o, 3GPP 2005p, 3GPP 2005q] for the complete description and details on these latest additions to the Parlay X Web Services specification set.

16.5 Summary

Parlay X Web Services have been designed with the objective of enabling large communities of Internet-savvy practitioners to design and build communications applications on complex network infrastructure. They are kept intentionally lightweight and follow the service mantra of the Internet whereby applications are developed using common IT toolkits and widespread protocol and middleware paradigms. Parlay X Web Services need to be powerful enough for skilled telecommunication protocol programmers, yet easy enough for students. And there is ample justification for this approach. A nimble interface, which does one thing and does it well, certainly deserves a chair on the sun deck with the more capable and functional interfaces. However, extreme care must be taken to keep the Parlay X Web Services at an even higher level of abstraction than the Parlay interfaces. Otherwise, although we have fully functional interfaces and more abstract interfaces in theory what we really have is both a belt and suspenders.

Bibliography

There are several books available on the topic of programmable networks and value-added applications. In what follows, we present a brief overview of the more relevant. Together with our book, it is hoped this provides a solid and comprehensive description of the field of play.

[Mueller 2002] offers an extensive survey of the protocols and APIs for voice services over converged networks and the software required to implement them. The book presents example protocol message exchanges and code fragments. The last chapter, Chapter 14, is dedicated to Parlay, taking the Parlay 2.1 Framework and Call Control APIs to walk through a good and detailed example of how a simple Call Processing application is set-up, and goes through the Framework handshake.

A deep and elaborate coverage of open network APIs is presented in [Jain 2004]. This book covers in detail the APIs defined by JTAPI, JAIN JCC and JCAT, and Parlay, and concludes with a description of XML programmability initiatives, including Parlay X and SPIRITS. The focus is on Call Processing and Call Models, with an in-depth coverage of the design patterns used to model the call processing behavior in the network. Running examples, code fragments and call flows are provided to expand the understanding of each of the technologies covered.

[Zuidweg 2002] describes the evolution of intelligence in the network that is available to create services and applications. The book show the evolution of IN and CAMEL and leads to interactions between IN and Internet (e.g. PINT and SPIRITS). The author then spends a chapter on Parlay and OSA. The focus of the book is on Service Logic and Service Creation.

In addition to the books described above, several papers have appeared in a number of journals and conference proceedings, either on Parlay in general or addressing dedicated topics of interest within Parlay. Some of the more prevalent general papers include [Stretch 2001,Unmehopa 2002b,Moerdijk 2003].

List of Abbreviations and Acronyms

2G	2nd Generation
3G	3rd Generation
3GPP	3rd Generation Partnership Project
3GPP2	3rd Generation Partnership Project 2
AAA	Authentication, Authorization and Accounting
ACG	Automatic Code Gapping
AFLT	Advanced Forward Link Trilateration
AM	Account Management
AMPU	Average Minutes Per User
ANSI	American National Standard Institute
API	Application Programming Interface
ARPU	Average Revenue Per User
AS	Application Server
ASP	Application Service Provider
ATM	Asynchronous Transfer Mode or Automated Teller Machine
BCSM	Basic Call State Model
BHCA	Busy Hour Call Attempt
BNF	Backus Naur Form
BOBO	Billing on behalf of
BSC	Base Station Controller
BSS	Base Station System
CA	Client Application
CAMEL	Customized Application for Mobile Enhanced Logic
CAP	CAMEL Application Part
CBC	Content Based Charging
CC	Call Control
CC/PP	Composite Capabilities/Preference Profile
CCC	Conference Call Control
CCCS	Conference Call Control Service
CCF	Call Control Function
CDMA	Code Division Multiple Access
CDR	Call Detail Record
CGI	Cell Global Identifier or Common Gateway Interface
CH	Charging
CHAM	Charging and Account Management
CHAP	Challenge Handshake Authentication Protocol
CIPID	Contact Information for Presence Information Data format

CN	Core Network
CO	Central Office
COM	Component Object Model
CORBA	Common Object Request Broker Architecture
CPIM	Common Profile for Instant Messaging
CPL	Call Processing Language
CPP	Common Presence Profile
CPU	Central Processing Unit
CPUTP	CPU Transaction Processing
CR	Change Request
CRM	Customer Relationship Management
CS	Capability Set
CSCF	Call Session Control Function
CSE	Camel Service Environment
CTD	Click-to-Dial
CUI	Call User Interaction
DCOM	Distributed Component Object Model
DFP	Distributed Functional Plane
DHCP	Dynamic Host Configuration Protocol
DIME	Direct Internet Message Encapsulation
DMZ	De-Militarized Zone
DP	Detection Point
DP-N	Detection Point – Notification
DP-R	Detection Point – Request
Dpx	Duration per transaction
DSA	Digital Signature Algorithm
DSC	Data Session Control
DSL	Digital Subscriber Line
DSP	Digital Signal Processor
DTD	Document Type Definition
DTMF	Dual Tone Multiple Frequency
E&M	Ear and Mouth
e2e	End-to-End
E911	Enhanced 911
EAI	Enterprise Application Integration
EDGE	Enhanced Data-rate for GSM Evolution
EDP	Event Detection Point
EDP-N	Event Detection Point – Notification
EDP-R	Event Detection Point – Request
EDR	Event Detail Record
EFLT	Enhanced Forward Link Trilateration
EH	Event Handling
EMS	Enhanced Message Service
ESC	Event State Compositor
ESME	External Short Message Entity
ESP	Encapsulating Security Payload
ETSI	European Telecommunication Standards Institute
EVDO	Evolution for Data Optimized
EVDV	Evolution for Data and Voice

FA	Foreign Agent
FCAPS	Fault Management, Configuration Management, Accounting Management, Performance Management, Security Management
FCC	Federal Communications Commission
FDMA	Frequency Division Multiple Access
FMO	Future Mode of Operation
FSM	Finite State Machine
FTP	File Transfer Protocol
FWK	Framework
GAA	Generic Authentication Architecture
GCC	Generic Call Control
GCCS	Generic Call Control Service
GERAN	GSM Enhanced Radio Access Network
GGSN	Gateway GPRS Support Node
GLMS	Group List Management Server
GMLC	Gateway Mobile Location Center
GMS	Generic Messaging Service
GPRS	General Packet Radio Service
GPS	Global Positioning System
GSM	Global System for Mobile communications
gsmSCF	GSM Service Control Function
GTP	GPRS Tunneling Protocol
GTT	Global Title Translation
GUI	Generic User Interaction
HA	High Availability or Home Agent
HLR	Home Location Register
HSS	Home Subscriber Server
HTML	HyperText Markup Language
HTTP	HyperText Transport Protocol
HTTPS	Hypertext Transfer Protocol over Transport Layer Security
I/O	Input/Output
ICAP	Internet Content Adaptation Protocol
I-CSCF	Interrogating Call Session Control Function
ICW	Internet Call Waiting
ID	Identifier
IDL	Interface Definition Language
IETF	Internet Engineering Task Force
iFC	Initial Filter Criteria
IM	Instant Messaging or IP Multimedia
IMAP	Internet Message Access Protocol
IMS	IP Multimedia Subsystem
IM-SSF	IP Multimedia – Service Switching Function
IN	Intelligent Networking
INAP	IN Application Protocol
IOR	Interoperable Object Reference
IP	Internet Protocol
IPCP	IP Control Protocol
IPsec	IP Security
IRTF	Internet Research Task Force

ISC	IMS Service Control
ISOC	Internet SOCiety
ISP	Internet Service Provider
ISUP	ISDN User Part
IT	Information Technology
ITU	International Telecommunication Union
ITU-T	International Telecommunications Union – Telecommunication Standardization Sector
IWF	InterWorking Function
J2EE	Java 2 platform, Enterprise Edition
J2SE	Java 2 platform, Standard Edition
JWG	Joint Working Group
KISS	Keep It Simple Stupid
LAI	Location Area Identifier
LAN	Local Area Network
LCS	Location Service
LIF	Location Interoperability Forum
LIF MLP	LIF Mobile Location Protocol
LPDP	Local Policy Decision Point
M4U	Movies For You
MAP	Mobile Application Part
MEP	Message Exchange Pattern
MG	Media Gateway
MGC	Media Gateway Controller
MIDL	Microsoft Interface Definition Language
MIME	Multipurpose Internet Mail Extensions
MIN	Mobile Identification Number
MLP	Mobile Location Protocol
MM	Mobility Management
MMCC	MultiMedia Call Control
MMCCS	MultiMedia Call Control Service
MMD	MultiMedia Domain
MMM	MultiMedia Messaging
MMS	Multimedia Message Service
MO-SM	Mobile Originated Short Message
MPC	Mobile Positioning Center
MPCC	MultiParty Call Control
MPCCS	MultiParty Call Control Service
MPEG	Motion Pictures Expert Group
ms	millisecond
MS	Mobile Station
MSC	Mobile Switching Center
MSISDN	Mobile Subscriber ISDN Number
MSRP	Message Session Relay Protocol
MTBF	Mean Time Between Failure
MT-SM	Mobile Terminated Short Message
MTTR	Mean Time Till Repair
MVE	Multi-Vendor Environment
MVNO	Mobile Virtual Network Operator

N/A	Not Applicable
N/W	Network
NAS	Network Access Server
NEBS	Network Equipment Building System
NNI	Network-to-Network Interface
OA&M	Operations, Administration and Maintenance
OAM&P	Operations, Administration, Maintenance and Provisioning
O-BCSM	Originating Basic Call State Model
OMA	Open Mobile Alliance™
OMG	Object Management Group
OO	Object Oriented
OPES	Open Pluggable End-Services
OPEX	Operational Expenses
OSA	Open Service Access
OSI	Open Systems Interconnection
OSS	Operations Support Systems
PA	Presence Agent
PAG	Presence and Availability Group
PAM	Presence and Availability Management
PC	Personal Computer
PCF	Packet Control Function
PCIM	Policy Core Information Model
P-CSCF	Proxy Call Session Control Function
PDA	Personal Digital Assistant
PDE	Position Determining Equipment
PDP	Packet Data Protocol or Policy Decision Point
PDSN	Packet Data Serving Node
PDU	Protocol Data Unit
PE	Policy Enforcer
PEEM	Policy Evaluation, Enforcement and Management
PEP	Policy Enforcement Point
PGA	Parental Guidance Application
PIC	Point In Call
PIN	Personal Identification Number or Policy Ignorant Node
PINT	PSTN/Internet Interworking
PLMN	Public Land Mobile Network
PM	Policy Management
PMO	Present Mode of Operation
PNA	Presence Network Agent
PoC	Push to talk over Cellular
POP	Point of Presence
POTS	Plain Old Telephony Service
PPM	Parlay Proxy Manager
PPP	Point to Point protocol
PPUA	Pay-Per-Use Application
PS	Presence Server
PSTN	Public Switched Telephone Network
PUA	Presence User Agent
QoS	Quality of Service

RADIUS	Remote Authentication Dial-In User Service
RFC	Request For Comments
RHS	Right Hand Side
RMI	Remote Method Invocation
RPC	Remote Procedure Call
RPID	Rich Presence Information Data format
RSA	Rivest-Shamir-Adelman
RTFM	Read The Forgotten Manual
RTP	Real-time Transport Protocol
RTT	Radio Transmission Technology or Round Trip Time
SAG	Subscription Assignment Group
SAP	Service Access Point
SCF	Service Capability Feature or Service Control Function
SCM	Service Combination Manager
SCP	Service Control Point
SCS	Service Capability Server
S-CSCF	Serving Call Session Control Function
SCTP	Stream Control Transmission Protocol
SDF	Service Data Function
SDK	Software Development Kit
SDP	Session Description Protocol
SDR	Service Detail Record
sFC	Subsequent Filter Criteria
SGSN	Serving GPRS Support Node
SIBB	Service Independent Building Block
SILM	Service Instance Lifecycle Manager
SIM	Service Interaction Manager or Subscriber Identity Module
SIMPLE	SIP for Instant Messaging & Presence Leveraging Extensions
SIP	Session Initiation Protocol
SIPPING	Session Initiation Protocol Project INvestiGation
SLA	Service Level Agreement
SMDPP	Short Message Delivery Point-to-Point protocol
SMF	Service Management Function
SMG	Service Mediation Gateway
SMPP	Short Message Peer to Peer protocol
SMS	Short Message Service
SMS GW	SMS Gateway
SMSC	SMS Service Center
SMTP	Simple Mail Transfer Protocol
SN	Service Node
SOAP	Simple Object Access Protocol
SPAN	Signaling and Protocols for Advanced Networks
SPIRITS	Services in PSTN/IN Requesting InTernet Services
SS7	Signaling System nr. 7
SSF	Service Switching Function
SSL	Secure Sockets Layer
SSO	Single Sign-On
SSP	Service Switching Point
STP	Signal Transfer Point

TACACS	Terminal Access Controller Access-Control System
TAM	Total Addressable Market
TAT	Turn Around Time
TC	Terminal Capabilities
TCAP	Transaction Capabilities Application Part
TCP	Transmission Control Protocol
TDM	Time-Division Multiplexing
TDMA	Time Division Multiple Access
TDP	Trigger Detection Point
TDP-N	Trigger Detection Point – Notification
TDP-R	Trigger Detection Point – Request
THIG	Topology Hiding Inter-network Gateway
TINA-C	Telecommunications Information Networking Architecture Consortium
TLS	Transport Layer Security
TPS	Transaction Per Second
TR	Technical Recommendation
TS	Technical Specification
TSAS	Telecommunications Service Access and Subscription
TUP	Telephone User Part
UDDI	Universal Description, Discovery and Integration
UDP	User Datagram Protocol
UI	User Interaction
UL	User Location
ULC	User Location Camel
ULE	User Location Emergency
ULP	Upper Layer Protocol
ULTr	Triggered User Location
UML	Unified Modeling Language
UMTS	Universal Mobile Telecommunications System
UNI	User-to-Network Interface
URI	Unique Resource Identifier
URL	Universal Resource Location
US	User Status
USSD	Unstructured Supplementary Service Data
UTRAN	UMTS Terrestrial Radio Access Network
VHE	Virtual Home Environment
VLR	Visitor Location Register
VNO	Virtual Network Operator
VoIP	Voice over IP
VPN	Virtual Private Network
W3C	World Wide Web Consortium
WAN	Wide Area Network
WAP	Wireless Application Protocol
WAP GW	WAP Gateway
WG	Working Group
WiFi	Wireless Fidelity
WRU	Where Are You
WSDL	Web Services Description Language
WSP	Web Service Provider or Wireless Service Provider or Wireless Session Protocol

WSR	Web Service Requester
WTLS	Wireless Transport Layer Security
WV	Wireless Village
XCAP	XML Configuration Access Protocol
XMI	XML Metadata Interchange
XML	eXtensible Markup Language
XMPP	Extensible Messaging and Presence Protocol
Xpd	Transaction per Duration
XSD	XML Schema Definition

References

Web References

[3GPP] http://www.3gpp.org/
[3GPP2] http://www.3gpp2.org/
[BlueTooth] https://www.bluetooth.org/
[ETSI] http://www.etsi.org/
[IETF] http://www.ietf.org/
[IRTF] http://www.irtf.org/
[ITU] http://www.itu.int/
[JAIN] http://java.sun.com/products/jain/
[OMA] http://www.openmobilealliance.org/
[OMG] http://www.omg.org/
[OPIUM] http://www.ist-opium.org/
[Parlay] http://www.parlay.org/
[Parlay@Wiley] http://www.wiley.com/go/parlay
[Skype] http://www.skype.com/
[TINA] http://www.tinac.com/
[Vonage] http://www.vonage.com/

Main References

[3GPP 2002a] 3GPP TS 22.228, 3rd Generation Partnership Project; Technical Specification Group Services and System Aspects; Service requirements for the IP Multimedia; Core Network Subsystem (Stage 1) (Release 5), Version 5.6.0 (June 2002), URL: http://www.3gpp.org/

[3GPP 2002b] 3GPP TS 22.105, 3rd Generation Partnership Project, 'Services and Service Capabilities', Version 5.2.0 (July 2002), URL: http://www.3gpp.org/

[3GPP 2002c] 3GPP TS 22.121, 3rd Generation Partnership Project, 'The Virtual Home Environment', Version 5.3.1 (July 2002), URL: http://www.3gpp.org/

[3GPP 2002d] 3GPP TS 22.127, 3rd Generation Partnership Project, 'Stage 1 Service Requirement for the Open Services Access (OSA)', Version 5.5.0 (December 2002), URL: http://www.3gpp.org/

[3GPP 2002e] 3GPP TS 23.127, 3rd Generation Partnership Project, 'Virtual Home Environment (VHE)/Open Service Access (OSA)', Version 5.2.0 (June 2002), URL: http://www.3gpp.org/

[3GPP 2002f] 3GPP TR 29.998-01, 3rd Generation Partnership Project, 'Open Service Access (OSA) Application Programming Interface (API) Mapping for OSA; Part 1: General Issues on API Mapping', Version 5.0.0 (June 2002), URL: http://www.3gpp.org/

[3GPP 2002g] 3GPP TR 29.998-04-1, 3rd Generation Partnership Project, 'Open Service Access (OSA) Application Programming Interface (API) Mapping for OSA; Part 4: Call Control Service Mapping; Subpart 1: API to CAP Mapping', Version 5.0.0 (June 2002), URL: http://www.3gpp.org/

[3GPP 2002h] 3GPP TR 29.998-05-1, 3rd Generation Partnership Project, 'Open Service Access (OSA) Application Programming Interface (API) Mapping for OSA; Part 5: User Interaction Service Mapping; Subpart 1: API to CAP Mapping', Version 5.0.0 (June 2002), URL: http://www.3gpp.org/

Parlay/OSA: From Standards to Reality Musa Unmehopa, Kumar Vemuri, Andy Bennett
Copyright © 2006 Lucent Technologies Inc. All Rights Reserved

[3GPP 2002i] 3GPP TR 29.99805-4, 3rd Generation Partnership Project, 'Open Service Access (OSA) Application Programming Interface (API) Mapping for OSA; Part 5: User Interaction Service Mapping; Subpart 4: API to SMS Mapping', Version 5.0.0 (June 2002), URL: http://www.3gpp.org/

[3GPP 2002j] 3GPP TR 29.998-06, 3rd Generation Partnership Project, 'Open Service Access (OSA) Application Programming Interface (API) Mapping for OSA; Part 6: User Location – User Status Service Mapping to MAP', Version 5.0.0 (June 2002), URL: http://www.3gpp.org/

[3GPP 2002k] 3GPP TR 29.998-08, 3rd Generation Partnership Project, 'Open Service Access (OSA) Application Programming Interface (API) Mapping for OSA; Part 8: Data Session Control Service Mapping to CAP', Version 5.0.0 (June 2002), URL: http://www.3gpp.org/

[3GPP 2003] 3GPP TS 23.002, 3rd Generation Partnership Project, 'Network Architecture', Version 5.12.0 (October 2003), URL: http://www.3gpp.org/

[3GPP 2004a] 3GPP TS 23.228, 3rd Generation Partnership Project; IP Multimedia Subsystem (IMS); Stage 2 (Release 5), Version 5.13.0 (December 2004), URL: http://www.3gpp.org/

[3GPP 2004b] 3GPP TS 24.228, 3rd Generation Partnership Project; Signalling Flows for the IP Multimedia Call Control Based on Session Initiation Protocol (SIP) and Session Description Protocol (SDP); Stage 3 (Release 5), Version 5.13.0 (June 2005), URL: http://www.3gpp.org/

[3GPP 2004c] 3G TS 29.002, 3rd Generation Partnership Project, 'Mobile Application Part (MAP) Specification', Version 5.10.0 (June 2004), URL: http://www.3gpp.org/

[3GPP 2004d] 3G TS 29.198-1, 3rd Generation Partnership Project, 'Open Service Access (OSA) Application Programming Interface (API); Part 1: Overview', Version 5.7.0 (September 2004), URL: http://www.3gpp.org/

[3GPP 2004e] 3G TS 29.198-2, 3rd Generation Partnership Project, 'Open Service Access (OSA) Application Programming Interface (API); Part 2: Common Data', Version 5.8.0 (September 2004), URL: http://www.3gpp.org/

[3GPP 2004f] 3G TS 29.198-3, 3rd Generation Partnership Project, 'Open Service Access (OSA) Application Programming Interface (API); Part 3: Framework', Version 5.8.0 (September 2004), URL: http://www.3gpp.org/

[3GPP 2004g] 3G TS 29.198-4-1, 3rd Generation Partnership Project, 'Open Service Access (OSA) Application Programming Interface (API); Part 4: Call Control; Subpart 1: Common Call Control Data Definitions', Version 5.7.0 (September 2004), URL: http://www.3gpp.org/

[3GPP 2004h] 3G TS 29.198-4-2, 3rd Generation Partnership Project, 'Open Service Access (OSA) Application Programming Interface (API); Part 4: Call Control; Subpart 2: Generic Call Control Data Service Capability Feature (SCF)', Version 5.8.0 (September 2004), URL: http://www.3gpp.org/

[3GPP 2004i] 3G TS 29.198-4-3, 3rd Generation Partnership Project, 'Open Service Access (OSA) Application Programming Interface (API); Part 4: Call Control; Subpart 3: Multi-party Call Control Data Service Capability Feature (SCF)', Version 5.8.0 (September 2004), URL: http://www.3gpp.org/

[3GPP 2004j] 3G TS 29.198-4-4, 3rd Generation Partnership Project, 'Open Service Access (OSA) Application Programming Interface (API); Part 4: Call Control; Subpart 4: Multimedia Call Control Service Capability Feature (SCF)', Version 5.8.0 (September 2004), URL: http://www.3gpp.org/

[3GPP 2004k] 3G TS 29.198-5, 3rd Generation Partnership Project, 3rd Generation Partnership Project, 'Open Service Access (OSA) Application Programming Interface (API); Part 5: Generic User Interaction', Version 5.8.0 (September 2004), URL: http://www.3gpp.org/

[3GPP 2004l] 3G TS 29.198-6, 3rd Generation Partnership Project, 'Open Service Access (OSA) Application Programming Interface (API); Part 6: Mobility', Version 5.6.0 (September 2004), URL: http://www.3gpp.org/

[3GPP 2004m] 3G TS 29.198-7, 3rd Generation Partnership Project, 'Open Service Access (OSA) Application Programming Interface (API); Part 7: Terminal capabilities', Version 5.7.0 (September 2004), URL: http://www.3gpp.org/

[3GPP 2004n] 3G TS 29.198-8, 3rd Generation Partnership Project, 'Open Service Access (OSA) Application Programming Interface (API); Part 8: Data Session Control', Version 5.7.0 (September 2004), URL: http://www.3gpp.org/

[3GPP 2004o] 3G TS 29.198-11, 3rd Generation Partnership Project, 'Open Service Access (OSA) Application Programming Interface (API); Part 11: Account Management', Version 5.6.0 (September 2004), URL: http://www.3gpp.org/

[3GPP 2004p] 3G TS 29.198-12, 3rd Generation Partnership Project, 'Open Service Access (OSA) Application Programming Interface (API); Part 12: Charging', Version 5.7.0 (September 2004), URL: http://www.3gpp.org/

[3GPP 2004q] 3G TS 29.198-13, 3rd Generation Partnership Project, 'Open Service Access (OSA) Application Programming Interface (API); Part 13: Policy management SCF', Version 5.6.0 (September 2004), URL: http://www.3gpp.org/

[3GPP 2004r] 3G TS 29.198-14, 3rd Generation Partnership Project, 'Open Service Access (OSA) Application Programming Interface (API); Part 14: Presence and Availability Management (PAM)', Version 5.7.0 (September 2004), URL: http://www.3gpp.org/

[3GPP 2004s] 3GPP TS 22.101, 3rd Generation Partnership Project, 'Service Principles', Version 5.13.0 (March 2004), URL: http://www.3gpp.org/

[3GPP 2004t] 3GPP TS 23.271, 3rd Generation Partnership Project, 'Functional Stage 2 Description of Location Services (LCS)', Version 5.13.0 (December 2004), URL: http://www.3gpp.org/

[3GPP 2004u] 3GPP TR 29.998-04-4, 3rd Generation Partnership Project, 'Open Service Access (OSA) Application Programming Interface (API) Mapping for Open Service Access; Part 4: Call Control Service Mapping; Subpart 4: Multiparty Call Control ISC', Version 5.0.3 (June 2004), URL: http://www.3gpp.org/

[3GPP 2004v] 3GPP TS 23.141, 3rd Generation Partnership Project, 'Presence Service; Architecture and Functional Description', Version 6.7.0 (September 2004), URL: http://www.3gpp.org/

[3GPP 2005a] 3GPP TS 23.218, 3rd Generation Partnership Project; IP Multimedia (IM) Session Handling; IM Call Model; Stage 2 (Release 5), Version 5.8.0 (March 2005), URL: http://www.3gpp.org/

[3GPP 2005b] 3GPP TS 24.229, 3rd Generation Partnership Project; IP Multimedia Call Control Protocol based on Session Initiation Protocol (SIP) and Session Description Protocol (SDP); Stage 3 (Release 5), Version 5.13.0 (June 2005), URL: http://www.3gpp.org/

[3GPP 2005c] 3G TS 29.198-15, 3rd Generation Partnership Project, 'Open Service Access (OSA) Application Programming Interface (API); Part 15: Multi-media Messaging (MM) Service Capability Feature', Version 6.2.0 (July 2005), URL: http://www.3gpp.org/

[3GPP 2005d] 3GPP TS 29.199-1, 3rd Generation Partnership Project, 'Open Service Access (OSA) Parlay X Web Services; Part 1: Common', Version 6.2.0 (June 2005), URL: http://www.3gpp.org/

[3GPP 2005e] 3GPP TS 29.199-2, 3rd Generation Partnership Project, 'Open Service Access (OSA) Parlay X Web Services; Part 2: Third Party Call', Version 6.1.0 (June 2005), URL: http://www.3gpp.org/

[3GPP 2005f] 3GPP TS 29.199-3, 3rd Generation Partnership Project, 'Open Service Access (OSA) Parlay X Web Services; Part 3: Call Notification', Version 6.1.0 (June 2005), URL: http://www.3gpp.org/

[3GPP 2005g] 3GPP TS 29.199-4, 3rd Generation Partnership Project, 'Open Service Access (OSA) Parlay X Web Services; Part 4: Short Messaging', Version 6.3.0 (June 2005), URL: http://www.3gpp.org/

[3GPP 2005h] 3GPP TS 29.199-5, 3rd Generation Partnership Project, '3rd Generation Partnership Project, Open Service Access (OSA) Parlay X Web Services; Part 5: Multimedia Messaging', Version 6.3.0 (June 2005), URL: http://www.3gpp.org/

[3GPP 2005i] 3GPP TS 29.199-6, 3rd Generation Partnership Project, 'Open Service Access (OSA) Parlay X Web Services; Part 6: Payment', Version 6.1.0 (June 2005), URL: http://www.3gpp.org/

[3GPP 2005j] 3GPP TS 29.199-7, 3rd Generation Partnership Project, 'Open Service Access (OSA) Parlay X Web Services; Part 7: Account Management', Version 6.1.0 (June 2005), URL: http://www.3gpp.org/

[3GPP 2005k] 3GPP TS 29.199-8, 3rd Generation Partnership Project, 'Open Service Access (OSA) Parlay X Web Services; Part 8: Terminal Status', Version 6.1.0 (June 2005), URL: http://www.3gpp.org/

[3GPP 2005l] 3GPP TS 29.199-9, 3rd Generation Partnership Project, 'Open Service Access (OSA) Parlay X Web Services; Part 9: Terminal Location', Version 6.2.0 (June 2005), URL: http://www.3gpp.org/

[3GPP 2005m] 3GPP TS 29.199-10, 3rd Generation Partnership Project, 'Open Service Access (OSA) Parlay X Web Services; Part 10: Call Handling', Version 6.1.0 (June 2005), URL: http://www.3gpp.org/

[3GPP 2005n] 3GPP TS 29.199-11, 3rd Generation Partnership Project, 'Open Service Access (OSA) Parlay X Web Services; Part 11: Audio Call', Version 6.1.0 (June 2005), URL: http://www.3gpp.org/

[3GPP 2005o] 3GPP TS 29.199-12, 3rd Generation Partnership Project, 'Open Service Access (OSA) Parlay X Web Services; Part 12: Multimedia Conference', Version 6.1.0 (June 2005), URL: http://www.3gpp.org/

[3GPP 2005p] 3GPP TS 29.199-13, 3rd Generation Partnership Project, 'Open Service Access (OSA) Parlay X Web Services; Part 13: Address List Management', Version 6.1.0 (June 2005), URL: http://www.3gpp.org/

[3GPP 2005q] 3GPP TS 29.199-14, 3rd Generation Partnership Project, 'Open Service Access (OSA) Parlay X Web Services; Part 14: Presence', Version 6.2.0 (June 2005), URL: http://www.3gpp.org/

[3GPP 2005r] 3GPP TS 23.218, 3rd Generation Partnership Project, 'IP Multimedia (IM) Session Handling; IM call model', Version 5.8.0 (March 2005), URL: http://www.3gpp.org/

[3GPP2 2002a] 3GPP2 P.S0001-B, 3rd Generation Partnership Project 2, 'Wireless IP Network Standard', Version 1.0.0 (October 2002), URL: http://www.3gpp2.org/

[3GPP2 2002b] 3GPP2 S.R0061-0, 3rd Generation Partnership Project 2, 'Wireless Immediate Messaging – Stage 1 Requirements', Version 1.0 (October 2002), URL: http://www.3gpp2.org/

[3GPP2 2002c] 3GPP2 S.R0062-0, 3rd Generation Partnership Project 2, 'Presence for Wireless Systems – Stage 1 Requirements', Version 1.0 (October 2002), URL: http://www.3gpp2.org/

[3GPP2 2003a] 3GPP2 X.S0013-000-0, 3rd Generation Partnership Project 2, 'All-IP Core Network Multimedia Domain – Overview', Version 1.0 (December 2003), URL: http://www.3gpp2.org/

[3GPP2 2003b] 3GPP2 X.S0013-002-0, 3rd Generation Partnership Project 2, 'All-IP Core Network Multimedia Domain – IP Multimedia Subsystem – Stage 2', Version 1.0 (December 2003), URL: http://www.3gpp2.org/

[3GPP2 2003c] 3GPP2 X.S0013-003-0, 3rd Generation Partnership Project 2, 'All-IP Core Network Multimedia Domain; IP Multimedia (IMS) Session Handling; IP Multimedia (IM) Call Model – Stage 2', Version 1.0 (December 2003), URL: http://www.3gpp2.org/

[3GPP2 2003d] 3GPP2 X.S0013-004-0, 3rd Generation Partnership Project 2, 'All-IP Core Network Multimedia Domain; IP Multimedia Call Control Protocol Based on SIP and SDP, Stage 3', Version 1.0 (December 2003), URL: http://www.3gpp2.org/

[3GPP2 2003e] 3GPP2 X.S0013-005-0, 3rd Generation Partnership Project 2, 'All-IP Core Network Multimedia Domain; IP Multimedia Subsystem Cx Interface Signaling Flows and Message Contents', Version 1.0 (December 2003), URL: http://www.3gpp2.org/

[3GPP2 2003f] 3GPP2 X.S0013-006-0, 3rd Generation Partnership Project 2, 'All-IP Core Network Multimedia Domain Cx Interface Based on the Diameter Protocol; Protocol Details', Version 1.0 (December 2003), URL: http://www.3gpp2.org/

[3GPP2 2003g] 3GPP2 X.S0013-007-0, 3rd Generation Partnership Project 2, 'All-IP Core Network Multimedia Domain; IP Multimedia Subsystem – Charging Architecture', Version 1.0 (December 2003), URL: http://www.3gpp2.org/

[3GPP2 2003h] 3GPP2 X.S0013-008-0, 3rd Generation Partnership Project 2, 'All-IP Core Network Multimedia Domain; IP Multimedia Subsystem – Accounting Information Flows and Protocol', Version 1.0 (December 2003), URL: http://www.3gpp2.org/

[3GPP2 2003i] 3GPP2 X.S0013-010-0, 3rd Generation Partnership Project 2, 'All-IP Core Network Multimedia Domain; IP Multimedia Subsystem Sh interface; Signaling flows and message contents – Stage 2', Version 1.0 (December 2003), URL: http://www.3gpp2.org/

[3GPP2 2003j] 3GPP2 X.S0013-011-0, 3rd Generation Partnership Project 2, 'All-IP Core Network Multimedia Domain; Sh Interface based on Diameter Protocols; Protocol Details – Stage 3', Version 1.0 (December 2003), URL: http://www.3gpp2.org/

[3GPP2 2003k] 3GPP2 S.R0037-0, 3rd Generation Partnership Project 2, 'IP Network Architecture Model for cdma2000 Spread Spectrum Systems', Version 3.0 (August 2003), URL: http://www.3gpp2.org/

[3GPP2 2003l] 3GPP2 X.S0017-0, 3rd Generation Partnership Project 2, 'Open Service Access (OSA) – Application Programming Interface (API) – OSA API', Version 1.0.0 (June 2003), URL: http://www.3gpp2.org/

[Andersson 2004] Andersson, J.K., 'Overload Control and Performance Evaluation in a Parlay/OSA Environment', Department of Communication Systems, Lund Institute of Technology, 2004, 82 pp.

[Bakker 2000] Bakker, J.-L., McGoogan, J.R., Opdyke, W.F. and Panken, F.J., 'Rapid Development and Delivery of Converged Services using APIs', *Bell Labs Technical Journal*, **5**(3), 2000, 12–29.

[Bakker 2002] Bakker, J.-L., Tweedie, D. and Unmehopa, M., 'Evolving Service Creation; New Developments in Network Intelligence', *Telektronikk* **98**(4), 2000, 58–68.

[Bennett 2003] Bennett, A.J., Grech, M.L.F., Unmehopa, M.R. and Vemuri, K.V., 'Service Mediation Standards', *Bell Labs Technical Journal*, **7**(4), 2003, 77–90, Published by Wiley Periodicals, Inc.

[Brenner 2005] Brenner, M.R., Grech, M.L.F., Torabi, M. and Unmehopa, M.R., 'The Open Mobile Alliance and Trends in Supporting the Mobile Services Industry', *Bell Labs Technical Journal*, **10**(1), 2005, 59–75, Published by Wiley Periodicals, Inc.

[Brooks 1995] Brooks, F.P., 'The Mythical Man-Month: Essays on Software Engineering', Addison-Wesley Professional, 1995, 336 p.

[Comer 1999] Comer, D.E. and Stevens, D.L. 'Internetworking with TCP/IP, Vol 2: Design, Implementation, and Internals', 3rd Edition, Prentice Hall, 1999, 660 pp.

[Comer 2000] Comer, D.E., 'Internetworking with TCP/IP Vol.1: Principles, Protocols, and Architecture, 4th Edition', Prentice Hall, 2000, 755 pp.

[Dobrowolski 2001] Dobrowolski, J. and Vemuri, K. 'Internet-based Service Creation and the Need for a VoIP Call Model', draft-dobrowolski-voip-cm-01.txt, IETF Internet Draft, Expired (May 2001). A copy can be downloaded from: http//www.potaroo.net/ietf/old-ids/draft-dobrowolski-voip-cm-01.txt

[ETSI 2005a] ETSI Standard ES 202 915-4-5, 'Open Service Access (OSA) Application Programming Interface (API); Part 4: Call Control; Sub-part 5: Conference Call Control SCF', Version 1.3.1 (March 2005), URL: http://portal.etsi.org/

[ETSI 2005b] ETSI Standard ES 202 915-9, 'Open Service Access (OSA) Application Programming Interface (API); Part 9: Generic Messaging SCF', Version 1.2.1 (March 2005), URL: http://portal.etsi.org/

[ETSI 2005c] ETSI Standard ES 202 915-10, 'Open Service Access (OSA) Application Programming Interface (API); Part 10: Connectivity Manager SCF', Version 1.3.1 (March 2005), URL: http://portal.etsi.org/

[Faynberg 1996] Faynberg, I., Gabuzda, L.R., Kaplan, M.P. and Shah, N.J., 'The Intelligent Network Standards: Their Application to Services', 1996, McGraw-Hill Professional, New York, 236 pp.

[Faynberg 2000] Faynberg, I., Gabuzda, L. and Lu, H., 'Converged Networks and Services: Internetworking IP and the PSTN', John Wiley & Sons, Inc. 2000, New York, 347 pp.

[FIW 2003] Seventh International Workshop on Feature Interactions in Telecommunication and Software Systems (FIW'03), URL: http://www.site.uottawa.ca/fiw03/

[Gurbani 2003] Gurbani, V.K., Brusilovsky, A., Faynberg, I., Lu, H-L., Sun, X-H. and Unmehopa, M., 'Internet Service Execution for Telephony Events', Proceedings of the 8th IEEE International Conference on Intelligence in Next Generation Networks, ICIN2003, Bordeaux, France, April 2003

[H.248.1 2002] ITU-T Recommendation H.248.1, Telecommunication Standardization Sector of ITU, Series H: Audiovisual and Multimedia Systems, 'Gateway Control Protocol: Version 2', May 2002

[H.323 2003] ITU-T Recommendation H.323, Telecommunication Standardization Sector of ITU, Series H: Audiovisual and Multimedia Systems, 'Packet-based Multimedia Communications Systems', July 2003

[Hanmer 2000] Hanmer, R., 'Real Time and Resource Overload Language', 7th Pattern Languages of Programs Conference (PLoP 2000), 13–16 August 2000, Monticello, Illinois, USA

[Henning 1999] Henning, M. and S. Vinoski, 'Advanced CORBA(R) Programming with C + +', Addison-Wesley Professional, 1999, 1120 pp.

[Holzmann 1991] Holzmann, G.J., 'Design and Validation of Computer Protocols', Prentice Hall PTR, 1991, Upper Saddle River, NJ, 512 pp.

[Hull 2004] Hull, R.B., Kumar, B.B. Qutub, S.S., Unmehopa, M.R. and Varney, D.W., 'Policy Enabling the Services Layer', *Bell Labs Technical Journal*, **9**(1), 2004, 5–18, Published by Wiley Periodicals, Inc.

[Hyde 1999] Hyde, P., 'Java Thread Programming', Sams Publishing, Indianapolis, Indiana, USA, 1999, 528 pp.

[IMSinOMA 2005] Open Mobile Alliance™, Utilization of IMS Capabilities Architecture Candidate Version 1.0 (2005), URL: http://www.openmobilealliance.org/

[Jain 2004] Jain, R., Bakker, J.-L. and Anjum, F., 'Programming Converged Networks: Call Control in Java, XML, and Parlay/OSA', Wiley-Interscience, 2004, 268 pp.

[JSR 116] JSR-000116 SIP Servlet API Specification 1.0 Final Release (2003), URL: http://www.jcp.org/en/jsr/detail?id=116

[Kozik 2000] Kozik, J., Faynberg, I. and Lu, H-L., 'On Opening PSTN to Enhanced Voice/data Services – The PINT Protocol Solution', *Bell Labs Technical Journal*, **5**(3), 2000, 153–165

[Kozik 2003] Kozik, J., Unmehopa, M.R. and Vemuri, K.V., 'A Parlay and SPIRITS-based Architecture for Service Mediation', *Bell Labs Technical Journal*, **7**(3), 2003, 105–122, Published by Wiley Periodicals, Inc.

[Lagerberg 2002] Lagerberg, K., Plas, D-J. and Wegdam, M., 'Web Services in Third-generation Service Platforms', *Bell Labs Technical Journal*, **7**(2), 2002, 167–183, Published by Wiley Periodicals, Inc.

[Liberty] Liberty Alliance Project, Liberty Alliance Specifications, URL: http://www.projectliberty.org/resources/specifications.php

[Lofthouse 2004] Lofthouse, H., Yates, M.J. and Stretch, R., 'Parlay X Web Services', *BT Technology Journal*, **22**(1), 2004, 81–86, Springer Science and Business Media.

[Mampaey 2000] Mampaey, M. and Couturier, A., 'Using TINA Concepts for IN Evolution', *IEEE Communications Magazine*, 2000, 94–99

[Miller 2002] Miller, M.A., 'Voice Over IP Technologies: Building the Converged Network', 2nd edition, John Wiley & Sons Ltd., 2002, 552 pp.

[Moerdijk 2003] Moerdijk, A.J. and Lucas Klostermann, 'Opening the Networks with Parlay/OSA: Standards and Aspects behind the APIs', *IEEE Network*, **17**(3), 2003, 58–64.

[Moore 2002] Moore, G.A., 'Crossing the Chasm', 2002, Collins (Revised Edition), 256 pp.

[Mouly 1992] Mouly, M. and Pautet, M-B., 'The GSM System for Mobile Communications', Telecom Publishing, 1992, 701 pp.

[Mueller 2002] Mueller, S.M., 'APIs and Protocols for Convergent Network Services', McGraw-Hill Professional, New York, 2002, 445 pp.

[Natsuno 2003] Natsuno, T., 'The i-mode Wireless Ecosystem', John Wiley & Sons Inc. (2003), NJ, US

[NEBS 2002] GR-63-CORE, 'Network Equipment-Building System (NEBS) Requirements: Physical Protection', Issue 02 (April 2002).

[Newcomer 2002] Newcomer, E., 'Understanding Web Services: XML, WSDL, SOAP, and UDDI', Addison-Wesley, Independent Technologies Guide (D. Chappell, series editor), 2002, 368 pp.

[Nielsen 2002] Nielsen, H.F., Sanders, H., Butek, R. and Nash, S., 'Direct Internet Message Encapsulation (DIME)', draft-nielsen-dime-02, IETF Internet Draft, Expired (June 2002)

[NIST 2005] National Institute of Standards and technology, NIST/SEMATECH e-Handbook of Statistical Methods [online], URL: http://www.itl.nist.gov/div898/handbook/ (Accessed July 2005). Please see, in particular, section 8.1.8.4, which explains 'R out of N' systems or N + K spared arrangements

[Norris 1998] Norris, J.R., 'Markov Chains', Cambridge Series in Statistical and Mathematical Probabilistics, Cambridge University Press, 1998, 237 pp.

[OMG 2000a] Object Management Group (OMG) Telecom Service Access & Subscription Specification, Version 1.0 (October 2002), URL: http://www.omg.org/

[OMG 2002a] Object Management Group (OMG) XML Metadata Interchange (XMI) Specification, Version 1.2 (January 2002), URL: http://www.omg.org/

[OMG 2003a] Object Management Group (OMG) XML Metadata Interchange (XMI) Specification, Version 2.0 (May 2003), URL: http://www.omg.org/

[Parlay 2002a] Parlay Web Services WSDL Style Guide, Version 1.0 (November 2002), URL: http://www.parlay.org/

[Parlay 2002b] Parlay Web Services – Application Deployment Infrastructure Version 1.0 (October 2002), URL: http://www.parlay.org/

[Parlay 2002c] Parlay Web Services Overview, Version 1.0 (October 2002), URL: http://www.parlay.org/

[Parlay 2003] Parlay 4.0 – Parlay X Web Services Specification, Version 1.0 (May 2003), URL: http://www.parlay.org/

[RFC 1305] Mills, D.L., 'Network Time Protocol (Version 3) Specification, Implementation and Analysis', IETF RFC 1305, March 1992

[RFC 1332] McGregor, G., 'The PPP Internet Protocol Control Protocol (IPCP)', IETF RFC 1332, May 1992

[RFC 1492] Finseth, C., 'An Access Control Protocol, Sometimes Called TACACS', IETF RFC 1492, July 1993

[RFC 1994] Simpson, W., 'PPP Challenge Handshake Authentication Protocol (CHAP)', IETF RFC 1994, August 1996

[RFC 2045] Freed, N. and Borenstein, N., 'Multipurpose Internet Mail Extensions (MIME) Part One: Format of Internet Message Bodies', IETF RFC 2045, November 1996

[RFC 2131] Droms, R., 'Dynamic Host Configuration Protocol', RFC 2131, March 1997

[RFC 2138] Rigney, C., Rubens, A., Simpson, W., and Willens, S., 'Remote Authentication Dial In User Service (RADIUS)', IETF RFC 2138, April 1997

[RFC 2401] Kent, S., and Atkinson, R., 'Security Architecture for the Internet Protocol', IETF RFC 2401, November 1998

[RFC 2848] Petrack, S. and Conroy, L. 'The PINT Service Protocol: Extensions to SIP and SDP for IP Access to Telephone Call Services', IETF RFC 2848, June 2000

[RFC 3050] Lennox, J., Schulzrinne, H. and Rosenberg, J., 'Common Gateway Interface for SIP', IETF RFC 3050, January 2001

[RFC 3261] Rosenberg, J., Schulzrinne, H., Camarillo, G., Johnston, A., Peterson, J., Sparks, R., Handley, M. and Schooler, E., 'SIP: Session Initiation Protocol', IETF RFC 3261, June 2002

[RFC 3507] Elson, J. and Cerpa, A., 'Internet Content Adaptation Protocol (ICAP)', IETF RFC 3507, April 2003

[RFC 3588] Calhoun, P., Loughney, J., Guttman, E., Zorn, G. and Arkko, J., 'Diameter Base Protocol', IETF RFC 3588, September 2003

[RFC 3835] Barbir, A., Penno, R., Chen, R., Hofmann, M. and Orman, H., 'An Architecture for Open Pluggable Edge Services (OPES)', IETF RFC 3835, August 2004

[RFC 3880] Lennox, J., Wu, X. and Schulzrinne, H., 'Call Processing Language (CPL): A Language for User Control of Internet Telephony Services', IETF RFC 3880, October 2004

[RFC 3880] Lennox, J., Wu, X. and Schulzrinne, H., 'Call Processing Language (CPL): A Language for User Control of Internet Telephony Services', IETF RFC 3880, October 2004

[RFC 3897] Barbir, A., 'Open Pluggable Edge Services (OPES) Entities and End Points Communication', IETF RFC 3897, September 2004

[RFC 3910] Gurbani, V., Brusilovsky, A., Faynberg, I., Gato, J., Lu, H. and Unmehopa, M., 'The SPIRITS (Services in PSTN requesting Internet Services) Protocol', IETF RFC 3910, October 2004

[Rising 2001] Rising, L., 'Design Patterns in Communication Software', Cambridge University Press, 2001, 548 pp.

[Robbins 2003] Robbins, K. and Robbins, S., 'Unix Systems Programming: Communication, Concurrency and Threads', Prentice Hall (2nd Edition), 2003, 912 pp.

[Russell 2002] Russell, T., 'Signaling System #7', 4th edition, McGraw-Hill Professional, 2002, 495 pp.

[Schmidt 2004] Schmidt, D.C., 'CORBA Tutorials' [online], URL: http://www.cs.wustl.edu/~schmidt/tutorials-corba.html (August 2004) (accessed July 2005)

[Shooman 1983] Shooman, M.L., 'Software Engineering: Design, Reliability, and Management', McGraw-Hill, Inc. New York, NY, USA, 1983, 704 pp.

[Sigtran] IETF Signaling Transport Working Group Charter, accessed July 2005, URL: http://www.ietf.org/html. charters/sigtran-charter.html

[Stretch 2001] Stretch, R., 'OSA and Other Related Issues', *BT Technology Journal*, **19**(1), 2001, 80–87, published by Kluwer Academic Publishers

[Tanenbaum 2003] Tanenbaum, A.S., 'Computer Networks', 4th edition, Prentice Hall PTR, Upper Saddle River, NJ, 2003, 912 pp.

[TINA 1995] Telecommunications Information Networking Architecture Consortium (TINA-C), Overall Concepts and Principles of TINA, Version 4.0 (Feb. 1995), URL: http://www.tinac.com/

[TINA 1997a] Telecommunications Information Networking Architecture Consortium (TINA-C), Service Architecture, Version 5.0 (June 1997), URL: http://www.tinac.com/

[TINA 1997b] Telecommunications Information Networking Architecture Consortium (TINA-C), TINA Business Model and Reference Points, Version 4.0 (May 1997), URL: http://www.tinac.com/

[TINA 1999] Telecommunications Information Networking Architecture Consortium (TINA-C), Ret Reference Point Specifications, Version 1.1 (April 1999), URL: http://www.tinac.com/

[Unmehopa 2002a] Unmehopa, M., Vemuri, K., Brusilovsky, A., Dacloush, E., Zaki, A., Haerens, F., Bakker, J.-L., Chatras, B., and Dobrowolski, J., 'On selection of IN Parameters to be carried by the SPIRITS Protocol', IETF Internet Draft draft-ietf-spirits-in-03.txt (July 2002, expired)

[Unmehopa 2002b] Unmehopa, M.R., Grech, M.L.F., Dobrowolski, J.A. and Stanaway, Jr., J.J., 'The Support of Mobile Internet Applications in UMTS Networks Through the Open Service Access', *Bell Labs Technical Journal*, **6**(2), 2002, 47–64, published by Wiley Periodicals, Inc.

[Vemuri 2000] K.V. Vemuri, 'SPHINX: A Study in Convergent Telephony and Advanced Scenarios for H.323-SIP Interoperation', Workshop on IP Telecom Service (IPTS), Georgia, USA, September 2000 URL: http://www.research.att.com/conf/ipts2000/

[Viterbi 1995] Viterbi, A.J., 'CDMA: Principles of Spread Spectrum Communication', Addison-Wesley Wireless Communications, 1995, 272 pp.

[W3C 2000] World Wide Web Consortium (W3C), SOAP Messages with Attachments, W3C Note (December 2000), URL: http://www.w3.org/TR/SOAP-attachments/

[W3C 2004] Voice eXtensible Markup Language (VoiceXML) W3C Recommendation, Version 2.0 (March 2004), URL: http://www.w3.org/TR/2004/REC-voicexml20-20040316/

[WAP] Open Mobile Alliance™, the WAP specifications. URL: http://www.openmobilealliance.org/

[WAP Push] WAP Push Architectural Overview, Version 03-July-2001, WAP-250-PushArchOverview-20010703-a

[WGS84] US Defense Mapping Agency (DMA) TR 8350.2, 'Department of Defense World Geodetic System 1984: Its Definition and Relationships with Local Geodetic Systems', 2nd Edition, 1991, 169 pp.

[WSI 2004] Web Services Interoperability Organization (WS-I.org) Basic Profile, Version 1.1 (August 2004), URL: http://www.ws-i.org/Profiles/BasicProfile-1.1.html

[Zuidweg 2002] Zuidweg, J., 'Next Generation Intelligent Networks', Artech House Telecommunications Library, Norwood, MA, USA, 2002, 366 pp.

Index